140 Derivation and Integration

This book is devoted to an invariant multi-dimensional process of recovering a function from its derivative. It considers additive functions defined on the family of all bounded BV sets that are continuous with respect to a suitable topology. A typical example is the flux of a continuous vector field. A very general Gauss-Green theorem follows from the sufficient conditions for the derivability of the flux.

Since the setting is invariant with respect to local lipeomorphisms, a standard argument extends the Gauss-Green theorem to the Stokes theorem on Lipschitz manifolds. In addition, the author proves the Stokes theorem for a class of top-dimensional normal currents — a first step toward solving a difficult open problem of derivation and integration in middle dimensions.

The book contains complete and detailed proofs of all new results, and of many known results for which the references are not easily available. It will provide valuable information to research mathematicians and advanced graduate students interested in geometric integration and related areas.

Washek F. Pfeffer is Professor Emeritus of Mathematics at the University of California, Davis. He was born in Prague, Czech Republic, where he studied mathematics at Charles University (1955–60). He immigrated to the United States in 1965, and in 1966 received his Ph.D. from the University of Maryland, College Park. Dr. Pfeffer has worked at the Czechoslovak Academy of Sciences in Prague and has taught at the Royal Institute of Technology in Stockholm, George Washington University, and the University of California, Berkeley. In 1994–95 he was a Fulbright Lecturer at Charles University. His primary research areas are analysis and topology. Dr. Pfeffer is a member of the American and Swedish Mathematical Societies. He serves on the Academic Board of the Center for Theoretical Study at Charles University. He has written the books *Integrals and Measures* (Marcel Dekker, 1977) and *The Riemann Approach to Integration* (Cambridge University Press, 1993).

CAMBRIDGE TRACTS IN MATHEMATICS

General Editors

B. BOLLOBAS, W. FULTON, A. KATOK, F. KIRWAN,
P. SARNAK

140 Derivation and Integration

Washek F. Pfeffer

University of California, Davis

Derivation and Integration

CAMBRIDGE
UNIVERSITY PRESS

CAMBRIDGE UNIVERSITY PRESS
Cambridge, New York, Melbourne, Madrid, Cape Town, Singapore,
São Paulo, Delhi, Dubai, Tokyo, Mexico City

Cambridge University Press
The Edinburgh Building, Cambridge CB2 8RU, UK

Published in the United States of America by Cambridge University Press, New York

www.cambridge.org
Information on this title: www.cambridge.org/9780521155656

First published 2001
First paperback edition 2010

A catalogue record for this publication is available from the British Library

Library of Congress Cataloguing in Publication Data

Pfeffer, Washek F.
    Derivation and integration / Washek F. Pfeffer
        p. cm. - (Cambridge tracts in mathematics ; 140)
    Includes bibliographical references and index.
    ISBN 0-521-79268-1
    1. Integrals, Generalized. I. Title. II. Series.

    QA312.P458 2001
    515'.4-dc21   00-065097

ISBN 978-0-521-79268-4 Hardback
ISBN 978-0-521-15565-6 Paperback

To my Italian friends and colleagues
with gratitude and affection

Contents

Preface

This book is devoted to a multi-dimensional version of a very classical problem: recovering a function from its derivative. An immediate application of our results yields the Gauss-Green and Stokes theorems of large generality.

The problem we consider has a long history. In dimension one, it was solved by Lebesgue for absolutely continuous functions, and by Denjoy and Perron (independently and by different means) for so called ACG$_*$ functions [75, Chapter 7, Section 8]. In higher dimensions, absolutely continuous functions become absolutely continuous measures, which can still be recovered from their Radon-Nikodym derivatives by means of the Lebesgue integral. A multi-dimensional analog of ACG$_*$ functions is more subtle, and has been defined only recently [18, 19].

There is no obvious extension of the Denjoy-Perron integral to higher dimensions. The early generalizations [4, 72] do not integrate partial derivatives of all differentiable functions, and give no indication how this can be achieved. Even the strikingly simple Riemannian definition of the Denjoy-Perron integral, obtained independently by Henstock [30] and Kurzweil [42], did not initially produce desirable results in higher dimensions [43, 52]. The first successful multi-dimensional generalization is due to Mawhin [50, 49], who modified the Henstock-Kurzweil definition so that the partial derivatives of each differentiable function are integrable and the Gauss-Green formula holds. Soon, more refined coordinate free generalizations were produced by several authors [35, 64, 44, 38], but none of these gave a useful descriptive definition of the integral: a simple necessary and sufficient condition that a function is differentiable and equal to the integral of its derivative. The ACG$_*$ functions, whose definition relies on the order structure of the reals, provided poor guidance for the multi-dimensional case. The correct point of view introduced in [63] has been fully developed only after Thomson managed to characterize the ACG_* functions in measure-theoretic terms [84].

There is a vast difference between one-dimensional and multi-dimensional settings, which is hard to overemphasize. In particular, the presence of geometry requires that in higher dimensions the classical real variable techniques are augmented by methods of geometric measure theory.

We present the most recent results connecting set functions with their derivates.* The set functions we consider are additive functions defined on the family of all bounded BV sets in $\mathbb{R}^m$ that are continuous with respect to a suitable topology. We call them charges, although they have little in common with charges employed in [70]. Aside from signed measures, a typical example of a charge is the flux of a continuous vector field. Derivates are calculated with respect to Lebesgue measure and a derivation base consisting of sequences of BV sets whose shapes and perimeters are controlled. If a charge is the flux of a vector field that is differentiable at some point, then the charge is derivable at this point and its derivate equals the divergence of the vector field. In view of this, the connection with the Gauss-Green theorem is imminent. In our setting we deal with discontinuous vector fields and place no topological restrictions on the sets of their singularities. Already within the context of Lebesgue integration (i.e., under the additional assumption that the divergence is Lebesgue integrable), our version of the Gauss-Green theorem complements those established previously [76, 24, 3].

While we investigate only the local situation, we show that charges and their derivates are invariant with respect to local lipeomorphisms, which need not be injective. Thus a standard argument can be used to extend the Gauss-Green theorem to the Stokes theorem on Lipschitz manifolds. In addition, we establish the Stokes theorem for a class of top-dimensional normal currents. It is our hope that this is a first step toward solving a difficult open problem of derivation and integration in middle dimensions. For the Lebesgue integrable derivates the problem was settled in [87].

The book is divided into six chapters, and each chapter contains several sections. Theorems, propositions, lemmas, etc., as well as various formulae, are numbered consecutively according to the chapter and

---

*Generally, we use *derivative* and *differentiation* for point functions, and *derivate* and *derivation* for set functions. The Radon-Nikodym derivative of a measure is a notable exception.

section in which they occur. Claims and formulae needed only for a particular proof or example are marked or numbered consecutively within that proof or example, respectively; as a rule, no outside references are made to them. Some material, considered supplementary to the main exposition, is presented in small print.

After necessary preliminaries presented in Chapter 1, we define charges and address the issue of their derivability in Chapter 2. The main result is an analog of Ward's theorem: if the lower and upper derivates of a charge are almost everywhere finite, then the charge is almost everywhere derivable. Using approximating partitions of BV sets, we show that a charge is nonnegative whenever its derivate is nonnegative outside a set whose codimension one Hausdorff measure is σ-finite.

In Chapter 3 we associate with each charge a Borel regular measure called the critical variation. We show that a charge is absolutely continuous if and only if its critical variation is absolutely continuous and locally finite. It turns out that charges whose critical variations are merely absolutely continuous, called AC_* charges,[†] retain many useful properties of absolutely continuous charges. In particular, they are derivable almost everywhere and are uniquely determined by their derivates — a property analogous to that of absolutely continuous Radon measures.

The duality between charges and BV functions is studied in Chapter 4. We prove that the linear space of all charges is dual to the linear space of all bounded BV functions with compact support, and that the latter space is pre-dual to any linear space of charges that contains all absolutely continuous charges. The algebra of all locally bounded locally BV functions acts naturally on the linear space of all charges; by multiplication, it also acts on their derivates. We show that the linear spaces of all AC_* charges and their derivates are invariant with respect to these actions, and that the actions commute with derivation.

Chapter 5 is devoted to the problem of recovering an AC_* charge from its derivate. We solve it by developing a Riemann type averaging process, called the R-integral, akin to the Henstock-Kurzweil integral and a fortiori to the classical Denjoy-Perron integral. The derivate of each AC_* charge is locally R-integrable, and the charge is the indefinite R-integral of its derivate. Applying this to the flux of a conti-

---

[†]There is no immediate connection between AC_* charges and AC_* functions introduced in [75, Chapter 7, Section 8]. However, AC_* charges are related to ACG_* functions defined ibid.

nuous pointwise Lipschitz vector field, we obtain the aforementioned Gauss-Green theorem. As the R-integral satisfies the area theorem for lipeomorphisms, it can be routinely lifted to Lipschitz manifolds where the Gauss-Green theorem becomes the Stokes theorem. Applying the results of Chapter 4 we prove that locally bounded locally BV functions are precisely the multipliers for the R-integral. This facilitates a change of variables theorem for local lipeomorphisms.

In Chapter 6 we extend the R-integral to a larger family of integrable functions. The extension is obtained by a transfinite iteration of improper integrals, and it preserves all desirable properties of the R-integral.

The book is not intended for beginners. The reader is expected to have a solid background in graduate level analysis, including some understanding of BV functions and BV sets. To this end, [22] and [88] are excellent references. On the other hand, we strive to provide complete and detailed proofs of every claim we make. A few arguments left to the reader are truly simple and straightforward. As usual, some auxiliary results, mainly in Chapter 1, are presented without proofs; however, in such cases we always provide precise references to the literature.

No attempt was made to trace systematically the origins of various results contained in the book. Occasional historical remarks are limited to situations witnessed directly by the author.

W.F.P.

Albion, California
November, 2000

Acknowledgements

Since retiring from the University of California, it has been my good fortune to spend time with colleagues interested in problems related to my work. I am indebted to my long term collaborators B. Bongiorno at the University of Palermo and J. Mawhin at the Catholic University of Louvain who periodically invited me to visit with them in their respective departments. The Italian Research Council and Catholic University of Louvain sponsored my visits. I am equally indebted to A. Volčič for his invitations to lecture at the Workshop on Measure Theory and Real Analysis in Grado. The interaction with other Workshop participants contributed significantly to my work. A Fulbright Grant supported my seven month stay in Prague, where I.M. Havel and J. Kurzweil kindly provided an office in the Mathematical Institute of the Czech Academy of Sciences and the Center for Theoretical Study of Charles University. J. Veselý organized a course I taught at Charles University; parts of this book originated from lectures I prepared for that course.

During the preparation of the text, I benefited greatly from discussions with Z. Buczolich, G.D. Chakerian, U. Darji, T. De Pauw, R.J. Gardner, G. Gruenhage, R.M. Hardt, J. Král, J. Malý, P. Mattila, D. Preiss, K. Prikry, A. Shibakov, D.P. Sullivan, B.S. Thomson, and W.P. Ziemer. In addition, T. De Pauw and B.S. Thomson read portions of the manuscript and made valuable comments and corrections. Consultations with K. Pitcoff improved the style and grammar at several places.

E. Howard carefully checked all my proofs and eliminated a multitude of mistakes, some quite serious. His contribution was invaluable. The responsibility for any errors which may not have been corrected is entirely mine.

The manuscript was prepared in $\mathcal{A}_{\mathcal{M}}\mathcal{S}$-LaTeX using Textures®2.0. I am grateful to K. Sharma for including me in the AppleMasters Program: using the computer equipment to which I was introduced at the AppleMasters Workshop minimized the time I spent typesetting. Throughout the typesetting process, I relied on the generous assistance of the American Mathematical Society T_EX-perts, B.N. Beeton and M.J. Downes.

Editorial help provided by the publisher was essential. I am particularly obliged to the Mathematics Editor, L. Cowles, who cooperated with me throughout the whole project with great patience and understanding. J. Carey deserves much credit for her careful copyediting of the manuscript. In the final stages of production I received aid and encouragement from C. Felgar.

Finally, I would like to acknowledge the nontechnical contribution of my friend Greg Gorman who lifted my life to unprecedented levels.

1

Preliminaries

We establish the notation and terminology, and present some basic facts that will be used throughout the book. Several well-known theorems are stated without proofs, however, those results for which we found no convenient references are proved in detail. In general, the reader is expected to have some prior knowledge of the concepts introduced in this chapter.

1.1. The setting

The set of all real numbers is denoted by $\mathbb{R}$, and the elements of $\mathbb{R}$ are referred to as *numbers* or *finite numbers*. The elements of $\mathbb{R} \cup \{\pm\infty\}$ are called the *extended real numbers*. In $\mathbb{R} \cup \{\pm\infty\}$ we consider the usual order and topology, and define the following algebraic operations:

$$a + \infty := +\infty + a := +\infty \quad \text{for } a > -\infty\,,$$

$$a - \infty := -\infty + a := -\infty \quad \text{for } a < +\infty\,,$$

$$a \cdot (\pm\infty) := \begin{cases} \pm\infty & \text{if } a > 0, \\ \mp\infty & \text{if } a < 0, \\ 0 & \text{if } a = 0. \end{cases}$$

Here and elsewhere, the meaning of $P := Q$ is that P is *defined* as equal to Q. Throughout, the symbol ∞ stands for $+\infty$.

The ambient space of this book is $\mathbb{R}^m$ where $m \geq 1$ is a fixed integer. For $x = (\xi_1, \ldots, \xi_m)$ and $y = (\eta_1, \ldots, \eta_m)$ in $\mathbb{R}^m$ we let

$$x \cdot y := \sum_{i=1}^{m} \xi_i \eta_i \quad \text{and} \quad |x| := \sqrt{x \cdot x}\,.$$

In $\mathbb{R}^m$ we shall use exclusively the usual Euclidean metric induced by the norm $|\cdot|$. For $x \in \mathbb{R}^m$ and $r \geq 0$, we let

$$B(x,r) := \{y \in \mathbb{R}^m : |x - y| < r\},$$
$$B[x,r] := \{y \in \mathbb{R}^m : |x - y| \leq r\};$$

if $r > 0$, we call the sets $B(x,r)$ and $B[x,r]$, respectively, the *open* and *closed balls* of radius r centered at x. The zero vector in $\mathbb{R}^m$ is denoted by 0, and we write $B(r)$ and $B[r]$ instead of $B(0,r)$ and $B[0,r]$, respectively.

The diameter, closure, interior, and boundary of a set $E \subset \mathbb{R}^m$ are denoted by $d(E)$, $\mathrm{cl}\, E$, $\mathrm{int}\, E$, and ∂E, respectively. If A and B are subsets of $\mathbb{R}^m$, we let

$$\mathrm{dist}\,(A,B) := \inf\{|x - y| : x \in A \text{ and } y \in B\},$$

and write $\mathrm{dist}\,(x,A)$ instead of $\mathrm{dist}\,(\{x\},A)$.

A *cell* is a compact nondegenerate subinterval of $\mathbb{R}^m$, i.e., the set

$$A := \prod_{i=1}^{m}[a_i, b_i]$$

where $a_i < b_i$ are real numbers for $i = 1, \ldots, m$. For $r > 0$, a *cube* of diameter $2r\sqrt{m}$ centered at the point $x := (\xi_1, \ldots, \xi_m)$ of $\mathbb{R}^m$ is the cell

$$K[x,r] := \prod_{i=1}^{m}[\xi_i - r, \xi_i + r];$$

we write $K[r] := [-r, r]^m$ instead of $K[0,r]$. A *dyadic cube* is the cell

$$\prod_{i=1}^{m}\left[\frac{k_i}{2^n}, \frac{k_i + 1}{2^n}\right]$$

where $n, k_1, \ldots, k_m$ are integers with $n \geq 0$. If A, B is a pair of dyadic cubes whose interiors meet, then $A \subset B$ or $B \subset A$. It follows any family $\mathcal{K}$ of dyadic cubes contains a subfamily $\mathcal{L}$ such that $\bigcup \mathcal{L} = \bigcup \mathcal{K}$ and $\mathcal{L}$ consists of cubes whose interiors are disjoint. Each dyadic cube C with $d(C) < \sqrt{m}$ is contained in a unique dyadic cube C^* for which $d(C^*) = 2d(C)$. The cube C^*, called the *mother* of C, is the smallest (with respect to inclusion) dyadic cube properly containing C. We say dyadic cubes B and C are *adjacent* whenever

$$d(B) = d(C) \quad \text{and} \quad B \cap C \neq \emptyset.$$

Each dyadic cube C is adjacent to 3^m dyadic cubes, including C itself.

Finite and countably infinite sets are called *countable*. For any pair of sets A and B, the set

$$A \triangle B := (A - B) \cup (B - A)$$

is called the *symmetric difference* of A and B.

Without additional attributes, functions are always real-valued. Occasionally, we refer to functions as real-valued or finite functions, to distinguish them from extended real-valued functions. When no confusion can arise, we use the same symbol to denote a function f defined on a set A and its restriction $f \upharpoonright B$ to a set $B \subset A$. If f is an extended real-valued function defined on a set $E \subset \mathbb{R}^m$, the sets

$$\{f = 0\} := \{x \in E : f(x) = 0\} \quad \text{and} \quad \operatorname{supp} f := \operatorname{cl}\{f \neq 0\}$$

are called, respectively, the *null set* and *support* of f; the set $\{f \neq 0\}$ is defined in the obvious way. We let

$$f^+ := \max\{f, 0\} \quad \text{and} \quad f^- := \max\{-f, 0\},$$

and note that $f = f^+ - f^-$ and $|f| = f^+ + f^-$. The *indicator* of a set $E \subset \mathbb{R}^m$ is the function χ_E defined by the formula

$$\chi_E(x) := \begin{cases} 1 & \text{if } x \in E, \\ 0 & \text{if } x \in \mathbb{R}^m - E. \end{cases}$$

The *zero extension* of an extended real-valued function f defined on a set $E \subset \mathbb{R}^m$ is the function $\overline{f}$ defined as follows

$$\overline{f}(x) := \begin{cases} f(x) & \text{if } x \in E, \\ 0 & \text{if } x \in \mathbb{R}^m - E. \end{cases}$$

Note that $\overline{f \upharpoonright E} = f\chi_E$ for each extended real-valued function f defined on $\mathbb{R}^m$.

Convention 1.1.1. Let $X(E)$ be a collection of extended real-valued functions defined on a set $E \subset \mathbb{R}^m$. We denote by $X_c(E)$ the family of all $f \in X(E)$ such that $\operatorname{supp} f$ is a compact subset of E. If $X(E)$ consists only of real-valued functions and $n \geq 1$ is an integer, we denote by $X(E; \mathbb{R}^n)$ the collection of all maps

$$x \mapsto \big(f_1(x), \dots, f_n(x)\big) : E \to \mathbb{R}^n$$

such that each f_i belongs to $X(E)$. The family $X_c(E; \mathbb{R}^n)$ is defined in the obvious way.

Convention 1.1.2. Let $X : U \to X(U)$ be a map assigning to each open set $U \subset \mathbb{R}^m$ a linear space $X(U)$ of functions defined on U so that

$$\{f \upharpoonright V : f \in X(U)\} \subset X(V)$$

for each open set $V \subset U$. Given an open set $\Omega \subset \mathbb{R}^m$, we denote by $X_{\mathrm{loc}}(\Omega)$ the linear space of all functions f defined on Ω such that $f \upharpoonright U$ belongs to $X(U)$ for each open set U whose closure is a compact subset of Ω. In view of the above inclusion, we have $X(\Omega) \subset X_{\mathrm{loc}}(\Omega)$.

In case the symbol X in Convention 1.1.1 or 1.1.2 has a superscript, such as X^{sup}, we adhere to the usual notation by writing X^{sup}_c and $X^{\mathrm{sup}}_{\mathrm{loc}}$ instead of $(X^{\mathrm{sup}})_c$ and $(X^{\mathrm{sup}})_{\mathrm{loc}}$, respectively.

Given a set $E \subset \mathbb{R}^m$, we denote by $C(E)$ the linear space of all continuous functions defined on E. If Ω is an open set and $k \geq 1$ is an integer, then $C^k(\Omega)$ denotes the linear space all functions defined on Ω that have continuous partial derivatives of order less than or equal to k. We let $C^\infty(\Omega) = \bigcap_{k=1}^{\infty} C^k(\Omega)$, and refer to elements of $C^k(\Omega)$ and $C^\infty(\Omega)$ as C^k and C^∞ functions on Ω, respectively. The spaces

$$C_c(E), \ C(E;\mathbb{R}^n), \ C_c(E;\mathbb{R}^n), \ C_c^k(\Omega), \ \text{etc.},$$

are defined according to Convention 1.1.1. Since continuity and differentiability are local properties, no new spaces are obtained by applying the subscript loc to $C(\Omega)$ or $C^k(\Omega)$ or $C^\infty(\Omega)$.

Throughout the book we use positive constants depending on certain parameters, such as the dimension m. If κ is a constant, we write $\kappa := \kappa(p_1, \ldots, p_k)$ to indicate κ depends *only* on parameters $p_1, \ldots, p_k$. When no confusion can arise, distinct constants appearing in different contexts may be denoted by the same letter.

1.2. Topology

All topological spaces in this book are Hausdorff. If $\mathcal{T}$ and $\mathcal{S}$ are two topologies in a set X, we say $\mathcal{S}$ is *larger* than $\mathcal{T}$, or equivalently $\mathcal{T}$ is *smaller* than $\mathcal{S}$, if and only if $\mathcal{T} \subset \mathcal{S}$. A closure of a subset E of a topological space $(X, \mathcal{T})$ is denoted by $\mathrm{cl}_{\mathcal{T}} E$. The *Borel σ-algebra* in $(X, \mathcal{T})$ is the smallest σ-algebra in X containing $\mathcal{T}$; its elements are called *Borel subsets* of X.

For a topological space $(X, \mathcal{T})$, the *subspace topology* in a set $Y \subset X$ is still denoted by $\mathcal{T}$. If $(X, \mathcal{T})$ and $(Y, \mathcal{S})$ are topological spaces, we denote by $\mathcal{T} \otimes \mathcal{S}$ the *product topology* in $X \times Y$.

A topological space X is called *sequential* if a set $A \subset X$ is closed whenever each sequence in A that converges in X converges to a point of A. Alternatively, X is sequential if a set $B \subset X$ is open whenever for each sequence $\{x_i\}$ in X that converges to an $x \in B$, all but finitely many points x_i belong to B. All closed and all open subspaces of a sequential space are sequential. A map f from a sequential space X to any topological space Y is continuous whenever

$$\lim f(x_i) = f(\lim x_i)$$

for each convergent sequence $\{x_i\}$ in X. Each first countable space is sequential, but not vice versa (cf. Proposition 1.7.7 below).

If $(X, \mathcal{T})$ is a sequential space, then the closure $\mathrm{cl}_{\mathcal{T}} A$ of a set $A \subset X$ is obtained by iterating transfinitely the process of taking limits. For each $E \subset X$, denote by $\mathrm{cl}_1 E$ the set consisting of all limits of sequences in E converging in X. Let $\mathrm{cl}_0 A = A$, and assuming $\mathrm{cl}_\beta A$ has been defined for each ordinal $\beta < \alpha \leq \omega_1$, let $\mathrm{cl}_\alpha A = \mathrm{cl}_1(\mathrm{cl}_\beta A)$ if $\alpha = \beta + 1$, and $\mathrm{cl}_\alpha A = \bigcup_{\beta < \alpha} \mathrm{cl}_\beta A$ if α is a limit ordinal. The set $\mathrm{cl}_{\omega_1} A$ contains A, and it is contained in any closed subset of X containing A. Observing $\mathrm{cl}_{\omega_1} A$ is a closed subset of X, we see that $\mathrm{cl}_{\mathcal{T}} A = \mathrm{cl}_{\omega_1} A$.

By a *linear space*, also referred to as a *vector space*, we always mean a linear space over $\mathbb{R}$. When no confusion is possible, 0 will denote the zero element of both $\mathbb{R}$ and the linear space under consideration. If A and B are subsets of a linear space X and $r \in \mathbb{R}$, we let

$$A + B := \{x + y : x \in A \text{ and } y \in B\} \quad \text{and} \quad rA := \{rx : x \in A\}.$$

For $x \in X$, we write $x + B$ instead of $\{x\} + B$, and we let $-A = (-1)A$. If $B \subset rA$ for an $r > 0$, we say that A *absorbs* B. A set $C \subset X$ is called

- *absorbing* if it absorbs $\{x\}$ for each $x \in X$;
- *convex* if $tx + (1 - t)y$ belongs to C for all x, y in C and each $t \in [0, 1]$;
- *symmetric* if $C = -C$.

Observe each convex symmetric set $C \subset X$ containing 0 has the property

$$|r| \leq |s| \implies rC \subset sC.$$

A topology $\mathcal{T}$ in a linear space X is called a *vector topology* if the maps

$$(x, y) \mapsto x + y : X \times X \to X \quad \text{and} \quad (t, x) \mapsto tx : \mathbb{R} \times X \to X$$

are continuous. If, in addition, $\mathcal{T}$ has a neighborhood base at 0 consisting of convex sets, we say $\mathcal{T}$ is a *locally convex topology* in X. A *topological vector space* is a pair $(X, \mathcal{T})$ where X is a linear space and $\mathcal{T}$ is a vector topology in X. If $\mathcal{T}$ is a locally convex topology in X, we call the pair $(X, \mathcal{T})$ a *locally convex space*.

In view of [21, Example 8.1.17], a vector topology $\mathcal{T}$ in a linear space X is induced by a uniformity. Thus $\mathcal{T}$ is completely regular by [21, Theorem 8.1.20].

Lemma 1.2.1. *Let $\mathcal{T}$ be a topology in a linear space X, and suppose $\mathcal{T}$ has a neighborhood base at 0 consisting of absorbing convex sets. Then $(X, \mathcal{T})$ is a locally convex space whenever the maps $z \mapsto x+z$ and $z \mapsto tz$ from X to X are continuous for each $x \in X$ and each $t \in \mathbb{R}$.*

PROOF. We only need to show $\mathcal{T}$ is a vector topology in X. By our assumptions, for each $x \in X$ and each $t \in \mathbb{R} - \{0\}$, the maps $z \mapsto x + z$ and $z \mapsto tz$ are homeomorphisms of X onto itself. Thus it suffices to prove the maps $(y, z) \mapsto y + z$ and $(t, z) \to tz$ are continuous at the points $(0, 0)$ and $(1, x)$, respectively.

We show first $\mathcal{T}$ has a neighborhood base consisting of absorbing convex and symmetric sets. Let U be an absorbing convex neighborhood of 0. Since $-U$ is a convex neighborhood of 0, the intersection V of the neighborhoods U and $-U$ is a symmetric convex neighborhood of 0. To show that V is also absorbing, select an $x \in X$ and find $r, s > 0$ so that rx and $s(-x)$ belong to U. As U is a convex set containing 0, it is easy to see that $tx \in V$ whenever $0 \le t \le \min\{r, s\}$.

Now let V be an absorbing convex and symmetric neighborhood of 0. Since

$$\tfrac{1}{2}V + \tfrac{1}{2}V \subset V,$$

the continuity of $(y, z) \mapsto y + z$ at the point $(0, 0)$ follows. Choose an $x \in X$, and find an $r > 0$ with $rx \in V$. Let $s = 1/(2 + r)$ and $|t - 1| < r/2$. If z is a point of $x + sV$, then $tz - x = t(z - x) + (t - 1)x$ belongs to

$$tsV + \tfrac{t-1}{r}V \subset \tfrac{1}{2}V + \tfrac{1}{2}V \subset V.$$

Thus tz belongs to $x + V$, which implies $(t, z) \mapsto tz$ is continuous at the point $(1, x)$. □

A subset E of a topological vector space $(X, \mathcal{T})$ is called *bounded* whenever it is absorbed by every neighborhood of zero. When $\mathcal{T}$ is induced by a metric, then a set $E \subset X$ is bounded if and only if the diameter of E is finite.

Proposition 1.2.2. *Let* $(X, \mathbf{S})$ *be a locally convex space, and suppose* $E_1 \subset E_2 \subset \cdots$ *are compact convex subsets of* X *such that*

(i) $0 \in E_1$ *and* $X = \bigcup_{n=1}^{\infty} E_n$;

(ii) *for each integer* $n \geq 1$, *each* $x \in X$, *and each* $t \in \mathbb{R}$ *there is an integer* $n(x,t) \geq 1$ *with* $x + E_n \subset E_{n(x,t)}$ *and* $tE_n \subset E_{n(x,t)}$.

If $\mathbf{T}$ *is the largest topology in* X *for which all inclusions*

$$\iota_n : (E_n, \mathbf{S}) \hookrightarrow (X, \mathbf{T})$$

are continuous, then the following claims hold.

(1) *A set* $A \subset X$ *is* $\mathbf{T}$-*closed if and only if* $A \cap E_n$ *is* $\mathbf{S}$-*closed for* $n = 1, 2, \ldots$. *In particular, the topologies* $\mathbf{S}$ *and* $\mathbf{T}$ *coincide in each* E_n.

(2) *A map* f *from* $(X, \mathbf{T})$ *to a topological space* Y *is continuous whenever each restriction* $(f \restriction E_n) : (E_n, \mathbf{S}) \to Y$ *is continuous.*

(3) *The topology* $\mathbf{T}$ *is locally convex.*

(4) *A sequence* $\{x_i\}$ *in* X $\mathbf{T}$-*converges to an* $x \in X$ *if and only if* $\{x_i\}$ *is a sequence in some* E_n *and* $\{x_i\}$ $\mathbf{S}$-*converges to* x.

(5) *A set* $A \subset X$ *is* $\mathbf{T}$-*bounded if and only if* A *is an* $\mathbf{S}$-*bounded subset of some* E_n.

(6) *The topology* $\mathbf{T}$ *is sequentially complete.*

(7) *If* $\mathbf{S}$ *is a sequential topology, then so is* $\mathbf{T}$.

PROOF. (1) Since $A \cap E_n = \iota_n^{-1}(A)$, the first claim is obvious.

(2) As $(f \restriction E_n)^{-1}(C) = f^{-1}(C) \cap E_n$ for each set $C \subset Y$, the second claim follows from claim (1).

(3) Claims (1) and (2) together with property (ii) imply the maps $z \mapsto x + z$ and $z \mapsto tz$ from $(X, \mathbf{T})$ into itself are continuous for each $x \in X$ and each $t \in \mathbb{R}$. Let U be a $\mathbf{T}$-neighborhood of 0, and let $x \in X$. Then x belongs to some E_n, and by claim (1), we can find a convex $\mathbf{S}$-neighborhood V of 0 so that $V \cap E_n \subset U \cap E_n$. As V is an absorbing convex set, there is a positive $t \leq 1$ with $tx \in V$. Now E_n is a convex set containing both 0 and x. Thus tx belongs to $V \cap E_n \subset U$. We conclude each $\mathbf{T}$ neighborhood of 0 is an absorbing set. In view of Lemma 1.2.1, it suffices to show that $\mathbf{T}$ has a neighborhood base at 0 consisting of convex sets.

To this end, select a $U \in \mathbf{T}$ containing 0. The set $K_0 := \{0\}$ is compact and convex. Find a $U_1 \in \mathbf{S}$ with $U \cap E_1 = U_1 \cap E_1$. As the topology $\mathbf{S}$ is regular, there is a convex set $V_1 \in \mathbf{S}$ containing 0 such that

$\operatorname{cl}_{\mathbf{S}}(K_0 + V_1) = \operatorname{cl}_{\mathbf{S}}V_1$ is a subset of U_1. The set $K_1 = E_1 \cap \operatorname{cl}_{\mathbf{S}}(K_0 + V_1)$ is compact and convex. Find a $U_2 \in \mathbf{S}$ with $U \cap E_2 = U_2 \cap E_2$. Since

$$K_1 \subset U_1 \cap E_1 = U \cap E_1 \subset U \cap E_2 = U_2 \cap E_2 \subset U_2,$$

there is a convex set $V_2 \in \mathbf{S}$ containing 0 such that $\operatorname{cl}_{\mathbf{S}}(K_1 + V_2)$ is a subset of U_2 [74, Theorem 1.10]. The set $K_2 = E_2 \cap \operatorname{cl}_{\mathbf{S}}(K_1 + V_2)$ is compact and convex. Now construct inductively convex sets $V_i \in \mathbf{S}$ containing 0 and convex compact sets K_i so that

$$K_{i-1} + V_i \subset K_i \quad \text{and} \quad E_i \cap (K_{i-1} + V_i) \subset U$$

for $i = 1, 2, \ldots$. The set

$$V := \bigcup_{i=1}^{\infty} \left[E_i \cap (K_{i-1} + V_i) \right],$$

being the union of an increasing sequence of convex sets, is convex. Clearly, $V \subset U$ and $0 \in V$. Since

$$E_n \cap V = E_n \cap \bigcup_{i=n}^{\infty} \left[E_i \cap (K_{i-1} + V_i) \right] = E_n \cap \bigcup_{i=n}^{\infty} (K_{i-1} + V_i)$$

is a relatively $\mathbf{S}$-open subset of E_n, it follows from claim (1) that V is $\mathcal{T}$-open.

(4) Let $\{x_i\}$ be a sequence in X that $\mathcal{T}$-converges to an $x \in X$. As $\mathbf{S}$ is smaller than $\mathcal{T}$, the sequence $\{x_i\}$ $\mathbf{S}$-converges to x. If $\{x_i\}$ is not a sequence in any E_n, then passing to a subsequence, still denoted by $\{x_i\}$, we may assume x_i is not in $E_i \cup \{x\}$ for $i = 1, 2, \ldots$. The set

$$U := \bigcap_{i=1}^{\infty} (X - \{x_i\})$$

contains x but no x_i, and

$$U \cap E_n = \bigcap_{i=1}^{\infty} (E_n - \{x_i\}) = \bigcap_{i=1}^{n-1} (E_n - \{x_i\})$$

is a relatively $\mathbf{S}$-open subset of E_n for $n = 1, 2, \ldots$. Claim (1) implies U is a $\mathcal{T}$ neighborhood of x, a contradiction. Since the topologies $\mathbf{S}$ and $\mathcal{T}$ coincide on each E_n, the converse is obvious.

(5) Assume $A \subset X$ is $\mathcal{T}$-bounded. Since $\mathbf{S}$ is smaller than $\mathcal{T}$, the set A is $\mathbf{S}$-bounded. It follows from property (ii), the sequence $\{E_n\}$ has a subsequence $\{E_{n_i}\}$ such that $iE_i \subset E_{n_i}$. Proceeding toward a contradiction, suppose A is not contained in any E_i. Then there is a

sequence $\{x_i\}$ in A such that $x_i \notin E_{n_i}$ for $i = 1, 2, \ldots$. According to [74, Theorem 1.30], the sequence $\{x_i / i\}$ $\mathcal{T}$-converges to 0. As x_i / i is not in E_i, this contradicts claim (4).

Conversely assume $A \subset E_n$ is $\mathcal{S}$-bounded, and select an arbitrary sequence $\{x_i\}$ in A, and a sequence $\{r_i\}$ in $\mathbb{R}$ converging to 0. Then $\{r_i x_i\}$ $\mathcal{S}$-converges to 0. Observe $r = \sup |r_i|$ is a real number. By property (ii), there is an integer $k \geq 1$ with $rE_n \cup (-rE_n) \subset E_k$. As E_k is a convex set containing 0, we see $\{r_i x_i\}$ is a sequence in E_k. According to claim (4), the sequence $\{r_i x_i\}$ $\mathcal{T}$-converges to 0. Applying [74, Theorem 1.30] again, we conclude A is $\mathcal{T}$-bounded.

(6) If $\{x_i\}$ is a $\mathcal{T}$-Cauchy sequence in X, it is $\mathcal{T}$-bounded. Claim (5) implies $\{x_i\}$ is a sequence in some E_n, and the set E_n is $\mathcal{T}$-compact. Thus $\{x_i\}$ contains a $\mathcal{T}$-convergent subsequence, and being $\mathcal{T}$-Cauchy, the sequence $\{x_i\}$ $\mathcal{T}$-converges.

(7) Suppose $\mathcal{S}$ is a sequential topology, and let $A \subset X$ be a set such that each $\mathcal{T}$-convergent sequence in A $\mathcal{T}$-converges to a point of A. Select an integer $n \geq 1$ and a sequence $\{x_i\}$ in $A \cap E_n$ that $\mathcal{S}$-converges to an $x \in X$. Since E_n is $\mathcal{S}$-closed, $x \in E_n$. According to what we just proved, $\{x_i\}$ $\mathcal{T}$-converges to x, and hence $x \in A$. As $\mathcal{S}$ is a sequential topology, the set $A \cap E_n$ is $\mathcal{S}$-closed. In view of claim (1), the set A is $\mathcal{T}$-closed. $\qquad\square$

Proposition 1.2.3. *Under the assumptions of Proposition 1.2.2, denote by $\mathcal{U}$ the largest topology in $X \times X$ for which all inclusions*

$$(E_n \times E_n, \mathcal{S} \otimes \mathcal{S}) \hookrightarrow (X \times X, \mathcal{U})$$

are continuous. Then $\mathcal{U} = \mathcal{T} \otimes \mathcal{T}$.

PROOF. To simplify the notation, denote by $\mathcal{S}^2$ and $\mathcal{T}^2$ the product topologies $\mathcal{S} \otimes \mathcal{S}$ and $\mathcal{T} \otimes \mathcal{T}$, respectively, and let $E^2 := E \times E$ for each set $E \subset X$. Observe $(A \cap B)^2 = A^2 \cap B^2$ for all $A, B \subset X$. In view of Proposition 1.2.2, both $(X^2, \mathcal{T}^2)$ and $(X^2, \mathcal{U})$ are topological vector spaces. Thus we only need to compare the $\mathcal{T}^2$ and $\mathcal{U}$ neighborhood bases at $0 \in X^2$.

Given $U \in \mathcal{T}$, find $V_n \in \mathcal{S}$ so that $U \cap E_n = V_n \cap E_n$ for $n = 1, 2, \ldots$. Observe V_n^2 belongs to $\mathcal{S}^2$ and $U^2 \cap E_n^2 = V_n^2 \cap E_n^2$. Proposition 1.2.2 implies U^2 belongs to $\mathcal{U}$, and we conclude $\mathcal{T}^2 \subset \mathcal{U}$.

Conversely, select a $W \in \mathcal{U}$ containing $0 \in X^2$, and find a W_1 in $\mathcal{S}^2$ so that $W \cap E_1^2 = W_1 \cap E_1^2$. Since $\mathcal{S}^2$ is a regular topology, there is a $U_1 \in \mathcal{S}$ containing $0 \in X$ such that $(\mathrm{cl}_{\mathcal{S}} U_1)^2$ is contained in W_1, and consequently

$$(\mathrm{cl}_{\,\mathbf{S}} U_1 \cap E_1)^2 \subset W_1 \cap E_1^2 \subset W \cap E_1^2 \,.$$

Suppose we have constructed $U_1, \ldots, U_k$ in $\mathbf{S}$ so that

$$\mathrm{cl}_{\,\mathbf{S}} U_{j-1} \cap E_{j-1} \subset U_j \quad \text{and} \quad (\mathrm{cl}_{\,\mathbf{S}} U_j \cap E_j)^2 \subset W \cap E_j^2$$

for $j = 2, \ldots, k$. Find a $W_{k+1} \in \mathbf{S}^2$ with $W_{k+1} \cap E_{k+1}^2 = W \cap E_{k+1}^2$, and observe $(\mathrm{cl}_{\,\mathbf{S}} U_k \cap E_k)^2 \subset W_{k+1}$. A simple compactness argument produces a $U_{k+1} \in \mathbf{S}$ such that

$$\mathrm{cl}_{\,\mathbf{S}} U_k \cap E_k \subset U_{k+1} \quad \text{and} \quad (\mathrm{cl}_{\,\mathbf{S}} U_{k+1})^2 \subset W_{k+1} \,.$$

The latter inclusion implies

$$(\mathrm{cl}_{\,\mathbf{S}} U_{k+1} \cap E_{k+1})^2 \subset W_{k+1} \cap E_{k+1}^2 = W \cap E_{k+1}^2 \,,$$

and the induction step is completed. Let $V = \bigcup_{i=1}^{\infty} (U_k \cap E_k)$. Clearly V^2 contains $0 \in X^2$, and as $\{U_k \cap E_k\}$ is an increasing sequence,

$$V^2 = \bigcup_{i=1}^{\infty} (U_k \cap E_k)^2 \subset W \,.$$

Moreover, for $n = 1, 2, \ldots$,

$$E_n \cap V = E_n \cap \bigcup_{k=n}^{\infty} (U_k \cap E_k) = E_n \cap \bigcup_{k=n}^{\infty} U_k \,.$$

Since each union $\bigcup_{k=n}^{\infty} U_k$ belongs to $\mathbf{S}$, the set V belongs to $\mathfrak{T}$. From this we infer $\mathcal{U} \subset \mathfrak{T}^2$. $\qquad\square$

Let X be a topological vector space. A *barrel* in X is any closed set $B \subset X$ that is symmetric, convex, and absorbing. If B is a compact barrel in X, then the sets nB, $n = 1, 2, \ldots$, satisfy conditions (i) and (ii) of Proposition 1.2.2. We say X is *barrelled* if it is locally convex and each barrel in X is a neighborhood of 0. It follows easily from the Baire category theorem [74, Theorem 2.2] every *Fréchet space* (i.e., a locally convex space which is completely metrizable) is barrelled; the converse is false, as Corollary 4.2.3, (ii) and Theorem 4.2.7 below demonstrate.

1.3. Measures

By a *measure* we always mean an outer measure in $\mathbb{R}^m$, i.e., a function μ defined on all subsets of $\mathbb{R}^m$ that satisfies the following conditions:

(i) $\mu(\emptyset) = 0$;

(ii) $\mu(B) \leq \mu(A)$ whenever $B \subset A \subset \mathbb{R}^m$;

(iii) $\mu(\bigcup_i A_i) \leq \sum_i \mu(A_i)$ for each countable collection $\{A_i\}$ of subsets of $\mathbb{R}^m$.

If μ is a measure and $E \subset \mathbb{R}^m$, we define a measure $\mu \, \llcorner \, E$ by letting

$$(\mu \, \llcorner \, E)(A) := \mu(A \cap E)$$

for each $A \subset \mathbb{R}^m$. If $\mu = \mu \, \llcorner \, E$, we say μ is a *measure in* E. A set $A \subset \mathbb{R}^m$ is called μ-*measurable* whenever

$$\mu(E) = \mu(E \cap A) + \mu(E - A)$$

for every set $E \subset \mathbb{R}^m$. The family of all μ-measurable sets is a σ-algebra in $\mathbb{R}^m$ denoted by $\mathfrak{M}_\mu$. A measure μ is called

- *locally finite* if $\mu(K) < \infty$ for each compact set K;
- σ-*finite* if $\mathbb{R}^m = \bigcup_{i=1}^\infty E_i$ and $\mu(E_i) < \infty$ for $i = 1, 2, \dots$;
- *Borel* if each Borel set is μ-measurable;
- *Borel regular* if it is Borel and each $E \subset \mathbb{R}^m$ is contained in a Borel set B with $\mu(B) = \mu(E)$;
- *Radon* if it is Borel regular and locally finite;
- *metric* if $\mu(A \cup B) = \mu(A) + \mu(B)$ for each pair of sets $A, B \subset \mathbb{R}^m$ with $\text{dist}(A, B) > 0$.

Each metric measure is Borel [22, Section 1.1, Theorem 5]. If μ is a Borel regular measure and A is a μ-measurable set with $\mu(A) < \infty$, then $\mu \, \llcorner \, A$ is a Radon measure [22, Section 1.1, Theorem 3].

The *support* of a measure μ is the closed set

$$\text{supp} \, \mu := \mathbb{R}^m - \bigcup \{ U \subset \mathbb{R}^m : U \text{ is open and } \mu(U) = 0 \} \,.$$

Since each subset of $\mathbb{R}^m$ has the Lindelöf property [21, Section 3.8 and Corollary 4.1.16], the countable subadditivity of μ, defined by property (iii) above, implies

$$\mu(\mathbb{R}^m - \text{supp} \, \mu) = 0 \,.$$

Let μ be a measure. A set $E \subset \mathbb{R}^m$ such that $\mu(E) = 0$ is called μ-*negligible*. Extended real-valued functions f and g defined on a set $E \subset \mathbb{R}^m$ are called μ-*equivalent* if they differ only on a μ-negligible subset of E. Subsets A and B of $\mathbb{R}^m$ are called μ-*equivalent* whenever their indicators are μ-equivalent, i.e., whenever the symmetric difference $A \triangle B$ is μ-negligible.

As a rule, we shall *not identify* an individual set $E \subset \mathbb{R}^m$, or an individual function f defined on E, with the μ-equivalence class determined by E, or f, respectively. On the other hand, spaces of sets or functions, are always viewed as spaces of the corresponding μ-equivalence classes. Given the context, the reader will have no difficulty to distinguish these viewpoints.

An extended real-valued function f defined on a Borel set $E \subset \mathbb{R}^m$ (or on a set $E \in \mathcal{M}_\mu$) is called *Borel measurable* (or *μ-measurable*) whenever for each Borel set $B \subset \mathbb{R}$, the set

$$f^{-1}(B) := \left\{ x \in E : f(x) \in B \right\}$$

is a Borel subset of $\mathbb{R}^m$ (or belongs to $\mathcal{M}_\mu$). The Borel measurability and μ-measurability of a map $\varphi : E \to \mathbb{R}^n$ are defined componentwise.

If f is a μ-measurable extended real-valued function defined on a set $E \in \mathcal{M}_\mu$, and the Lebesgue integral of f over E with respect to μ exists (finite or infinite), we denoted it by the usual symbol $\int_E f \, d\mu$. A μ-measurable function f defined on a set $E \in \mathcal{M}_\mu$ with

$$\int_E |f| \, d\mu < \infty$$

is called *μ-integrable* in E. The linear space of all μ-integrable functions in E is denoted by $L^1(E, \mu)$. If μ is a Borel measure and $\Omega \subset \mathbb{R}^m$ is an open set, the linear space $L^1_{\mathrm{loc}}(\Omega, \mu)$ of all *locally μ-integrable functions* in Ω is defined by Convention 1.1.2. Convention 1.1.1 defines the linear spaces $L^1(E, \mu; \mathbb{R}^n)$ and $L^1_{\mathrm{loc}}(\Omega, \mu; \mathbb{R}^n)$.

Let μ be a Radon measure, and let $h := (h_1, \ldots, h_n)$ be a map in $L^1(\mathbb{R}^m, \mu; \mathbb{R}^n)$. It follows from [22, Section 1.1, Lemma 1, (ii)] that for an arbitrary set $A \subset \mathbb{R}^m$, there is a Borel set B such that $A \subset B$ and $\mu(B) = \mu(A)$. Letting

$$\nu(A) := \left(\int_B h_1 \, d\mu, \ldots, \int_B h_n \, d\mu \right),$$

it is easy to see that $\nu(A)$ does not depend on the choice of B. Thus ν is a well defined map from the family of all subsets of $\mathbb{R}^m$ to $\mathbb{R}^n$. We call it a *vector-valued measure*, denoted by $\mu \llcorner h$. If $n = 1$, it is common to call $\mu \llcorner h$ a *signed measure*. The signed measures $\nu_i := \mu \llcorner h_i$ are called the *components* of the vector-valued measure ν. If B is a Borel set and $v : B \to \mathbb{R}^n$ is a μ-measurable map such that the integral $\int_B v \cdot h \, d\mu$

exists, we let

$$\int_B v \cdot d\nu := \int_B v \cdot h \, d\mu ,$$

and call this extended real number the integral of v over B with respect to the vector-valued measure ν. The function

$$|h| := \sqrt{h_1^2 + \cdots + h_m^2}$$

is μ-integrable, and it is easy to verify that $\mu \, \mathsf{L} \, |h|$ is a finite Radon measure, called the *total variation* of ν, denoted by $\|\nu\|$. For $i = 1, \dots, n$, let s_i be a function on $\mathbb{R}^m$ such that

$$s_i(x) = \frac{h_i(x)}{|h(x)|}$$

whenever $|h(x)| > 0$. Observe the map $s = (s_1, \dots, s_n)$ is $\|\nu\|$-integrable, $|s| = 1$ $\|\nu\|$-almost everywhere, and

$$\nu = \|\nu\| \, \mathsf{L} \, s . \tag{1.3.1}$$

Following techniques of [73, Chapter 6], it is not difficult to show that both $\|\nu\|$ and the $\|\nu\|$-equivalence class of s are uniquely determined by ν. Since $\|\nu\|$ is a Radon measure, we may assume that s is Borel measurable. The formula (1.3.1) is referred to as the *polar representation* of ν. Clearly, $\nu_i = \|\nu\| \, \mathsf{L} \, s_i$ for $i = 1, \dots, n$.

Remark 1.3.1. If μ is a Radon measure and $h \in L^1_{\mathrm{loc}}(\mathbb{R}^m, \mu; \mathbb{R}^n)$, then $\nu(A)$ can still be defined for all bounded subsets of $\mathbb{R}^m$. The signed measures ν_i, $i = 1, \dots, n$, are defined for all subsets of $\mathbb{R}^m$, and their values are extended real numbers. Thus Radon measures can be regarded as nonnegative signed measures. The concepts and notation associated with ν for $h \in L^1(\mathbb{R}^m, \mu; \mathbb{R}^n)$ remain meaningful when $h \in L^1_{\mathrm{loc}}(\mathbb{R}^m, \mu; \mathbb{R}^n)$; in particular, $\|\nu\|$ is a Radon measure, which need not be finite. Observe

$$\mu \, \mathsf{L} \, E = \mu \, \mathsf{L} \, \chi_E$$

for each μ-measurable set E.

Let μ be a Radon measure, and let ν be a signed measure. We say ν is *absolutely continuous* with respect to μ whenever $\nu(E) = 0$ for each μ-negligible set E. According to the *Radon-Nikodym-Lebesgue theorem* [22, Section 1.6.2, Theorem 3], there are a μ-negligible Borel set E and a Borel measurable function $h \in L^1_{\mathrm{loc}}(\mathbb{R}^m, \mu)$, unique up to μ-equivalence, such that

$$\nu = \nu \, \mathsf{L} \, E + \mu \, \mathsf{L} \, h .$$

We call $\nu \, \mathsf{L} \, E$ and $\mu \, \mathsf{L} \, h$ the *singular* and *absolutely continuous* parts of ν with respect to μ, respectively.

The *Lebesgue measure* in $\mathbb{R}^m$ is denoted by $\mathcal{L}^m$, however, for a set $E \subset \mathbb{R}^m$, we usually write $|E|$ instead of $\mathcal{L}^m(E)$. We say sets $A, B \subset \mathbb{R}^m$ *overlap* whenever $|A \cap B| > 0$.

Lebesgue measure plays an important role in our presentation. Unless specified otherwise, the words *measure, measurable, integrable,* or *Lebesgue integrable* for emphasis, *negligible, equivalent, singular,* and *absolutely continuous,* as well as the expressions *almost all* and *almost everywhere,* always refer to the measure $\mathcal{L}^m$. When no confusion is possible, we write $\int_A f(x)\,dx$ instead of $\int_A f\,d\mathcal{L}^m$.

Suppose E and Ω are, respectively, a measurable and an open subset of $\mathbb{R}^m$. We let

$$L^1(E) := L^1(E, \mathcal{L}^m) \quad \text{and} \quad L^1_{\mathrm{loc}}(\Omega) := L^1_{\mathrm{loc}}(\Omega, \mathcal{L}^m)\,.$$

The linear space of all bounded measurable functions defined on E is denoted by $L^\infty(E)$; the linear space $L^\infty_{\mathrm{loc}}(\Omega)$ is defined by Convention 1.1.2. Note the meanings of ∞ in L^∞ and C^∞ (see Section 1.1) are completely different. In $L^1(E)$ and $L^\infty(E)$, we use the usual norms

$$|f|_1 := \int_E |f(x)|\,dx \qquad \text{and} \qquad |f|_\infty := \operatorname*{ess\,sup}_{x \in E} |f(x)|\,,$$

respectively. The topologies in $L^1(E)$ and $L^\infty(E)$ induced by the norms $|\cdot|_1$ and $|\cdot|_\infty$ are called the L^1-*topology* and L^∞-*topology*, respectively. Aside from the commonly used spaces defined above, it will be convenient to introduce the linear space

$$L^1_\ell(E) := \left\{ f \upharpoonright E : f \in L^1_{\mathrm{loc}}(\mathbb{R}^m) \right\}\,. \tag{1.3.2}$$

Note a function f defined on E belongs to $L^1_\ell(E)$ if and only if f is integrable in each bounded measurable subset of E, or equivalently, if and only if the zero extension $\bar{f}$ of f belongs to $L^1_{\mathrm{loc}}(\mathbb{R}^m)$. If $\Omega \subset \mathbb{R}^m$ is an open set, then $L^1_{\mathrm{loc}}(\Omega) \subset L^1_\ell(\Omega)$ and this inclusion is proper if and only if $\emptyset \neq \Omega \neq \mathbb{R}^m$.

Given an integer $n \geq 1$, it is easy to verify the linear space $L^p(E; \mathbb{R}^n)$ where $p = 1$ or $p = \infty$ consists of all maps $\varphi := (f_1, \ldots, f_n)$ from E into $\mathbb{R}^n$ such that each f_i is measurable and the function $|\varphi| : x \mapsto |\varphi(x)|$ belongs to $L^p(E)$. The L^p norm of $\varphi \in L^p(E; \mathbb{R}^n)$, still denoted by $|\varphi|_p$, is defined as the L^p norm of $|\varphi|$. Explicitly,

$$|\varphi|_1 := \int_E \sqrt{f_1(x)^2 + \cdots + f_n(x)^2}\, dx\,,$$

$$|\varphi|_\infty := \operatorname*{ess\,sup}_{x \in E} \sqrt{f_1(x)^2 + \cdots + f_n(x)^2}\,.$$

In the rest of this section we recall the most basic facts about Hausdorff measures, including their definition. An extensive treatment of Hausdorff measures can be found in [46]; for a brief introduction we recommend [22, Chapter 2]. Given $s \geq 0$, let

$$\Gamma(s) := \int_0^\infty t^{s-1} e^{-t}\, dt \quad \text{and} \quad \alpha(s) := \frac{\Gamma\left(\frac{1}{2}\right)^s}{\Gamma\left(\frac{s}{2} + 1\right)}\,.$$

The map $s \mapsto \Gamma(s)$ is the usual gamma function, and the geometric meaning of the constant $\alpha(m)$ is given by the equality

$$\bigl|B(r)\bigr| = \alpha(m) r^m \tag{1.3.3}$$

which holds for each $r \geq 0$ [24, Sections 2.7.16 and 3.2.13]. Given $E \subset \mathbb{R}^m$ and a $\delta > 0$, let

$$\mathcal{H}^s_\delta(E) := \inf_{\mathcal{C}} \sum_{C \in \mathcal{C}} \alpha(s) \left[\frac{d(C)}{2}\right]^s$$

where $\mathcal{C}$ is a countable family of subsets of $\mathbb{R}^m$ such that $E \subset \bigcup \mathcal{C}$ and $d(C) < \delta$ for each $C \in \mathcal{C}$; here we let $0^0 := 1$ and $d(\emptyset)^s := 0$. The s-dimensional *Hausdorff measure* of E is the extended real number

$$\mathcal{H}^s(E) := \sup_{\delta > 0} \mathcal{H}^s_\delta(E)\,.$$

The map $\mathcal{H}^s : E \mapsto \mathcal{H}^s(E)$ is a Borel regular measure that is invariant with respect to translations and orthogonal transformations. The measure $\mathcal{H}^0$ is the usual *counting measure* in $\mathbb{R}^m$, and $\mathcal{H}^m = \mathcal{L}^m$.

The Hausdorff measures will be used extensively throughout the whole book. In particular, we shall need a lemma, which follows from [23, Theorem 5.1].

Lemma 1.3.2. *Let $E \subset \mathbb{R}^m$, and for an $s \geq 0$, let $\mathcal{H}^s(E) < c$. There is a constant $\kappa := \kappa(m) > 1$ such that for each $\delta > 0$, we can find a family $\mathcal{K}$ of dyadic cubes of diameters less than δ with*

$$E \subset \operatorname{int} \bigcup \mathcal{K} \quad \text{and} \quad \sum_{K \in \mathcal{K}} d(K)^s < \kappa c\,.$$

1.4. Covering theorems

It follows immediately from Zorn's lemma [36, Chapter 1, Section 5, Theorem 16] each family of sets contains a maximal (with respect to inclusion) disjoint subfamily. Recall each disjoint family of open subsets of $\mathbb{R}^m$ is countable. In particular, any nonoverlapping family of closed balls or cubes is countable. Throughout this section, for $B := B[x, r]$, we let $B^\bullet := B(x, 5r)$.

Theorem 1.4.1. *Let* $\mathcal{B}$ *be a family of closed balls with*

$$\sup\{d(B) : B \in \mathcal{B}\} < \infty .$$

There is a disjoint family $\mathcal{C} \subset \mathcal{B}$ *such that for each* $B \in \mathcal{B}$ *we can find a* $C \in \mathcal{C}$ *with* $B \cap C \neq \emptyset$ *and* $B \subset C^\bullet$. *In particular,*

$$\bigcup \mathcal{B} \subset \bigcup \{C^\bullet : C \in \mathcal{C}\} .$$

PROOF. Let $d := \sup\{d(B) : B \in \mathcal{B}\}$, and let

$$\mathcal{B}_k := \left\{ B \in \mathcal{B} : \frac{d}{2^k} < d(B) \leq \frac{d}{2^{k-1}} \right\}$$

for $k = 1, 2, \ldots$. Find a maximal disjoint family $\mathcal{C}_1 \subset \mathcal{B}_1$, and assuming the families $\mathcal{C}_2 \subset \mathcal{B}_2, \ldots, \mathcal{C}_{k-1} \subset \mathcal{B}_{k-1}$ have been defined, let $\mathcal{C}_k$ be the maximal disjoint subfamily of

$$\left\{ B \in \mathcal{B}_k : B \cap \bigcup_{i=1}^{k-1} \left(\bigcup \mathcal{C}_i \right) = \emptyset \right\} .$$

Now let $\mathcal{C} := \bigcup_{k=1}^\infty \mathcal{C}_k$, and choose a $B \in \mathcal{B}$. Then $B \in \mathcal{B}_k$ for a unique integer $k \geq 1$, and it follows from the maximality of $\mathcal{C}_k$ that B meets a $C \in \bigcup_{i=1}^k \mathcal{C}_i$. Since $d(B) \leq d2^{-k+1} < 2d(C)$, it is easy to see $B \subset C^\bullet$. $\square$

Corollary 1.4.2. *Let* $E \subset \mathbb{R}^m$ *have finite measure, and let* $\mathcal{B}$ *be a family of closed balls such that for each* $x \in E$ *and each* $\eta > 0$ *there is a* $B \in \mathcal{B}$ *with* $x \in B$ *and* $d(B) < \eta$. *Given* $\varepsilon > 0$, *there is a finite disjoint family* $\mathcal{C} \subset \mathcal{B}$ *with*

$$\left| E - \bigcup \mathcal{C} \right| < \varepsilon .$$

PROOF. Select an open set $U \subset \mathbb{R}^m$ so that $E \subset U$ and $|U| < \infty$. With no loss of generality, we may assume $\bigcup \mathcal{B} \subset U$. Let $\{C_1, C_2, \ldots\}$ be an enumeration of the disjoint family $\mathcal{C} \subset \mathcal{B}$ whose existence was established in Theorem 1.4.1. Select an integer $k \geq 1$, and an x in

$E - \bigcup_{i=1}^{k} C_i$. Since $\bigcup_{i=1}^{k} C_i$ is a closed set, x is contained in some ball $B \in \mathcal{B}$ which does not meet $\bigcup_{i=1}^{k} C_i$. By Theorem 1.4.1, there is a ball C_i with $i > k$ such that $B \cap C_i \neq \emptyset$, and $B \subset C_i^\bullet$. Thus

$$E - \bigcup_{i=1}^{k} C_i \subset \bigcup_{i>k} C_i^\bullet \qquad (*)$$

for $k = 1, 2, \ldots$. Now if $\mathcal{C}$ is finite, then $E - \bigcup \mathcal{C}$ is empty and the corollary is proved. If $\mathcal{C}$ is infinite, then

$$\sum_{i=1}^{\infty} |C_i| = \left| \bigcup_{i=1}^{\infty} C_i \right| \leq |U| < \infty.$$

Thus given $\varepsilon > 0$, there is an integer $k \geq 1$ with $\sum_{i>k} |C_i| < \varepsilon/5^m$, and inclusion $(*)$ implies

$$\left| E - \bigcup_{i=1}^{k} C_i \right| \leq \sum_{i>k} |C_i^\bullet| = 5^m \sum_{i>k} |C_i| < \varepsilon. \qquad \square$$

The *shape* of a set $C \subset \mathbb{R}^m$ is the number

$$s(C) := \begin{cases} \frac{|C|}{d(C)^m} & \text{if } d(C) > 0, \\ 0 & \text{otherwise.} \end{cases}$$

In view of (1.3.3), the shape of a ball is $\alpha(m)/2^m$. The *isodiametric inequality* states

$$s(C) \leq \frac{\alpha(m)}{2^m} \qquad (1.4.1)$$

for every set $C \subset \mathbb{R}^m$ [24, Corollary 2.10.33]. Note inequality (1.4.1) says that among all sets of a given diameter, the ball has the largest measure.

A family $\mathcal{C}$ of closed subsets of $\mathbb{R}^m$ is called a *Vitali cover* of a set $E \subset \mathbb{R}^m$ if there is a positive function γ defined on E such that given $x \in E$ and $\eta > 0$, we can find a set $C \in \mathcal{C}$ with

$$x \in C, \quad d(C) < \eta, \quad \text{and} \quad s(C) > \gamma(x).$$

Theorem 1.4.3 (Vitali). *Let $\mathcal{C}$ be a Vitali cover of a set $E \subset \mathbb{R}^m$. There is a countable disjoint family $\mathcal{D} \subset \mathcal{C}$ with*

$$\left| E - \bigcup \mathcal{D} \right| = 0.$$

Case 1. Assume $0 < |E| < \infty$, and $s(C) > \varepsilon\alpha(m)$ for each $C \in \mathbf{C}$ and a positive $\varepsilon < 1$. In each $C \in \mathbf{C}$ select a point x_C, and let $C^* := B[x_C, d(C)]$. Observe $C \subset C^*$ and

$$|C| = s(C)d(C)^m = \frac{s(C)}{\alpha(m)}|C^*|. \qquad (*)$$

Select an open set $U \subset \mathbb{R}^m$ so that $E \subset U$ and $|U| < (1+\varepsilon)|E|$, and let $\mathbf{C}^* := \{C^* \subset U : C \in \mathbf{C}\}$. By Corollary 1.4.2, there is a disjoint subcollection $\{C_1^*, \ldots, C_p^*\}$ of $\mathbf{C}^*$ such that

$$\left| E - \bigcup_{i=1}^{p} C_i^* \right| < \varepsilon^3 |E|.$$

If $\mathbf{C}_1 := \{C_1, \ldots, C_p\}$ and $\theta := 1 - \varepsilon^2 + \varepsilon^3$, then equality $(*)$ implies

$$\left| E - \bigcup \mathbf{C}_1 \right| \leq \left| E - \bigcup_{i=1}^{p} C_i^* \right| + \sum_{i=1}^{p} |C_i^* - C_i|$$

$$< \varepsilon^3 |E| + (1-\varepsilon) \sum_{i=1}^{p} |C_i^*|$$

$$\leq \varepsilon^3 |E| + (1-\varepsilon)|U| < \theta |E|.$$

The set $\bigcup \mathbf{C}_1$ is closed. Applying the previous argument to $E - \bigcup \mathbf{C}_1$ and the family of all sets in $\mathbf{C}$ disjoint from $\bigcup \mathbf{C}_1$, we obtain a finite collection $\mathbf{C}_2$ such that $\mathbf{C}_1 \cup \mathbf{C}_2$ is a disjoint subfamily of $\mathbf{C}$ and

$$\left| E - \left(\bigcup \mathbf{C}_1 \right) \cup \left(\bigcup \mathbf{C}_2 \right) \right| = \left| \left(E - \bigcup \mathbf{C}_1 \right) - \bigcup \mathbf{C}_2 \right|$$

$$< \theta \left| E - \bigcup \mathbf{C}_1 \right| < \theta^2 |E|.$$

Proceeding inductively, we construct finite collections $\mathbf{C}_1, \mathbf{C}_2, \ldots$ such that $\bigcup_{k=1}^{\infty} \mathbf{C}_k$ is a disjoint subfamily of $\mathbf{C}$ and

$$\left| E - \bigcup_{k=1}^{n} \left(\bigcup \mathbf{C}_k \right) \right| < \theta^n |E| \qquad (**)$$

for $n = 1, 2, \ldots$. Since $0 < \theta < 1$, we see that $\lim \theta^n = 0$. Note that inequality $(**)$ holds trivially if $|E| = 0$.

Case 2. Given any set $E \subset \mathbb{R}^m$, let γ be a positive function on E associated with the Vitali cover $\mathbf{C}$, and let

$$E_k := \left\{ x \in E \cap B(k) : \gamma(x) > \frac{\alpha(m)}{k} \right\}$$

for $k = 2, 3, \ldots$. Applying Case 1 to the set E_2 and the family

$$\left\{ C \in \mathcal{C} : C \subset B(2) \text{ and } s(C) > \frac{\alpha(m)}{2} \right\},$$

inequality $(\ast\ast)$ implies there is a finite disjoint collection $\mathcal{D}_2 \subset \mathcal{C}$ such that $|E_2 - \bigcup \mathcal{D}_2| < 1/2$. As $\bigcup \mathcal{D}_2$ is a closed set, we can apply Case 1 to the set $E_3 - \bigcup \mathcal{D}_2$ and the family

$$\left\{ C \in \mathcal{C} : C \subset B(3) - \bigcup \mathcal{D}_2 \text{ and } s(C) > \frac{\alpha(m)}{3} \right\}.$$

As before, we obtain a finite collection $\mathcal{D}_2$ such that $\mathcal{D}_1 \cup \mathcal{D}_2$ is a disjoint subfamily of $\mathcal{C}$, and

$$\left| E_3 - \left(\bigcup \mathcal{D}_2 \right) \cup \left(\bigcup \mathcal{D}_3 \right) \right| = \left| \left(E_3 - \bigcup \mathcal{D}_2 \right) - \bigcup \mathcal{D}_3 \right| < \frac{1}{3}.$$

Proceeding inductively, construct finite collections $\mathcal{D}_2, \mathcal{D}_3, \ldots$ such that $\mathcal{D} := \bigcup_{k=2}^{\infty} \mathcal{D}_k$ is a disjoint subfamily of $\mathcal{C}$, and

$$\left| E_n - \bigcup_{k=2}^{n} \left(\bigcup \mathcal{D}_k \right) \right| < \frac{1}{n}$$

for $n = 2, 3, \ldots$. Since E is the union of the increasing sequence $\{E_n\}$, we conclude

$$\left| E - \bigcup \mathcal{D} \right| = \lim_n \left| E_n - \bigcup \mathcal{D} \right| \leq \lim_n \left| E_n - \bigcup_{k=2}^{n} \left(\bigcup \mathcal{D}_k \right) \right| = 0. \qquad \square$$

For completeness we state another result, which generalizes the classical covering theorem of Besicovitch [22, Section 1.5.2]. Its proof is appreciably more involved than the proofs presented in this section. The interested reader is referred to [24, Theorem 2.8.14].

Theorem 1.4.4. *Let C be a compact convex symmetric set with nonempty interior, and let r be a bounded positive function defined on a set $E \subset \mathbb{R}^m$. For each $x \in E$, let*

$$C_x := x + r(x)C.$$

There are a positive integer $N := N(m, C)$ and subsets $E_1, \ldots, E_N$ of E such that each family $\mathcal{C}_i := \{ C_x : x \in E_i \}$ is disjoint and

$$E \subset \bigcup_{i=1}^{N} \left(\bigcup \mathcal{C}_i \right).$$

The following theorem, which rests on Theorem 1.4.4, is a Vitali type result that holds for any Borel measure. We shall apply it in the proof of Proposition 4.5.8 below.

Theorem 1.4.5. *Let C be a compact convex symmetric set with nonempty interior, and let $E \subset \mathbb{R}^m$. For each $x \in E$, let T_x be a set of positive real numbers with $\inf T_x = 0$. If μ is a Borel measure, then the family*

$$\mathcal{C} := \{x + tC : x \in E \text{ and } t \in T_x\}$$

contains a disjoint subfamily $\mathcal{D}$ such that

$$\mu\left(E - \bigcup \mathcal{D}\right) = 0.$$

PROOF. We prove only the special case when $0 < \mu(E) < \infty$; the general case follows as in the prove of Vitali's theorem. There is a bounded positive function r_1 defined on E with $C_x := x + r_1(x)C$ in $\mathcal{C}$ for each $x \in E$. By Theorem 1.4.4, there are an integer $N \geq 1$ and subsets $E_1, \dots, E_N$ of E such that each family $\mathcal{C}_i := \{C_x : x \in E_i\}$ is disjoint and $E \subset \bigcup_{i=1}^{N} (\bigcup \mathcal{C}_i)$. Thus

$$\mu(E) \leq \sum_{i=1}^{N} \mu\left(E \cap \bigcup \mathcal{C}_i\right),$$

and there is an integer k with $1 \leq k \leq N$ and $\mu(E)/N \leq \mu(E \cap \bigcup \mathcal{C}_k)$. The family $\mathcal{C}_k$ contains a finite subfamily $\mathcal{D}_1$ for which

$$\frac{1}{N+1}\mu(E) < \mu\left(E \cap \bigcup \mathcal{D}_1\right).$$

Since the set $D_1 := \bigcup \mathcal{D}_1$ is closed and since μ is a Borel measure, we have

$$\mu(E) = \mu(E \cap D_1) + \mu(E - D_1) > \frac{1}{N+1}\mu(E) + \mu(E - D_1),$$

and hence $\mu(E - D_1) < c\mu(E)$ where $c = N/(N+1)$.

If $\mu(E - D_1) = 0$, we are done. If $\mu(E - D_1) > 0$, proceed as follows. There is a bounded positive function r_2 defined on $E - D_1$ such that for each $x \in E - D_1$, the set $x + r_2(x)C$ belongs to $\mathcal{C}$ and is disjoint from the closed set D_1. Arguing as above, find a finite disjoint family $\mathcal{D}_2 \subset \mathcal{C}$ such that the closed set $D_2 := \bigcup \mathcal{D}_2$ is disjoint from D_1, and

$$\mu\big[E - (D_1 \cup D_2)\big] = \mu\big[(E - D_1) - D_2\big] < c\mu(E - D_1) < c^2\mu(E).$$

Now construct inductively finite subfamilies $\mathcal{D}_1, \mathcal{D}_2, \dots$ of $\mathcal{C}$ so that the family $\mathcal{D} := \bigcup_{i=1}^{\infty} \mathcal{D}_i$ is disjoint, and

$$\mu\left(E - \bigcup \mathcal{D}\right) \leq \mu\left[E - \bigcup_{i=1}^{p}\left(\bigcup \mathcal{D}_i\right)\right] < c^p\mu(E)$$

for $p = 1, 2, \dots$. As $0 < c < 1$, we conclude $\mu(E - \bigcup \mathcal{D}) = 0$. $\square$

1.5. Densities

Let $E \subset \mathbb{R}^m$. Given $x \in \mathbb{R}^m$, we call the numbers

$$\underline{\Theta}(E, x) := \liminf_{r \to 0+} \frac{|E \cap B[x, r]|}{|B[x, r]|}$$

and

$$\overline{\Theta}(E, x) := \limsup_{r \to 0+} \frac{|E \cap B[x, r]|}{|B[x, r]|}$$

the *lower* and *upper density* of E at x, respectively. The sets

$$\mathrm{int}_* E := \left\{ x \in \mathbb{R}^m : \underline{\Theta}(E, x) = 1 \right\},$$

$$\mathrm{cl}_* E := \left\{ x \in \mathbb{R}^m : \overline{\Theta}(E, x) > 0 \right\},$$

and $\partial_* E := \mathrm{cl}_* E - \mathrm{int}_* E$, which depend only on the equivalence class of E, are called the *essential interior, essential closure,* and *essential boundary* of E, respectively. Points of $\mathrm{int}_* E$ and $\mathbb{R}^m - \mathrm{cl}_* E$ are called, respectively, *density* and *dispersion points* of E. Clearly,

$$\mathrm{int}\, E \subset \mathrm{int}_* E \subset \mathrm{cl}_* E \subset \mathrm{cl}\, E \quad \text{and} \quad \partial_* E \subset \partial E, \tag{1.5.1}$$

and $\mathrm{int}_*(\mathbb{R}^m - E) = \mathbb{R}^m - \mathrm{cl}_* E$ whenever E is a measurable set.

Proposition 1.5.1. *If $E \subset \mathbb{R}^m$, then $\mathrm{cl}_* E$, $\mathrm{int}_* E$, and $\partial_* E$ are Borel sets.*

PROOF. The function

$$\varphi : (x, r) \mapsto \frac{|E \cap B[x, r]|}{|B[x, r]|}$$

is continuous on $\mathbb{R}^m \times (0, \infty)$. Let $r_1, r_2, \ldots$ be an enumeration of all positive rationals, and let $\varphi_i(x) = \varphi(x, r_i)$ for each $x \in \mathbb{R}^m$ and each integer $i \geq 1$. Since

$$\underline{\Theta}(E, x) = \lim_{n \to \infty} \inf_{r_i < 1/n} \varphi_i(x),$$

$$\overline{\Theta}(E, x) = \lim_{n \to \infty} \sup_{r_i < 1/n} \varphi_i(x)$$

for each $x \in \mathbb{R}^m$, the maps $x \mapsto \underline{\Theta}(E, x)$ and $x \mapsto \overline{\Theta}(E, x)$ are Baire functions, and the proposition follows. $\square$

The next result, called the *density theorem*, is very important. True to its name, it shows that almost all points of a set $E \subset \mathbb{R}^m$ are density points of E.

Theorem 1.5.2. *For each $E \subset \mathbb{R}^m$, the set $E - \text{int}_* E$ is negligible.*

PROOF. Assume first $|E| < \infty$, and for $0 < \theta \leq 1$, let

$$E_\theta := \left\{ x \in E : \underline{\Theta}(E, x) < \theta \right\}.$$

Choose an $\varepsilon > 0$ and find an open set $U \subset \mathbb{R}^m$ with $E_\theta \subset U$ and $|U| < |E_\theta| + \varepsilon$. Applying Corollary 1.4.2 to the family $\mathcal{B}$ of all balls $B[x, r] \subset U$ for which $x \in E_\theta$ and $\left| E \cap B[x, r] \right| < \theta \left| B[x, r] \right|$, we obtain disjoint balls $B_1, \ldots, B_k$ in $\mathcal{B}$ such that $\left| E_\theta - \bigcup_{i=1}^{k} B_i \right| < \varepsilon$. Thus

$$|E_\theta| \leq \sum_{i=1}^{k} |E_\theta \cap B_i| + \left| E_\theta - \bigcup_{i=1}^{k} B_i \right|$$

$$< \theta \sum_{i=1}^{k} |B_i| + \varepsilon \leq \theta |U| + \varepsilon < \theta |E_\theta| + (1 + \theta)\varepsilon,$$

and letting $\varepsilon \to 0$ yields $|E_\theta| = 0$ whenever $\theta < 1$. Since

$$E_1 = \bigcup_{n=1}^{\infty} E_{n/(n+1)} \,,$$

the proposition holds for any set E of finite measure.

Now let $E \subset \mathbb{R}^m$ be arbitrary, and for $n = 1, 2, \ldots$, let $B_n = B(n)$. By the previous argument,

$$\underline{\Theta}(E, x) = \underline{\Theta}(E \cap B_n, x) = 1$$

for almost all $x \in E \cap B_n$. As $E = \bigcup_{n=1}^{\infty} (E \cap B_n)$, the proposition holds in general.   □

Corollary 1.5.3. *If E is a measurable set, then*

$$|E \bigtriangleup \text{int}_* E| = |E \bigtriangleup \text{cl}_* E| = 0.$$

PROOF. As E is measurable, the density theorem implies $|E - \text{int}_* E| = 0$ and also

$$|\text{cl}_* E - E| = \left| (\mathbb{R}^m - E) - (\mathbb{R}^m - \text{cl}_* E) \right|$$
$$= \left| (\mathbb{R}^m - E) - \text{int}_* (\mathbb{R}^m - E) \right| = 0.$$

As $\text{int}_* E \subset \text{cl}_* E$, the corollary follows.   □

To see that the measurability assumption in Corollary 1.5.3 is essential, consider a subset E of a cell K such that both E and $K - E$ meet each uncountable compact subset of K. Employing the axiom of choice, such a set E, called the *Bernstein set*, is easy to construct by transfinite induction [58, Chapter 5]. If B is a Borel subset of K containing either E or $K - E$, then $K - B$ contains no uncountable compact set. Thus $|K - B| = 0$ by [22, Section 1.1, Theorem 4], and from the Borel regularity of Lebesgue measure, we infer $|E| = |K - E| = |K|$. It follows the set E is not measurable. Since the density theorem implies that $|\text{int}_* E| = |K|$, we have

$$0 < |K| = |K - E| \le |K - \text{int}_* E| + |\text{int}_* E - E| = |\text{int}_* E - E|.$$

Proposition 1.5.4. *Let $E \subset \mathbb{R}^m$. There is a Borel set B such that $E \subset B$ and $|E \cap C| = |B \cap C|$ for each measurable set C. In particular,*

$$\underline{\Theta}(E, x) = \underline{\Theta}(B, x) \quad and \quad \overline{\Theta}(E, x) = \overline{\Theta}(B, x)$$

for each $x \in \mathbb{R}^m$.

PROOF. The set E is the union of the sets $E_n := E \cap K[n]$, $n = 1, 2, \dots$, and there are bounded Borel sets A_n with $E_n \subset A_n$ and $|A_n| = |E_n|$. Now $B_n := \bigcap_{k=n}^{\infty} A_k$ is a bounded Borel set containing E_n. If C is a measurable set, the inequality

$$|E_n| = |E_n \cap C| + |E_n - C| \le |B_n \cap C| + |B_n - C|$$
$$= |B_n| \le |A_n| = |E_n|$$

involves only real numbers, and so $|E_n \cap C| = |B_n \cap C|$. The Borel set $B := \bigcup_{n=1}^{\infty} B_n$ contains E, and since the sequences $\{E_n\}$ and $\{B_n\}$ are increasing, we have $|E \cap C| = |B \cap C|$. $\qquad\square$

The following corollary is an immediate consequence of Proposition 1.5.4 and Corollary 1.5.3.

Corollary 1.5.5. *Let $E \subset \mathbb{R}^m$. There is a Borel set B containing E such that*

$$\text{int}_* E = \text{int}_* B \quad and \quad \text{cl}_* E = \text{cl}_* B.$$

In particular, $\partial_* E = \partial_* B$ is a negligible set.

Lemma 1.5.6. *Let $x \in \mathbb{R}^m$, and let $\{C_i\}$ be a sequence of measurable sets of positive measure such that*

$$\lim d(C_i \cup \{x\}) = 0 \quad and \quad \liminf s(C_i \cup \{x\}) > 0.$$

Then for each set $E \subset \mathbb{R}^m$ and each $x \in \mathbb{R}^m$,

$$\lim \frac{|E \cap C_i|}{|C_i|} = \begin{cases} 1 & \text{if } x \in \text{int}_* E, \\ 0 & \text{if } x \notin \text{cl}_* E. \end{cases}$$

PROOF. In view of Proposition 1.5.4 and Corollary 1.5.5, we may assume that E is a Borel set. Suppose $x \notin \mathrm{cl}_* E$, and choose a positive $s < \liminf s(C_i \cup \{x\})$. If $d_i := d(C_i \cup \{x\})$ and $B_i := B(x, d_i)$, then $\liminf |C_i|/|B_i| \geq s/\alpha(m)$. Thus

$$
0 = \lim \frac{|E \cap B_i|}{|B_i|} \geq \limsup \left(\frac{|E \cap C_i|}{|C_i|} \cdot \frac{|C_i|}{|B_i|} \right)
$$

$$
\geq \frac{s}{\alpha(m)} \limsup \frac{|E \cap C_i|}{|C_i|} \geq 0 \,,
$$

and we have $\lim \big(|E \cap C_i|/|C_i| \big) = 0$.

Now if $x \in \mathrm{int}_* E$ then $x \notin \mathrm{cl}_* (\mathbb{R}^m - E)$, and the first part of our proof implies

$$
0 = \lim \frac{|(\mathbb{R}^m - E) \cap C_i|}{|C_i|} = \lim \frac{|C_i| - |E \cap C_i|}{|C_i|} = 1 - \lim \frac{|E \cap C_i|}{|C_i|} \,.
$$

$\square$

Corollary 1.5.7. *If A is a measurable set and $B \subset \mathbb{R}^m$, then*

$$
\mathrm{cl}_* A \cap \mathrm{int}_* B \subset \mathrm{cl}_* (A \cap B) \,,
$$

$$
\mathrm{int}_* A \cap \mathrm{int}_* B = \mathrm{int}_* (A \cap B) \,.
$$

PROOF. Select an $x \in \mathrm{cl}_* A \cap \mathrm{int}_* B$. As $x \in \mathrm{cl}_* A$, there is a sequence $\{r_i\}$ of positive numbers such that $\lim r_i = 0$ and

$$
\frac{|A \cap B_i|}{|B_i|} \geq c > 0
$$

for $B_i := B[x, r_i]$ and $i = 1, 2, \ldots$. Thus

$$
s\big[(A \cap B_i) \cup \{x\} \big] \geq \frac{|A \cap B_i|}{(2r_i)^m} = \frac{\alpha(m)}{2^m} \cdot \frac{|A \cap B_i|}{|B_i|} > \frac{\alpha(m)}{2^m} c
$$

for $i = 1, 2, \ldots$. Since

$$
\frac{|(A \cap B) \cap B_i|}{|B_i|} = \frac{|B \cap (A \cap B_i)|}{|A \cap B_i|} \cdot \frac{|A \cap B_i|}{|B_i|} \,,
$$

Lemma 1.5.6 implies $x \in \mathrm{cl}_* (A \cap B)$; moreover, $x \in \mathrm{int}_* (A \cap B)$ whenever $x \in \mathrm{int}_* A$. As $\mathrm{int}_* (A \cap B) \subset \mathrm{int}_* A \cap \mathrm{int}_* B$, the corollary is proved. $\square$

We say a set $E \subset \mathbb{R}^m$ is *essentially closed* or *essentially open* when $\mathrm{cl}_* E$ is closed or $\mathrm{int}_* E$ is open, respectively. A subset of $\mathbb{R}^m$ that is simultaneously essentially closed and essentially open is called *essentially clopen*.

Observation 1.5.8. *Let E be an essentially clopen set such that*

$$\text{int}_* E \subset E \subset \text{cl}_* E.$$

Then $\text{cl}\, E = \text{cl}_* E$, $\text{int}\, E = \text{int}_* E$, *and* $\partial E = \partial_* E$.

PROOF. Since E is essentially clopen, we have

$$\text{int}_* E \subset \text{int}\, E \subset \text{cl}\, E \subset \text{cl}_* E,$$

and the observation follows from inclusions (1.5.1). □

Proposition 1.5.9. *A set E is essentially clopen if and only if $\partial_* E$ is closed.*

PROOF. In view of Corollary 1.5.5, it suffices to prove the proposition for a measurable set E. As the converse is obvious, assume $\partial_* E$ is closed. Working toward a contradiction, suppose there is an $x \in \text{int}_* E$ that is not an interior point of $\text{int}_* E$. Choose an $s_0 > 0$ so that $B_0 := B[x, s_0]$ does not meet $\partial_* E$, and find a $y \in B(x, s_0)$ that is a dispersion point of E. Let $r_1 := |y - x|$, and select a positive $s_1 \leq \min\{s_0 - r_1, 1\}$ with

$$\frac{|E \cap B[x, s_1]|}{|B[x, s_1]|} > \frac{2}{3} \quad \text{and} \quad \frac{|E \cap B[y, s_1]|}{|B[y, s_1]|} < \frac{1}{3}.$$

Since the function $z \mapsto |E \cap B[z, s_1]|/|B[z, s_1]|$ is continuous, the connected set $B[x, r_1]$ contains a point z_1 such that

$$\frac{|E \cap B[z_1, s_1]|}{|B[z_1, s_1]|} = \frac{1}{2}.$$

The closed ball $B_1 := B[z_1, s_1]$ is contained in B_0. From the density theorem and

$$|B(z_1, s_1) \cap E| = |B(z_1, s_1) - E| = \tfrac{1}{2}|B(z_1, s_1)| > 0$$

we infer that the open ball $B(z_1, s_1)$ contains a density point u and a dispersion point v. Let $r_2 := \max\{|u - z_1|, |v - z_1|\}$, and choose a positive $s_2 \leq \min\{s_1 - r_2, 1/2\}$ with

$$\frac{|E \cap B[u, s_2]|}{|B[u, s_2]|} > \frac{2}{3} \quad \text{and} \quad \frac{|E \cap B[v, s_2]|}{|B[v, s_2]|} < \frac{1}{3}.$$

As the function $z \mapsto |E \cap B[z, s_2]|/|B[z, s_2]|$ is continuous, the connected set $B[z_1, r_2]$ contains a point z_2 with

$$\frac{|E \cap B[z_2, s_2]|}{|B[z_2, s_2]|} = \frac{1}{2}.$$

The closed ball $B_2 := B[z_2, s_2]$ is contained in B_1. Inductively, we obtain a decreasing sequence $\{B_i\}$ of closed balls such that $\lim d(B_i) = 0$ and $|E \cap B_i|/|B_i| = 1/2$ for $i = 1, 2, \ldots$. It follows from Lemma 1.5.6 the unique point z common to all closed balls B_i belongs to $B_0 \cap \partial_* E$. This contradiction shows that $\mathrm{int}_* E$ is an open set. Now $\partial_*(\mathbb{R}^m - E) = \partial_* E$, and so

$$\mathbb{R}^m - \mathrm{cl}_* E = \mathrm{int}_*(\mathbb{R}^m - E)$$

is also an open set. $\qquad\square$

Proposition 1.5.10. *Let $\mathcal{B}$ be a family of closed subsets of $\mathbb{R}^m$. If $\mathrm{cl}_* B = B$ for each $B \in \mathcal{B}$, then $E := \bigcup \mathcal{B}$ is a measurable set.*

PROOF. For each point x in E, find a set B_x in $\mathcal{B}$ so that $x \in B_x$, and let $\gamma(x) := \overline{\Theta}(B_x, x)/2$. Observe $\gamma(x) > 0$, since $B_x = \mathrm{cl}_* B_x$. The family

$$\mathcal{C} := \big\{ B_x \cap B[x, r] : x \in E \text{ and } r > 0 \big\}$$

consists of closed subsets of E. Given $x \in E$, there is a sequence $\{r_k\}$ of positive numbers converging to zero such that for $k = 1, 2, \ldots$,

$$\big| B_x \cap B[x, r_k] \big| > \gamma(x) \big| B[x, r_k] \big|$$

and consequently

$$s\big(B_x \cap B[x, r_k]\big) \geq \frac{\big| B_x \cap B[x, r_k] \big|}{(2r_k)^m} > \frac{\alpha(m)}{2^m} \gamma(x) \, .$$

It follows $\mathcal{C}$ is a Vitali cover of E. By Vitali's theorem, there is a disjoint countable family $\mathcal{D} \subset \mathcal{C}$ with $\big| E - \bigcup \mathcal{D} \big| = 0$. As $\bigcup \mathcal{D}$ is a measurable subset of E, the set E is measurable. $\qquad\square$

The last theorem of this section summarizes the main density properties of the measure $\mathcal{H}^s$. Its proof can be found [46, Theorem 6.2].

Theorem 1.5.11. *Let $s \geq 0$. If $E \subset \mathbb{R}^m$ and $\mathcal{H}^s(E) < \infty$, then*

$$2^{-s} \leq \limsup_{r \to 0+} \frac{\mathcal{H}^s\big(E \cap B[x, r]\big)}{r^s} \leq 1$$

for $\mathcal{H}^s$-almost all $x \in E$. If, in addition, E is $\mathcal{H}^s$-measurable, then

$$\lim_{r \to 0+} \frac{\mathcal{H}^s\big(E \cap B[x, r]\big)}{r^s} = 0$$

for $\mathcal{H}^s$-almost all $x \in \mathbb{R}^m - E$.

1.6. Lipschitz maps

Let $n \geq 1$ be an integer, and let $E \subset \mathbb{R}^m$. A map $\phi : E \to \mathbb{R}^n$ is called *Lipschitz* if there is a real number c such that

$$|\phi(x) - \phi(y)| \leq c|x - y|$$

for all $x, y \in E$. The least number c for which the above inequality holds is called the *Lipschitz constant* of ϕ, denoted by $\mathrm{Lip}(\phi)$. A bijective Lipschitz map $\phi : E \to \mathbb{R}^n$ is called a *lipeomorphism* whenever the inverse map $\phi^{-1} : \phi(E) \to \mathbb{R}^m$ is Lipschitz. Given a Lipschitz map $\phi : E \to \mathbb{R}^n$, it is easy to verify that

$$\mathcal{H}^s\big[\phi(E)\big] \leq \mathrm{Lip}(\phi)^s \, \mathcal{H}^s(E) \tag{1.6.1}$$

for each $s \geq 0$ [22, Section 2.4, Theorem 1].

Theorem 1.6.1 (Kirszbraun). *Let $E \subset \mathbb{R}^m$, and let $\phi : E \to \mathbb{R}^n$ be a Lipschitz map. There is a Lipschitz map $\psi : \mathbb{R}^m \to \mathbb{R}^n$ such that $\psi \upharpoonright E = \phi$ and $\mathrm{Lip}(\psi) = \mathrm{Lip}(\phi)$.*

Theorem 1.6.2 (Rademacher). *Each Lipschitz map $\phi : \mathbb{R}^m \to \mathbb{R}^n$ is differentiable almost everywhere.*

Proofs of the Kirszbraun and Rademacher theorems can be found in [24, Theorems 2.10.43 and 3.1.6]. A weaker version of Kirszbraun's theorem, which suffices for many applications and is easier to proof, can be found in [22, Section 3.1., Theorem 1]. An alternative proof of Rademacher's theorem is given in [22, Section 3.1.2, Theorem 2].

Let $E \subset \mathbb{R}^m$ be a measurable set, let $\phi : E \to \mathbb{R}^m$ be a Lipschitz map, and let $\psi = (f_1, \ldots, f_m)$ be a Lipschitz extension of ϕ to $\mathbb{R}^m$ (Kirszbraun's theorem). By Rademacher's theorem, the matrix

$$D\psi := \begin{pmatrix} \frac{\partial f_1}{\partial x_1} & \cdots & \frac{\partial f_1}{\partial x_m} \\ \vdots & \cdots & \vdots \\ \frac{\partial f_m}{\partial x_1} & \cdots & \frac{\partial f_m}{\partial x_m} \end{pmatrix}$$

is defined almost everywhere, and we let

$$J_\phi(x) := \big|\det D\psi(x)\big|$$

for each $x \in E$ at which $D\psi(x)$ is defined. The next lemma shows that almost everywhere in E the function J_ϕ is uniquely determined by the map ϕ and does not depend on the extension ψ.

Lemma 1.6.3. *Let E be a measurable subset of an open set $U \subset \mathbb{R}^m$, and let f and g be measurable functions defined on U that have partial derivatives at each $x \in E$. If $f \restriction E = g \restriction E$, then*

$$\frac{\partial f}{\partial x_i} = \frac{\partial g}{\partial x_i}, \quad i = 1, \ldots, m,$$

almost everywhere in E.

PROOF. Observe the set

$$N := \left\{ x \in E : \frac{\partial f}{\partial x_1}(x) \neq \frac{\partial g}{\partial x_1}(x) \right\}$$

is measurable. Assuming $|N| > 0$ and applying Fubini's theorem [22, Section 1.4, Theorem 1], we can find a $\xi \in \mathbb{R}^{m-1}$ for which the set $N^\xi := \left\{ t \in \mathbb{R} : (t, \xi) \in N \right\}$ is uncountable. It follows that there is a $t \in N^\xi$ and a sequence $\{t_k\}$ in $N^\xi - \{t\}$ with $\lim t_k = t$. Thus

$$\frac{\partial f}{\partial x_1}(t, \xi) = \lim \frac{f(t_k, \xi) - f(t, \xi)}{t_k - t}$$

$$= \lim \frac{g(t_k, \xi) - g(t, \xi)}{t_k - t} = \frac{\partial g}{\partial x_1}(t, \xi),$$

a contradiction. The lemma follows by symmetry. $\qquad\qquad\square$

Theorem 1.6.4. *Let $\phi : E \to \mathbb{R}^m$ be a Lipschitz map of a measurable set $E \subset \mathbb{R}^m$. The set $\phi(E)$ is measurable, and if f is a nonnegative measurable function on E, then*

$$\int_E f(x) J_\phi(x)\, dx = \int_{\phi(E)} \sum \left\{ f(x) : x \in \phi^{-1}(y) \right\} dy.$$

In particular, $\int_E J_\phi(x)\, dx = \big|\phi(E)\big|$ whenever ϕ is injective.

Theorem 1.6.4, known as the *area theorem*, follows immediately from [22, Section 3.3.3, Theorem 2].

Theorem 1.6.5. *If $E \subset \mathbb{R}^m$ and $\phi : E \to \mathbb{R}^m$ is a lipeomorphism, then ϕ has a unique extension to a lipeomorphism $\bar\phi$ from $\operatorname{cl} E$ onto $\operatorname{cl} \phi(E)$. The equalities*

$$\operatorname{Lip}(\bar\phi) = \operatorname{Lip}(\phi) \quad and \quad \operatorname{Lip}(\bar\phi^{-1}) = \operatorname{Lip}(\phi^{-1})$$

hold, and if E is measurable, then

$$\bar\phi(\operatorname{cl}_* E) = \operatorname{cl}_* \phi(E), \quad \bar\phi(\operatorname{int}_* E) = \operatorname{int}_* \phi(E), \quad \bar\phi(\partial_* E) = \partial_* \phi(E).$$

The first three claims of Theorem 1.6.5 are easy to verify, and so is the equality $\bar{\phi}(\mathrm{cl}_* E) = \mathrm{cl}_* \phi(E)$ [64, Lemma 6.5]. Among the remaining two equalities, each implies the other, but neither is easy to prove. The interested reader is referred to [13] or [27, Theorem A.9].

A map $\phi : E \to \mathbb{R}^n$ is called *Lipschitz at a point* $x \in E$ if there are numbers c_x and $\delta_x > 0$ such that

$$\left| \phi(x) - \phi(y) \right| \le c_x |x - y|$$

for each $y \in E \cap B(x, \delta_x)$. A map that is Lipschitz at each point of its domain need not be Lipschitz, even if the domain is compact. Indeed, let f be a continuous function defined on $[0, 1/4]$ that is linear in the cells

$$[2^{-n-1}, 2^{-n} - 2^{-2n}] \quad \text{and} \quad [2^{-n} - 2^{-2n}, 2^{-n}],$$

and for which

$$f(2^{-n}) := 2^{-n} \quad \text{and} \quad f(2^{-n} - 2^{-2n-2}) := \tfrac{3}{4} \cdot 2^{-n}$$

where $n = 2, 3, \dots$. Then f is strictly increasing, Lipschitz at each $x \in [0, 1/4]$, but not Lipschitz.

Theorem 1.6.6 (Stepanoff). *Suppose $\Omega \subset \mathbb{R}^m$ is an open set, and let $\phi : \Omega \to \mathbb{R}^n$ be a map that is Lipschitz at each point of a set $E \subset \Omega$. Then ϕ is differentiable almost everywhere in E.*

Theorem 1.6.7 (Whitney). *Suppose $A \subset \mathbb{R}^m$ be a closed set, and let $\phi \in C(A; \mathbb{R}^n)$ be Lipschitz at each point of a set $E \subset A$. There is a $\psi \in C(\mathbb{R}^m; \mathbb{R}^n)$ such that $\psi \restriction A = \phi$ and ψ is Lipschitz at each point of E. In particular, ψ is differentiable almost everywhere in E.*

Stepanoff's theorem is proved in [24, Theorem 3.1.9], and the proof of Whitney's theorem is an easy modification of the proof presented in [80, Chapter 6, Theorem 3].

1.7. BV functions

Let Ω be an open subset of $\mathbb{R}^m$. The *variation* of a function $f \in L^1_{\mathrm{loc}}(\Omega)$ is the extended real number

$$\|f\| := \sup_v \int_\Omega f(x) \mathrm{div}\, v(x) \, dx \qquad (1.7.1)$$

where $v \in C_c^1(\Omega; \mathbb{R}^m)$ and $|v|_\infty \leq 1$. A function $f \in L^1(\Omega)$ such that $\|f\| < \infty$ is called a *function of bounded variation* in Ω, or simply a *BV function* in Ω. Since

$$\|f + g\| \leq \|f\| + \|g\| \quad \text{and} \quad \|cf\| = |c| \cdot \|f\| \qquad (1.7.2)$$

for all $f, g \in L_{\text{loc}}^1(\Omega)$ and each $c \in \mathbb{R}$, the family of all BV functions in Ω is a linear space, denoted by $BV(\Omega)$. If f is a BV function in Ω, then $f \restriction U$ is a BV function in U for each open set $U \subset \Omega$. Hence using Convention 1.1.2, we can define the space $BV_{\text{loc}}(\Omega)$ whose elements are called *locally BV functions* in Ω.

The following theorem summarizes important elementary properties of BV functions. Its proof can be found in [22, Section 5.2].

Theorem 1.7.1. *For an open set $\Omega \subset \mathbb{R}^m$, the following is true.*

(1) *Let $f \in L_{\text{loc}}^1(\Omega)$, and let $\{f_i\}$ be a sequence in $L_{\text{loc}}^1(\Omega)$ such that*

$$\lim \int_K |f_i - f|\, d\mathcal{L}^m = 0$$

for each compact set $K \subset \Omega$. Then $\|f\| \leq \liminf \|f_i\|$.

(2) *For each $f \in BV(\Omega)$, there is a sequence $\{f_i\}$ in $BV(\Omega) \cap C^\infty(\Omega)$ with*

$$\lim |f_i - f|_1 = 0 \quad \text{and} \quad \lim \|f_i\| = \|f\|.$$

(3) *Let $\{f_i\}$ be a sequence in $BV(\Omega)$ for which*

$$\bigcup_{i=1}^{\infty} \operatorname{supp} f_i \subset \Omega \quad \text{and} \quad \sup\left(|f_i|_1 + \|f_i\|\right) < \infty.$$

If Ω is bounded, then there is a subsequence $\{f_{i_j}\}$ of $\{f_i\}$ and an $f \in BV(\Omega)$ such that $\lim |f_{i_j} - f|_1 = 0$.

Given an open set $\Omega \subset \mathbb{R}^m$, a function $f \in L_{\text{loc}}^1(\Omega)$ belongs to $BV_{\text{loc}}(\Omega)$ if and only if there is a vector-valued measure Df such that the total variation $\|Df\|$ of Df is a Radon measure in Ω, and

$$\int_\Omega f \operatorname{div} v\, d\mathcal{L}^m = -\int_\Omega v \cdot d(Df) \qquad (1.7.3)$$

for each $v \in C_c^1(\Omega; \mathbb{R}^m)$ [22, Section 5.1, Theorem 1]. The vector-valued measure Df, called the *distributional gradient* of f, is uniquely determined by f; the components of Df are denoted by $D_1 f, \ldots, D_m f$. The

Radon measure $\|Df\|$ is called the *variational measure* of f, and the equality

$$\|f\| = \|Df\|(\Omega) \tag{1.7.4}$$

is easy to establish. In view of Remark 1.3.1 and the Radon-Nikodym theorem, there is a negligible set E and a Borel measurable locally integrable map

$$\nabla f := \left(\frac{\partial f}{\partial x_1}, \ldots, \frac{\partial f}{\partial x_m} \right)$$

from $\mathbb{R}^m$ to $\mathbb{R}^m$, unique up to equivalence, such that

$$Df = Df \llcorner E + \mathcal{L}^m \llcorner \nabla f \quad \text{and} \quad D_i f = D_i f \llcorner E + \mathcal{L}^m \llcorner \frac{\partial f}{\partial x_i}$$

for $i = 1, \ldots, m$. If $f \in BV_{\mathrm{loc}}(\Omega) \cap C^1(\Omega)$, then $E = \emptyset$ and $\nabla f \upharpoonright \Omega$ is the usual gradient of f.

Remark 1.7.2. Let $m = 1$. If $\Omega \subset \mathbb{R}$ is an open set, then a function $f : \Omega \to \mathbb{R}$ belongs to $BV_{\mathrm{loc}}(\Omega)$ if and only if f is equivalent to a function $g : \Omega \to \mathbb{R}$ whose *classical variation* [37, Section 5.4] is finite on each subcell of Ω. This is a consequence of [24, Theorem 4.5.9, (23)], partially quoted in Theorem 4.4.10 below.

The next theorem follows from [22, Section 6.1.1, Theorem 1] by a routine application of Hölder's inequality [73, Theorem 3.5] if $m \geq 2$, and from Remark 1.7.2 if $m = 1$. It shows that ∇f displays a gradient-like behavior for any function $f \in BV_{\mathrm{loc}}(\Omega)$.

Theorem 1.7.3. *If $\Omega \subset \mathbb{R}^m$ is an open set and $f \in BV_{\mathrm{loc}}(\Omega)$, then*

$$\lim_{r \to 0+} \frac{1}{r^{m+1}} \int_{B(x,r)} |f(y) - f(x) - \nabla f(x) \cdot (y - x)| \, dy = 0$$

for almost all $x \in \Omega$.

Let $\Omega \subset \mathbb{R}^m$ be an open set. Using Theorem 1.7.1, (1) and (2), it is easy to verify

$$\|fg\| \leq \|f\| \cdot |g|_\infty + \|g\| \cdot |f|_\infty \tag{1.7.5}$$

for each pair f, g of bounded functions in $BV(\Omega)$. Consequently, the linear spaces

$$BV^\infty(\Omega) := BV(\Omega) \cap L^\infty(\Omega),$$
$$BV_{\mathrm{loc}}^\infty(\Omega) := BV_{\mathrm{loc}}(\Omega) \cap L_{\mathrm{loc}}^\infty(\Omega) \tag{1.7.6}$$

are algebras, and so is the linear space $BV_c^\infty(\Omega)$ defined according to Convention 1.1.1. Note the symbol $BV_{loc}^\infty(\Omega)$ is in agreement with Convention 1.1.2.

A bounded open set $\Omega \subset \mathbb{R}^m$ is called a *Lipschitz domain* if for each $x \in \partial\Omega$ there is an $\varepsilon > 0$ and a Lipschitz function φ on $\mathbb{R}^{m-1}$ such that, after a suitable rotation about z and permutation of coordinates,

$$\Omega \cap K[x,\varepsilon] = \big\{(\xi,t) \in \mathbb{R}^m : t < \varphi(\xi)\big\} \cap K[x,\varepsilon].$$

Each Lipschitz domain Ω is an essentially clopen set with $int_*\Omega = \Omega$; hence $cl_*\Omega = cl\,\Omega$ and $\partial_*\Omega = \partial\Omega$ by Observation 1.5.8.

Theorems 1.7.4 through 1.7.6 below are examples of a special behavior exhibited by BV functions in Lipschitz domains; their proofs can be found in [22, Section 5.3 and 5.4].

Theorem 1.7.4. *Let $\Omega \subset \mathbb{R}^m$ be such that either Ω or $\mathbb{R}^m - \Omega$ is a Lipschitz domain. If f is a BV function in Ω, then a finite limit*

$$\operatorname{Tr} f(x) := \lim_{r \to 0+} \frac{1}{|\Omega \cap B[x,r]|} \int_{\Omega \cap B[x,r]} f(y)\,dy$$

exists for $\mathcal{H}^{m-1}$-almost all $x \in \partial\Omega$, and the function

$$\operatorname{Tr} f : x \mapsto \operatorname{Tr} f(x),$$

defined $\mathcal{H}^{m-1}$-almost everywhere on $\partial\Omega$, belongs to $L^1(\partial\Omega, \mathcal{H}^{m-1})$.

The function $\operatorname{Tr} f$ defined in Theorem 1.7.4 is called the *trace* of f on $\partial\Omega$. We view $\operatorname{Tr} f$ as the *boundary value* of f on $\partial\Omega$.

Theorem 1.7.5. *Let Ω be a Lipschitz domain, and let f be a function defined on $\mathbb{R}^m$. If the restrictions $g = f \restriction \Omega$ and $h = f \restriction (\mathbb{R}^m - cl\,\Omega)$ are BV functions in Ω and $\mathbb{R}^m - cl\,\Omega$, respectively, then f is a BV function in $\mathbb{R}^m$ and*

$$\|f\| = \|g\| + \|h\| + \int_{\partial\Omega} |\operatorname{Tr} g - \operatorname{Tr} h|\,d\mathcal{H}^{m-1}.$$

Theorem 1.7.6. *Let ν_Ω be the unit exterior normal of a Lipschitz domain Ω, and let v belong to $C^1(\mathbb{R}^m; \mathbb{R}^m)$. Then*

$$\int_\Omega f \operatorname{div} v\,d\mathcal{L}^m = \int_{\partial\Omega} (\operatorname{Tr} f)v \cdot \nu_\Omega\,d\mathcal{H}^{m-1} - \int_\Omega v \cdot d(Df)$$

for each BV function f in Ω.

Our next task is to topologize the algebra $BV_c^\infty(\mathbb{R}^m)$. According to Theorem 1.7.1, (3), the sets

$$BV_n := \left\{ f \in BV_c^\infty(\mathbb{R}^m) : \operatorname{supp} f \subset B[n] \text{ and } \|f\| + |f|_\infty \leq n+1 \right\},$$

$n = 1, 2, \ldots$, are compact in the L^1-topology. Denote by $\mathcal{T}$ the largest topology in $BV_c^\infty(\mathbb{R}^m)$ for which all inclusion maps

$$\left(BV_n, |\cdot|_1 \right) \hookrightarrow \left(BV_c^\infty(\mathbb{R}^m), \mathcal{T} \right)$$

are continuous. The following proposition summarizes the main features of the topology $\mathcal{T}$.

Proposition 1.7.7. *The topology $\mathcal{T}$ in $BV_c^\infty(\mathbb{R}^m)$ is a locally convex, sequential, and sequentially complete topology which has the following properties.*

(1) *A set $E \subset BV_c^\infty(\mathbb{R}^m)$ is $\mathcal{T}$-closed if and only if $E \cap BV_n$ is closed in the L^1-topology for $n = 1, 2, \ldots$. In particular, the topology $\mathcal{T}$ coincides with the L^1-topology in each BV_n.*

(2) *A map f from $\left(BV_c^\infty(\mathbb{R}^m), \mathcal{T} \right)$ to a topological space Y is continuous whenever each restriction $f \restriction BV_n$ is continuous with respect to the L^1-topology.*

(3) *The product map $(f, g) \mapsto fg$ from $BV_c^\infty(\mathbb{R}^m) \times BV_c^\infty(\mathbb{R}^m)$ to $BV_c^\infty(\mathbb{R}^m)$ is $(\mathcal{T} \otimes \mathcal{T}, \mathcal{T})$-continuous.*

(4) *A sequence $\{f_i\}$ in $BV_c^\infty(\mathbb{R}^m)$ $\mathcal{T}$-converges to 0 if and only if $\{f_i\}$ is a sequence in some BV_n and $\lim |f_i|_1 = 0$.*

(5) *A set $E \subset BV_c^\infty(\mathbb{R}^m)$ is $\mathcal{T}$ bounded if and only if E is a subset of some BV_n and $\sup_{f \in E} |f|_1 < \infty$.*

(6) *$\mathcal{T}$ is not first countable. In particular, $\mathcal{T}$ is strictly larger than the L^1-topology in $BV_c^\infty(\mathbb{R}^m)$.*

PROOF. Observe that $\{BV_n\}$ is an increasing sequence of compact convex sets containing 0, and $BV_c^\infty(\mathbb{R}^m) = \bigcup_{n=1}^\infty BV_n$. Moreover,

$$BV_i + BV_j \subset BV_{i+j}$$

for all positive integers i and j, and if $j \geq i|r|$ for an $r \in \mathbb{R}$, then $rBV_i \subset BV_j$. Thus it follows from Propositions 1.2.2 and 1.2.3 that $\mathcal{T}$ is a locally convex, sequential, and sequentially complete topology which satisfies claims (1)–(5).

We prove claim (6) only for $m = 1$, leaving the easy modification for arbitrary dimension m to the reader. For $i, k = 1, 2, \ldots$, select numbers

$a_{i,k} < b_{i,k}$ in the open interval $\big(1/(i+1), 1/i\big)$ so that

$$\lim_{k \to \infty} a_{i,k} = \frac{1}{i+1} \quad \text{and} \quad \lim_{k \to \infty} b_{i,k} = \frac{1}{i}.$$

Denote by f_i and $f_{i,k}$ the indicators of the sets

$$A_i := \big[1/(2i), 1/i\big] \quad \text{and} \quad A_{i,k} := \bigcup_{j=i}^{2i-1} [a_{j,k}, b_{j,k}],$$

respectively. Since $\|f_i\| = 2$ and $\|f_{i,k}\| = 2i$ by [22, Section 5.10, Theorem 1], we see that $\{f_i\}$ and $\{f_{i,k}\}$ are sequences in $BV_c^\infty(\mathbb{R})$. As

$$\lim |f_i|_1 = 0 \quad \text{and} \quad \lim_{k \to \infty} |f_{i,k} - f_i|_1 = 0$$

for $i = 1, 2, \ldots$, claim (4) implies the sequence $\{f_i\}$ $\mathcal{T}$-converges to 0, and each sequence $\{f_{i,k}\}$ $\mathcal{T}$-converges to f_i as $k \to \infty$. On the other hand, again by claim (4), there is no sequence in the countable family $\{f_{i,k} : i, k = 1, 2, \ldots\}$ that $\mathcal{T}$-converges to 0. An elementary argument shows this precludes the first countability of $\mathcal{T}$. $\square$

1.8. BV sets

The *perimeter* of a measurable set $E \subset \mathbb{R}^m$ is the extended real number

$$\|E\| := \sup_v \int_E \operatorname{div} v(x) \, dx$$

where $v \in C_c^1(\mathbb{R}^m; \mathbb{R}^m)$ and $|v|_\infty \leq 1$. Clearly $\|E\| = \|\chi_E\|$. We say $E \subset \mathbb{R}^m$ is, respectively, a *BV set*, or a *locally BV set*, in $\mathbb{R}^m$ whenever χ_E is a BV function, or a locally BV function, in $\mathbb{R}^m$. For a locally BV set E in $\mathbb{R}^m$, we write $\|\mathcal{D}E\|$ instead of $\|D\chi_E\|$.

Convention 1.8.1. When the ambient space $\mathbb{R}^m$ is clearly understood, we usually say BV set instead BV set in $\mathbb{R}^m$. A similar convention applies to locally BV sets, BV functions, and locally BV functions.

Let E be a locally BV set, and let $D\chi_E = \|\mathcal{D}E\| \mathsf{L} s$ be the polar representation of $D\chi_E$. Then

$$\nu_E : x \mapsto -s(x) : \partial_* E \to \mathbb{R}^m$$

is a Borel measurable map defined $\mathcal{H}^{m-1}$-almost everywhere on $\partial_* E$, called the *unit exterior normal* of E. The *Gauss-Green formula*

$$\int_E \operatorname{div} v \, d\mathcal{L}^m = \int_{\partial_* E} v \cdot \nu_E \, d\mathcal{H}^{m-1} \tag{1.8.1}$$

holds for each $v \in C_c^1(\mathbb{R}^m; \mathbb{R}^m)$. This, as well as Theorem 1.8.2 below, follows from [22, Sections 5.7, 5.8, and 5.11].

Theorem 1.8.2. *The following statements are true.*

(1) *If E is a measurable set, then $\|E\| = \mathcal{H}^{m-1}(\partial_* E)$.*

(2) *A measurable set E is a BV set if and only if $|E|$ and $\mathcal{H}^{m-1}(\partial_* E)$ are finite numbers.*

(3) *A measurable set E is a locally BV set if and only if $\mathcal{H}^{m-1} \llcorner \partial_* E$ is a locally finite measure, in which case $\|\mathcal{D}E\| = \mathcal{H}^{m-1} \llcorner \partial_* E$.*

Corollary 1.8.3. *Each Lipschitz domain Ω is a BV set.*

PROOF. Since $\partial_* \Omega = \partial \Omega$ is a compact set, it is easy to deduce from estimate (1.6.1) that $\mathcal{H}^{m-1}(\partial_* \Omega) < \infty$. $\qquad \square$

Proposition 1.8.4. *If A and B are measurable sets, then*

$$\max\{\|A \cup B\|, \|A \cap B\|, \|A - B\|\} \leq \|A\| + \|B\|.$$

In particular, if A and B are BV sets, then so are the sets $A \cup B$, $A \cap B$, and $A - B$.

PROOF. Since Corollary 1.5.7 implies

$$\partial_*(A \cup B) \cup \partial_*(A \cap B) \cup \partial_*(A - B) \subset \partial_* A \cup \partial_* B,$$

the proposition follows from Theorem 1.8.2, (1) and (2). $\qquad \square$

Proposition 1.8.5. *If A and B are locally BV sets, then*

$$\|A \cap B\| \leq \|\mathcal{D}A\|(\text{int}_* B) + \|\mathcal{D}B\|(\text{cl}_* A),$$

$$\|A \cap B\| \geq \|\mathcal{D}A\|(\text{int}_* B) + \|\mathcal{D}B\|(\text{int}_* A).$$

In particular, the intersection of a locally BV set and a bounded BV set is a BV set.

PROOF. Since $\partial_* A$ and $\text{int}_* A$ are disjoint sets, Corollary 1.5.7 yields

$$\text{int}_* B \cap \partial_* A \subset \text{int} B \cap \text{cl}_* A - (\text{int}_* A \cap \text{int}_* B)$$
$$\subset \text{cl}_*(A \cap B) - \text{int}_*(A \cap B) = \partial_*(A \cap B).$$

By symmetry, $\text{int}_* A \cup \partial_* B \subset \partial_*(A \cap B)$. Choose an $x \in \partial_*(A \cap B)$. As $\partial_*(A \cap B)$ is contained in both $\text{cl}_* A \cap \text{cl}_* B$ and $\partial_* A \cup \partial_* B$, the point x is either in $\partial_* A \cap \text{int}_* B$ or in $(\partial_* B) \cap (\text{cl}_* A)$, and consequently

$$\partial_*(A \cap B) \subset (\partial_* A \cap \text{int}_* B) \cup (\partial_* B \cap \text{cl}_* A).$$

Now it suffices to observe that $\text{int}_* A \cap \partial_* B$ and $\text{int}_* B \cap \partial_* A$ are disjoint sets, and apply Theorem 1.8.2, (3). $\qquad \square$

Theorem 1.8.6. *Let E be a measurable set, and let $\phi : A \to \mathbb{R}^m$ be a lipeomorphism. Then*

$$|\phi(E)| \leq \operatorname{Lip}(\phi)^m |E| \quad and \quad \|\phi(E)\| \leq \operatorname{Lip}(\phi)^{m-1} \|E\| .$$

Moreover, if E is a BV set, or locally BV set, then so is $\phi(E)$, respectively.

PROOF. From inequality (1.6.1) and Theorems 1.8.2, (1) and 1.6.5, we obtain

$$|\phi(E)| = \mathcal{H}^m\big(\phi(E)\big) \leq \operatorname{Lip}(\phi)^m \mathcal{H}^m(E) = \operatorname{Lip}(\phi)^m |E|$$

and

$$\begin{aligned}\|\phi(E)\| &= \mathcal{H}^{m-1}\big[\partial_*\phi(E)\big] = \mathcal{H}^{m-1}\big[\bar\phi(\partial_* E)\big] \\ &\leq \operatorname{Lip}(\bar\phi)^{m-1} \mathcal{H}^{m-1}(\partial_* E) = \operatorname{Lip}(\phi)^{m-1} \|E\| .\end{aligned}$$

According to Theorem 1.8.2, (2), the set $\phi(E)$ is BV whenever E is.

Now let E be a locally BV set, and let $B \subset \partial_*\phi(E)$ be a bounded set. By Theorem 1.6.5, the set $A := \phi^{-1}(B)$ is a bounded subset of $\partial_* E$, and inequality (1.6.1) yields

$$\mathcal{H}^{m-1}(B) = \mathcal{H}^{m-1}\big[\phi(A)\big] \leq \operatorname{Lip}(\phi)^{m-1} \mathcal{H}^{m-1}(A) < \infty .$$

It follows from Theorem 1.8.2, (3) that $\phi(E)$ is a locally BV set. $\quad\square$

The next theorem establishes important relationships between the Lebesgue measure and perimeter of a BV set. It follows from [45, Section 2.2] and [88, Theorem 5.12.7 applied to the proof of Theorem 5.4.3]. The first inequality in Theorem 1.8.7 below is called the *isoperimetric inequality*, the second is called the *relative isoperimetric inequality*. Observe the isoperimetric inequality says that among all bounded BV sets of a given perimeter, the ball has the largest measure.

Theorem 1.8.7. *Let $m \geq 2$, and let E and Ω be a bounded BV set and a Lipschitz domain, respectively. Then*

$$|E|^{\frac{m-1}{m}} \leq m^{-1}\alpha(m)^{-\frac{1}{m}} \|E\| ,$$

$$\min\{|\Omega \cap E|, |\Omega - E|\}^{\frac{m-1}{m}} \leq \beta\|DE\|(\Omega) ,$$

where $\beta := \beta(m,\Omega)$. If $\Omega' = \phi(r\Omega)$ where $r > 0$ and ϕ is an affine bijection of $\mathbb{R}^m$, then $\beta(m,\Omega') = \beta(m,\Omega)$.

If f is a function defined on a set Ω and $t \in \mathbb{R}$, we call

$$\{f > t\} := \big\{x \in \Omega : f(x) > t\big\} \quad \text{and} \quad \{f < t\} := \big\{x \in \Omega : f(x) < t\big\}$$

the *level sets* of f. Theorem 1.8.8 below, proved in [22, Section 5.5], links the variation of an integrable function with the perimeters of its level sets. It is referred to as the *coarea theorem*.

Theorem 1.8.8. *Let $\Omega \subset \mathbb{R}^m$ be an open set, and let $f \in L^1(\Omega)$. The extended real-valued function $t \mapsto \mathcal{H}^{m-1}\big(\Omega \cap \partial_*\{f > t\}\big)$ defined on $\mathbb{R}$ is $\mathcal{L}^1$-measurable, and*

$$\int_{\mathbb{R}} \mathcal{H}^{m-1}\big(\Omega \cap \partial_*\{f > t\}\big)\, dt = \|f\|.$$

Corollary 1.8.9. *If f is a BV function, then for $\mathcal{L}^1$-almost all $t \in \mathbb{R}$, the level set $\{f > t\}$ is a locally BV set and $\big\|\{f > t\}\big\| < \infty$.*

Corollary 1.8.9 follows immediately from the coarea theorem and Theorem 1.8.2, (3).

Proposition 1.8.10. *If f is a BV function, then*

$$\int_{\mathbb{R}} \big\|\mathcal{D}\{f > t\}\big\|(B)\, dt = \|Df\|(B)$$

for every Borel set $B \subset \mathbb{R}^m$.

PROOF. Observe $\|Df\| < \infty$ by our assumption, and $\big\|\mathcal{D}\{f > t\}\big\| < \infty$ for $\mathcal{L}^1$-almost all $t \in \mathbb{R}$ by Corollary 1.8.9. In view of the coarea theorem and Theorem 1.8.2, (3),

$$\int_{\mathbb{R}} \big\|\mathcal{D}\{f > t\}\big\|(U)\, dt = \|Df\|(U)$$

for each open set $U \subset \mathbb{R}^m$. A G_δ set B is the intersection of a decreasing sequence $\{U_i\}$ of open sets. Since $\lim \|Df\|(U_i) = \|Df\|(B)$, and

$$\lim \big\|\mathcal{D}\{f > t\}\big\|(U_i) = \big\|\mathcal{D}\{f > t\}\big\|(B)$$

for $\mathcal{L}^1$-almost all $t \in \mathbb{R}$, the monotone convergence theorem yields

$$\int_{\mathbb{R}} \big\|\mathcal{D}\{f > t\}\big\|(B)\, dt = \lim \int_{\mathbb{R}} \big\|\mathcal{D}\{f > t\}\big\|(U_i)\, dt$$

$$= \lim \|Df\|(U_i) = \|Df\|(B).$$

It follows the family $\mathcal{B}$ of all sets $B \subset \mathbb{R}^m$ for which the proposition holds contains all G_δ sets. In particular, $\mathcal{B}$ contains the algebra generated by all open sets. By the monotone convergence theorem, $\mathcal{B}$ is closed with respect to the unions and intersections of monotone sequences, and we conclude that $\mathcal{B}$ contains all Borel sets [60, Exercise 7-6]. □

Proposition 1.8.11. *If f is a measurable function defined on $\mathbb{R}^m$, then $\partial_*\{f < t\} = \partial_*\{f > t\}$ for all but countably many $t \in \mathbb{R}$. In particular, if f is a BV function, then*

$$\|\mathcal{D}\{f < t\}\| = \|\mathcal{D}\{f > t\}\| < \infty$$

for $\mathcal{L}^1$-almost all $t \in \mathbb{R}$.

PROOF. As the measurable sets

$$\{f = t\} = \{f \geq t\} - \{f > t\} = \left(\mathbb{R}^m - \{f < t\}\right) \triangle \{f > t\}$$

are disjoint, it is easy to see that all but countably many of them are negligible. Consequently, for all but countably many $t \in \mathbb{R}$, the sets $\{f > t\}$ and $\mathbb{R}^m - \{f < t\}$ are equivalent. It follows from Section 1.5 that for those $t \in \mathbb{R}$, we have

$$\partial_*\{f < t\} = \partial_*\left(\mathbb{R}^m - \{f < t\}\right) = \partial_*\{f > t\}. \qquad (*)$$

Now assume f is a BV function. By Corollary 1.8.9, for $\mathcal{L}^1$-almost all $t \in \mathbb{R}$, the level set $\{f > t\}$ is a locally BV set with $\|\{f > t\}\| < \infty$. Selecting such a $t \in \mathbb{R}$, equality $(*)$ and Theorem 1.8.2, (3) imply that the level set $\{f < t\}$ is a locally BV set, and that

$$\|\mathcal{D}\{f < t\}\| = \mathcal{H}^{m-1} \mathsf{L} \, \partial_*\{f < t\}$$
$$= \mathcal{H}^{m-1} \mathsf{L} \, \partial_*\{f > t\} = \|\mathcal{D}\{f > t\}\|$$

is a finite Radon measure. □

Theorem 1.8.12. *If f is a BV function, then so are $|f|$, f^-, f^+, and*

$$\| \, |f| \, \| \leq \|f^-\| + \|f^+\| = \|f\|.$$

PROOF. From inequality (1.7.2) and Proposition 1.8.11 we obtain

$$\| \, |f| \, \| \leq \|f^-\| + \|f^+\| = \int_0^\infty \|\{-f > t\}\| \, dt + \int_0^\infty \|\{f > t\}\| \, dt$$
$$= \int_{-\infty}^0 \|\{f < t\}\| \, dt + \int_0^\infty \|\{f > t\}\| \, dt$$
$$= \int_{\mathbb{R}} \|\{f > t\}\| \, dt = \|f\|. \qquad \square$$

1.9. Slices of BV sets

The perimeter of a measurable set E depends only on the equivalence class of E. Each BV set E in $\mathbb{R}$ is $\mathcal{L}^1$-equivalent to a set $\bigcup_{i=1}^{p}(a_i, b_i)$ where

$$a_1 < b_1 < \cdots < a_p < b_p$$

are real numbers; in this case $\|E\| = 2p$. We show how this observation can be used to estimate and calculate perimeters of BV sets in $\mathbb{R}^m$. We state first a Fubini type result for the perimeters of BV sets.

Let $\mathbf{u}_1, \ldots, \mathbf{u}_m$ be an orthonormal base in $\mathbb{R}^m$. For a positive integer $k \le m$, denote by $\langle \mathbf{u}_{i_1} \ldots \mathbf{u}_{i_k} \rangle$ the linear span of the vectors $\mathbf{u}_{i_1}, \ldots, \mathbf{u}_{i_k}$. Fix a positive integer $s < m$, and let

$$V := \langle \mathbf{u}_1, \ldots, \mathbf{u}_s \rangle \quad \text{and} \quad W := \langle \mathbf{u}_{s+1}, \ldots, \mathbf{u}_m \rangle.$$

Since $\mathbb{R}^m = V \oplus W$, each point of $\mathbb{R}^m$ can be written uniquely as (x, y) where $x \in V$ and $y \in W$.

Let $E \subset \mathbb{R}^m$ be a measurable set. For $\mathcal{H}^{m-s}$-almost all $y \in W$,

$$E^y := \big\{ x \in V : (x, y) \in E \big\}$$

is an $\mathcal{H}^s$-measurable subset of W. Making the obvious identification between V and $\mathbb{R}^s$, we denote by $\|E^y\|$ the perimeter of E^y in V. The extended real number

$$\|E\|_V := \sup_v \int_E \operatorname{div} v \, d\mathcal{L}^m,$$

where $v \in C_c^1(\mathbb{R}^m; V)$ and $|v|_\infty \le 1$, is called the *partial perimeter* of E in the direction of V. Given a vector $\mathbf{e} \in \partial B(1)$, we write $\|E\|_{\mathbf{e}}$ instead of $\|E\|_{\langle \mathbf{e} \rangle}$, and if

$$\mathbf{e}_i := (0, \ldots, \underset{i\text{-th place}}{1}, \ldots, 0)$$

we write $\|E\|_i$ instead of $\|E\|_{\mathbf{e}_i}$. A direct verification reveals

$$\max\big\{ \|E\|_V, \|E\|_W \big\} \le \|E\| \le \|E\|_V + \|E\|_W, \tag{1.9.1}$$

and by induction, we obtain the inequality

$$\max_{i=1,\ldots,m} \|E\|_i \le \|E\| \le \sum_{i=1}^{m} \|E\|_i. \tag{1.9.2}$$

Employing the orthogonal transformation of $\mathbb{R}^m$ that maps the base $\mathbf{u}_1, \ldots, \mathbf{u}_m$ onto the standard base $\mathbf{e}_1, \ldots, \mathbf{e}_m$, the next theorem follows from [45, Section 2.2.1, Theorem 2].

Theorem 1.9.1. *Let $E \subset \mathbb{R}^m$ be a measurable set. The extended real valued function $y \mapsto \|E^y\|$ defined on W is $\mathcal{H}^{m-s}$-measurable, and*

$$\int_W \|E^y\| \, d\mathcal{H}^{m-s}(y) = \|E\|_V \, .$$

If E is a BV set in $\mathbb{R}^m$, then E^y is a BV in V for $\mathcal{H}^{m-s}$-almost all points $y \in W$.

Denote by $\Pi_{\mathbf{e}}$ the linear subspace of $\mathbb{R}^m$ perpendicular to $\mathbf{e} \in \partial B(1)$, and for each $x \in \Pi_{\mathbf{e}}$, denote by l_x the line passing through x and parallel to $\mathbf{e}$. By Theorem 1.9.1,

$$\int_{\Pi_{\mathbf{e}}} \|l_x \cap E\| \, d\mathcal{H}^{m-1}(x) = \|E\|_{\mathbf{e}} \qquad (1.9.3)$$

for each measurable set E.

Proposition 1.9.2. *If C is a BV set and $\{B_n\}$ is a sequence of BV sets with $\lim \|B_n\| = 0$, then*

$$\lim \|B_n - C\| = \lim \|B_n \cap C\| = 0 \, .$$

PROOF. Select a vector $\mathbf{e} = \mathbf{e}_i$, and observe

$$E_n := \big\{ x \in \Pi_{\mathbf{e}} : \|l_x \cap B_n\| > 0 \big\} = \big\{ x \in \Pi_{\mathbf{e}} : \|l_x \cap B_n\| \geq 2 \big\}.$$

Note $\lim \mathcal{H}^{m-1}(E_n) = 0$, since formulae (1.9.3) and (1.9.2) imply

$$2\mathcal{H}^{m-1}(E_n) \leq \int_{E_n} \|l_x \cap B_n\| \, d\mathcal{H}^{m-1}(x) = \|B_n\|_i \leq \|B_n\| \, .$$

Now $\big\| l_x \cap (B_n - C) \big\| = 0$ for each x in $\Pi_{\mathbf{e}} - E_n$. Thus

$$\|B_n - C\|_i = \int_{E_n} \big\| l_x \cap (B_n - C) \big\| \, d\mathcal{H}^{m-1}(x)$$

$$\leq \int_{E_n} \|l_x \cap B_n\| \, \mathcal{H}^{m-1}(x) + \int_{E_n} \|l_x \cap C\| \, \mathcal{H}^{m-1}(x)$$

$$\leq \|B_n\| + \int_{E_n} \|l_x \cap C\| \, \mathcal{H}^{m-1}(x)$$

according to equation (1.9.3) and Proposition 1.8.4. As

$$\int_{\Pi_{\mathbf{e}}} \|l_x \cap C\| \, \mathcal{H}^{m-1}(x) = \|C\|_i \leq \|C\| < \infty,$$

$\lim \|B_n - C\|_i = 0$ for $i = 1, \ldots, m$. Hence $\lim \|B_n - C\| = 0$ by inequality (1.9.2). This and Proposition 1.8.4 yield

$$\lim \|B_n \cap C\| = \lim \|B_n - (B_n - C)\|$$
$$\leq \lim \|B_n\| + \lim \|B_n - C\| = 0. \qquad \Box$$

Theorem 1.9.3. *If E is a BV set, then the function $\mathbf{e} \mapsto \|E\|_{\mathbf{e}}$ defined on $\partial B(1)$ is $\mathcal{H}^{m-1}$-measurable and*

$$\|E\| = \frac{1}{2\alpha(m-1)} \int_{\partial B(1)} \|E\|_{\mathbf{e}} \, d\mathcal{H}^{m-1}(\mathbf{e}) .$$

PROOF. According to [86, §2, Section 6], for $\mathcal{H}^{m-1}$-almost all $x \in \Pi_{\mathbf{e}}$, the intersection $l_x \cap \mathrm{int}_* E$ is the union of finitely many open intervals whose closures are disjoint, and

$$\partial(l_x \cap \mathrm{int}_* E) = l_x \cap \partial_* E .$$

Hence $l_x \cap \mathrm{int}_* E$ is a BV set in l_x whose perimeter $\|l_x \cap \mathrm{int}_* E\|$ is equal to the number of elements in $l_x \cap \partial_* E$.

Let $\beta := \left[2\alpha(m-1) \right]^{-1}$, and let $\mathcal{J}^{m-1}$ be the $(m-1)$-dimensional *integral-geometric measure* defined in [56, Section 2.4]. For a detailed construction of $\mathcal{J}^{m-1}$ see [24, Section 2.10.5], where this measure is denoted by $\mathcal{J}^{m-1}_1$ — cf. [24, Theorem 2.10.15].

As $\partial_* E$ is a Borel set by Proposition 1.5.1, equality (1.9.3) implies the function $\mathbf{e} \mapsto \|E\|_{\mathbf{e}}$ is $\mathcal{H}^{m-1}$-measurable and

$$\mathcal{J}^{m-1}(\partial_* E) = \beta \int_{\partial B(1)} \left[\int_{\Pi_{\mathbf{e}}} \|l_x \cap \mathrm{int}_* E\| \, d\mathcal{H}^{m-1}(x) \right] d\mathcal{H}^{m-1}(\mathbf{e})$$
$$= \beta \int_{\partial B(1)} \|\mathrm{int}_* E\|_{\mathbf{e}} \, d\mathcal{H}^{m-1}(\mathbf{e})$$
$$= \beta \int_{\partial B(1)} \|E\|_{\mathbf{e}} \, d\mathcal{H}^{m-1}(\mathbf{e}) .$$

In view of [88, Theorem 5.7.3 and Lemma 5.9.5], the essential boundary $\partial_* E$ is $(\mathcal{H}^{m-1}, m-1)$ rectifiable [24, Section 3.2.14]. Thus by [24, Theorem 3.2.26], the measures $\mathcal{J}^{m-1}$ and $\mathcal{H}^{m-1}$ coincide on $\partial_* E$, and Theorem 1.8.2, (1) implies $\|E\| = \mathcal{J}^{m-1}(\partial_* E)$. $\qquad \Box$

Corollary 1.9.4. *Let C be a convex set. Then $\|A \cap C\| \leq \|A\|$ for each BV set A. In particular, C is a locally BV set and $\|C\| \leq 2m\, d(C)^{m-1}$.*

PROOF. The set C is measurable, since $|\partial C| = 0$ according to Fubini's theorem. Select a BV set A, and observe

$$\|l \cap C\| \le 2 \quad \text{and} \quad \|l \cap A \cap C\| \le \|l \cap A\|$$

for each line l in $\mathbb{R}^m$. Equality (1.9.3) yields $\|A \cap C\|_{\mathbf{e}} \le \|A\|_{\mathbf{e}}$ for every vector $\mathbf{e} \in \partial B(1)$. Thus $A \cap C$ is a BV set with $\|A \cap C\| \le \|A\|$ by inequality (1.9.2) and Theorem 1.9.3. Now it follows from Theorem 1.8.2, (3) and Proposition 1.8.5 that C is a locally BV set. Formulae (1.9.3) and (1.9.2) imply $\|C\| \le 2m\, d(C)^{m-1}$. $\qquad\square$

1.10. Approximating BV sets

Denote by $\mathcal{BV}(\mathbb{R}^m)$ the family of all bounded BV sets in $\mathbb{R}^m$, and let

$$\mathcal{BV}(E) := \{B \in \mathcal{BV}(\mathbb{R}^m) : B \subset E\}$$

for each locally BV set E in $\mathbb{R}^m$. We give $\mathcal{BV}(\mathbb{R}^m)$ the unique topology $\mathfrak{T}$ for which the injection

$$\chi : B \mapsto \chi_B : \big(\mathcal{BV}(\mathbb{R}^m), \mathfrak{T}\big) \to \big(BV_c^\infty(\mathbb{R}^m), \mathfrak{J}\big)$$

is a homeomorphism. Since $\chi\big[\mathcal{BV}(\mathbb{R}^m)\big]$ is a closed subset of the space $\big(BV_c^\infty(\mathbb{R}^m), \mathfrak{J}\big)$, the space $\big(\mathcal{BV}(\mathbb{R}^m), \mathfrak{T}\big)$ is sequential and sequentially complete. Note that in establishing Proposition 1.7.7, (7) we actually proved that the space $\big(\mathcal{BV}(\mathbb{R}^m), \mathfrak{T}\big)$ is not first countable. Let

$$\mathcal{BV}_n := \{B \in \mathcal{BV}(\mathbb{R}^m) : B \subset B[n] \text{ and } \|B\| \le n\}$$

for $n = 1, 2, \ldots$, and observe $\chi(\mathcal{BV}_n) \subset BV_n$. The topology $\mathfrak{T}$ on $\mathcal{BV}_n$ is compact and induced by the metric $(B, C) \mapsto |B \triangle C|$.

A sequence $\{B_i\}$ of subsets of $\mathbb{R}^m$ is called *bounded* if $\bigcup_{i=1}^\infty B_i$ is a bounded set. It follows from Proposition 1.7.7, (4) that a sequence $\{B_i\}$ in $\mathcal{BV}(\mathbb{R}^m)$ $\mathfrak{T}$-*converges* to a set $B \in \mathcal{BV}(\mathbb{R}^m)$ if and only if

  (i) $\{B_i\}$ is a bounded sequence,
  (ii) $\lim |B_i \triangle B| = 0$ and $\sup \|B_i\| < \infty$.

If a sequence $\{B_i\}$ in $\mathcal{BV}(\mathbb{R}^m)$ $\mathfrak{T}$-converges to a set $B \in \mathcal{BV}(\mathbb{R}^m)$, we write $\{B_i\} \to B$.

Using conditions (i) and (ii) above, we can define $\{B_i\} \to B$ for a sequence $\{B_i\}$ in $\mathcal{BV}(\mathbb{R}^m)$ and any set $B \subset \mathbb{R}^m$. However, this leads to no new concept, since by Theorem 1.7.1, (1) each set $B \subset \mathbb{R}^m$ for which there is a sequence $\{B_i\}$ in $\mathcal{BV}(\mathbb{R}^m)$ with $\{B_i\} \to B$ already belongs to $\mathcal{BV}(\mathbb{R}^m)$.

The following theorem is the main approximation result for bounded BV sets. For its proof we refer to [26, Theorem 1.24].

Theorem 1.10.1 (De Giorgi). *Let B be a bounded BV set, and let U be a bounded open subset of $\mathbb{R}^m$ containing* cl B. *There is a sequence $\{B_i\}$ of compact BV sets contained in U such that*

- *each ∂B_i is a C^∞ submanifold of $\mathbb{R}^m$,*
- $\lim |B_i \triangle B| = 0$ *and* $\lim \|B_i\| = \|B\|$.

In particular, $\{B_i\} \to B$ and each B_i is a bounded essentially clopen BV set with $B_i = \mathrm{cl}_* B_i$.

Note the boundedness of the open set U in Theorem 1.10.1 is assumed only for convenience: it guarantees that $\{B_i\} \to B$. The same is true for Proposition 1.10.3 below.

A finite (possibly empty) union of cells is called a *figure*, and the family of all figures in $\mathbb{R}^m$ is denoted by $\mathcal{F}(\mathbb{R}^m)$. A *dyadic figure* is a figure that is the union of finitely many dyadic cubes; the family of all dyadic figures in $\mathbb{R}^m$ is denoted by $\mathcal{F}_d(\mathbb{R}^m)$. Clearly each figure is a bounded BV set, and we show that $\mathcal{F}_d(\mathbb{R}^m)$, and hence $\mathcal{F}(\mathbb{R}^m)$, is dense in $(\mathcal{BV}(\mathbb{R}^m), \mathfrak{T})$. When no confusion is possible we write $\mathcal{BV}$, $\mathcal{F}$, and $\mathcal{F}_d$ instead of $\mathcal{BV}(\mathbb{R}^m)$, $\mathcal{F}(\mathbb{R}^m)$, and $\mathcal{F}_d(\mathbb{R}^m)$, respectively.

Lemma 1.10.2. *If A is a bounded essentially clopen BV set, then there is a sequence $\{A_i\}$ of dyadic figures contained in* int$_* A$ *such that*

$$\lim |A - A_i| = 0 \quad and \quad \sup \|A_i\| \leq \beta \|A\|$$

where $\beta := \beta(m) > 1$. In particular, $\{A_i\} \to A$.

PROOF. As the lemma is clear for $m = 1$, suppose $m \geq 2$. If $\|A\| = 0$, then $|A| = 0$ by the isoperimetric inequality, and it suffices to let $A_i = \emptyset$ for $i = 1, 2, \ldots$. Thus assume $\|A\| = \mathcal{H}^{m-1}(\partial_* A)$ is a positive number. Fix an integer $i \geq 1$, and let κ be the constant from Lemma 1.3.2. As $\mathcal{H}^{m-1}(\partial_* A) < 2\|A\|$, Lemma 1.3.2 implies there are dyadic cubes $K_1, K_2, \ldots$ of diameters less than $1/i$ such that

$$\partial_* A \subset \mathrm{int} \bigcup_{j=1}^{\infty} K_j \quad \text{and} \quad \sum_{j=1}^{\infty} d(K_j)^{m-1} < 2\kappa \|A\|.$$

For $j = 1, 2, \ldots$, let L_j be the union of all dyadic cubes adjacent to the cube K_j. Since the interiors of the cubes L_j cover the compact set $\partial_* A$, there is an integer $p \geq 1$ such that $\partial_* A$ is contained in the interior of the figure $B = \bigcup_{j=1}^{p} L_j$. Observe

$$\|B\| \leq \sum_{j=1}^{p} \|L_j\| < 2m \sum_{j=1}^{p} d(L_j)^{m-1} < 4m \cdot 3^{m-1} \kappa \|A\|,$$

$$|B| \leq \sum_{j=1}^{p} |L_j| < \frac{3}{i} \sum_{j=1}^{p} d(L_j)^{m-1} < \frac{2}{i} \cdot 3^m \kappa \|A\|.$$

Let $d := \min\{d(K_1), \ldots, d(K_p)\}$, and denote by $\mathcal{K}$ the family of all dyadic cubes of diameter d which do not overlap B. Suppose there is a $K \in \mathcal{K}$ that meets both $\mathrm{int}_* A$ and $\mathbb{R}^m - \mathrm{cl}_* A$. As $0 < |K \cap A| < |K|$, the relative isoperimetric inequality implies K meets $\partial_* A$, a contradiction. Thus for each $K \in \mathcal{K}$, either $K \subset \mathrm{int}_* A$ or $K \cap \mathrm{int}_* A = \emptyset$. It follows

$$A_i := \bigcup \{K \in \mathcal{K} : K \cap \mathrm{int}_* A \neq \emptyset\}$$

is a subset of $\mathrm{int}_* A$, and as $\mathrm{int}_* A$ is bounded, A_i is a figure. Since $\|A_i\| \leq \|B\|$ and $|A - A_i| \leq |B|$, it suffices to let $\beta = 4m \cdot 3^{m-1} \kappa$. $\quad\square$

Proposition 1.10.3. *Let A be a bounded BV set, and let $U \subset \mathbb{R}^m$ be a bounded open set containing $\mathrm{cl}\, A$. There is a sequence $\{A_i\}$ of dyadic figures contained in U such that*

$$\lim |A_i \bigtriangleup A| = 0 \quad and \quad \limsup \|A_i\| \leq \beta \|A\|$$

where β is the constant from Lemma 1.10.2. In particular, $\{A_i\} \to A$.

PROOF. Using De Giorgi's theorem, obtain a sequence $\{B_i\}$ of essentially clopen bounded BV sets contained in U for which

$$\lim |B_i \bigtriangleup A| = 0 \quad and \quad \lim \|B_i\| = \|A\|.$$

By Lemma 1.10.2, there are dyadic figures $C_i \subset \mathrm{int}_* B_i$, $i = 1, 2, \ldots$, such that $|B_i - C_i| < 1/i$ and $\|C_i\| \leq \beta \|B_i\|$. The proposition follows. $\quad\square$

Corollary 1.10.4. *The space $(\mathcal{BV}, \mathfrak{T})$ is the sequential completion of the space $(\mathcal{F}_\mathrm{d}, \mathfrak{T})$, and a fortiori, of the space $(\mathcal{F}, \mathfrak{T})$.*

In contrast to Lemma 1.10.2, approximations in De Giorgi's theorem and in Proposition 1.10.3 are neither from *inside* nor from *outside*. More precisely, given a bounded BV set A, there may be no sequence $\{A_i\}$ of figures, or of compact BV sets whose boundaries are C^∞ submanifolds of $\mathbb{R}^m$, such that $\{A_i\} \to A$ and $A_i \subset A$ or $A \subset A_i$ for $i = 1, 2, \ldots$ [26,

Remark 1.27]. However, we still have a more modest approximation from inside established in [20] and [81].

Proposition 1.10.5. *Let A be a bounded BV set. There are essentially closed BV sets $A_i \subset A$ such that*

$$\|A_i\| \leq \|A\| \quad and \quad |A - A_i| \leq \frac{1}{i}\|A\|$$

for $i = 1, 2, \ldots$. In particular, $\{A_i\} \to A$.

PROOF. Since all BV sets in $\mathbb{R}$ are equivalent to figures, the proposition is trivially true for $m = 1$. Thus assume $m \geq 2$, fix an integer $i \geq 1$, and define a function φ on $\mathcal{BV}(A)$ by the formula

$$\varphi(C) := \|C\| - i|C|.$$

Note $c := \inf\{\varphi(C) : C \in \mathcal{BV}(A)\}$ is a real number, and find a sequence $\{C_k\}$ in $\mathcal{BV}(A)$ with $\lim \varphi(C_k) = c$. Since

$$\sup_k \|C_k\| = \sup_k [\varphi(C_k) + i|C_k|] \leq \sup_k \varphi(C_k) + i|A| < \infty,$$

it follows from Theorem 1.7.1, (3) there is a set $E \in \mathcal{BV}(A)$ and a subsequence of $\{C_k\}$, still denoted by $\{C_k\}$, such that $\lim |E \triangle C_k| = 0$. By Theorem 1.7.1, (1),

$$c \leq \varphi(E) \leq \lim \varphi(C_k) = c,$$

and we conclude $\|E\| - i|E| = c \leq \|A\| - i|A|$. As the last inequality implies

$$\|E\| \leq \|A\| \quad and \quad i|A - E| \leq \|A\|,$$

it suffices to show E is essentially closed. To this end, we establish a stronger result, namely $\mathrm{cl}_* E = \mathrm{cl}\, E$.

Replacing E by $E \cap \mathrm{cl}_* E$, we may assume $|B(x, t) \cap E| > 0$ for each $x \in \mathrm{cl}\, E$ and each $t > 0$. Select an $x \in \mathrm{cl}\, E$, and to simplify the notation, let $B_t := B(x, t)$. By the isoperimetric inequality, for each $t > 0$,

$$0 < \gamma \leq \|E \cap B_t\| \cdot |E \cap B_t|^{\frac{1}{m} - 1} \tag{1}$$

where $\gamma := m\, \alpha(m)^{\frac{1}{m}}$. As E is minimal, $\varphi(E) \leq \varphi(E - B_t)$ and hence

$$\|E\| \leq \|E - B_t\| + i|E \cap B_t|. \tag{2}$$

Throughout the remainder of this proof, all relations involving t hold for $\mathcal{L}^1$-almost all $t > 0$. According to [51, Section 6.2.3, Lemma 4] and

Theorem 1.8.2, (2),

$$\|E \cap B_t\| = \mathcal{H}^{m-1}(\partial_* E \cap B_t) + \mathcal{H}^{m-1}(E \cap \partial B_t),$$
$$\|E - B_t\| = \mathcal{H}^{m-1}(\partial_* E - B_t) + \mathcal{H}^{m-1}(E \cap \partial B_t).$$

Adding these equalities and using (2), we obtain

$$\|E \cap B_t\| \le 2\mathcal{H}^{m-1}(E \cap \partial B_t) + i|E \cap B_t|,$$

and in view of (1),

$$\gamma \le 2\mathcal{H}^{m-1}(E \cap \partial B_t)|E \cap B_t|^{\frac{1}{m}-1} + i|E \cap B_t|^{\frac{1}{m}}. \tag{3}$$

Applying [22, Section 3.4.4, Proposition 1], it is easy to verify

$$|E \cap B_t| = \int_0^t \mathcal{H}^{m-1}(E \cap \partial B_s)\, ds$$

and hence

$$\mathcal{H}^{m-1}(E \cap \partial B_t) = \frac{d}{dt}|E \cap B_t|.$$

Rewriting inequality (3) in the form

$$\gamma \le 2m\frac{d}{dt}\left(|E \cap \partial B_t|^{\frac{1}{m}}\right) + i|E \cap B_t|^{\frac{1}{m}},$$

and integrating the latter inequality over the interval $(0,\varepsilon)$ yields

$$\gamma \le \frac{2m}{\varepsilon} \int_0^\varepsilon \frac{d}{dt}\left(|E \cap \partial B_t|^{\frac{1}{m}}\right) dt + i|E \cap B_t|^{\frac{1}{m}}\, dt$$
$$\le \frac{2m}{\varepsilon}|E \cap \partial B_t|^{\frac{1}{m}} + i|E \cap B_\varepsilon|^{\frac{1}{m}}$$
$$= 2m\left(\frac{|E \cap B_\varepsilon|}{\varepsilon^m}\right)^{\frac{1}{m}} + i|E \cap B_\varepsilon|^{\frac{1}{m}}$$

for each $\varepsilon > 0$. From this we infer $x \in \mathrm{cl}_* E$. $\square$

The isoperimetric inequality shows that for $m \ge 2$, the following result is stronger than Proposition 1.10.5. For its proof we refer to [82].

Theorem 1.10.6. *For each bounded BV set A there is a sequence $\{A_i\}$ consisting of essentially closed BV sets contained in A such that*

$$\lim \|A - A_i\| = 0.$$

2

Charges

The notion of charge, which originated from the ideas of Mařík [47, 40], lies at the basis of our work. Intuitively, charges are multi-dimensional generalizations of additive functions in the real line induced by continuous functions. A major portion of this chapter is devoted to derivation of charges.

2.1. The definition and examples

Let $\mathcal{C}$ be a family of sets. An *additive function* on $\mathcal{C}$ is a function F defined on $\mathcal{C}$ such that

$$F(A \cup B) = F(A) + F(B)$$

for each pair A, B of disjoint sets from $\mathcal{C}$ for which $A \cup B$ belongs to $\mathcal{C}$.

Definition 2.1.1. A *charge* in $\mathbb{R}^m$ is an additive function F on the family $\mathcal{BV}(\mathbb{R}^m)$ that is $\mathcal{T}$-continuous.

When the ambient space $\mathbb{R}^m$ is clearly agreed on, a charge in $\mathbb{R}^m$ is called simply a charge.

Proposition 2.1.2. *An additive function F on $\mathcal{BV}$ is a charge if and only if either of the following conditions is satisfied.*

(i) *Given $\varepsilon > 0$, there is an $\eta > 0$ such that $|F(C)| < \varepsilon$ for each BV set $C \subset B(1/\varepsilon)$ with $\|C\| < 1/\varepsilon$ and $|C| < \eta$.*

(ii) $\lim F(A_i) = 0$ *for each sequence $\{A_i\}$ in $\mathcal{BV}$ with $\{A_i\} \to \emptyset$.*

PROOF. It is clear that each charge satisfies (i), and that (i) implies (ii). Suppose (ii) holds, and let $\{A_i\}$ be a sequence in $\mathcal{BV}$ with $\{A_i\} \to A$

for an $A \in \mathcal{BV}$. Then $\lim F(A_i) = F(A)$, since $\{A_i \bigtriangleup A\} \to \emptyset$ and

$$|F(A_i) - F(A)| = |F(A_i - A) - F(A - A_i)|$$
$$\leq |F(A_i - A)| + |F(A - A_i)|.$$

As $(\mathcal{BV}, \mathfrak{T})$ is a sequential space, F is a charge. $\square$

By continuity, a charge F vanishes on all bounded negligible sets, and by additivity, $F(E)$ depends only on the equivalence class of a set $E \in \mathcal{BV}$. It follows

$$F(A \cup B) = F(A) + F(B)$$

whenever $A, B \in \mathcal{BV}$ *do not overlap*. Since an additive continuous functions F on $(\mathcal{F}, \mathfrak{T})$ or $(\mathcal{F}_d, \mathfrak{T})$ is uniformly continuous, Corollary 1.10.4 implies F has a unique extension to a charge, still denoted by F.

Let E be a locally BV set. Given a function F defined on $\mathcal{BV}$, let

$$(F \, \llcorner \, E)(A) := F(A \cap E)$$

for each $A \in \mathcal{BV}$, and call $F \, \llcorner \, E$ the *reduction* of F to E. If F is additive or $\mathfrak{T}$-continuous, then so is the reduction $F \, \llcorner \, E$, respectively. In particular, $F \, \llcorner \, E$ is a charge whenever F is a charge. If F is a charge and $F = F \, \llcorner \, E$, we say that F is a *charge in* E. The linear space of all charges in a locally BV set E is denoted by $CH(E)$. When no confusion is possible, we write CH instead of $CH(\mathbb{R}^m)$.

Example 2.1.3. Let E be a locally BV set, and let $f \in L_\ell^1(E)$. By definition (1.3.2), the function f is integrable in each bounded BV set $B \subset E$. The absolute continuity of the Lebesgue integral implies

$$F : A \mapsto \int_{A \cap E} f \, d\mathcal{L}^m : \mathcal{BV} \to \mathbb{R}$$

is a charge in E. The charge F is called the *indefinite integral* of f, denoted by $\int f \, d\mathcal{L}^m$ or $\int f(x) \, dx$.

Example 2.1.4. If E is a locally BV set and $v \in C(\mathrm{cl}\, E; \mathbb{R}^m)$, then

$$F : A \mapsto \int_{\partial_*(A \cap E)} v \cdot \nu_{A \cap E} \, d\mathcal{H}^{m-1} : \mathcal{BV} \to \mathbb{R}$$

is an additive function and $F = F \, \llcorner \, E$. To show the function F is $\mathfrak{T}$-continuous, choose positive numbers ε and θ, and find a vector field $w \in C^1(\mathbb{R}^m; \mathbb{R}^m)$ so that $|v(x) - w(x)| < \theta$ for each $x \in B[1/\varepsilon] \cap \mathrm{cl}\, E$.

Let $B := B(1/\varepsilon)$. Given a BV set $A \subset B$ with $\|A\| < 1/\varepsilon$, the Gauss-Green formula and Proposition 1.8.4 imply

$$
\begin{aligned}
|F(A)| &\leq \int_{\partial_*(A \cap E)} |v - w| \, d\mathcal{H}^{m-1} + \int_{A \cap E} |\operatorname{div} w| \, d\mathcal{L}^m \\
&\leq \theta \|A \cap E\| + |A \cap E| \cdot |\operatorname{div} w|_\infty \\
&\leq \theta \big(\|A\| + \|B \cap E\| \big) + |A| \cdot |\operatorname{div} w|_\infty \\
&< \theta \left(\frac{1}{\varepsilon} + \|B \cap E\| \right) + |A| \cdot |\operatorname{div} w|_\infty .
\end{aligned}
$$

Choosing θ sufficiently small, we obtain

$$
|F(A)| < \frac{\varepsilon}{2} + |A| \cdot |\operatorname{div} w|_\infty .
$$

It follows that $|F(A)| < \varepsilon$ whenever $|A|$ is sufficiently small. We conclude the function F is a charge in E, called the *flux* of v and denoted by $\int v \cdot \nu \, d\mathcal{H}^{m-1}$.

Remark 2.1.5. If $m = 1$, then every bounded BV set A is equivalent to a unique figure $\bigcup_{i=1}^{p}[a_i, b_i]$ where $a_1 < b_1 < \cdots < a_p < b_p$ are real numbers. It follows that each additive function F on $\mathcal{BV}$ is the flux of a vector field $v : \mathbb{R} \to \mathbb{R}$. Explicitly

$$
F(A) = \int_{\partial_* A} v \cdot \nu_A \, d\mathcal{H}^0 = \sum_{i=1}^{p} \big[v(b_i) - v(a_i) \big],
$$

and it is easy to see F is a charge if and only if v is continuous. In other words, each one-dimensional charge is the flux of a continuous vector field.

We show (Example 2.1.10 below) that in higher dimensions a charge need not be the flux of a continuous vector field. To this end, we need some preliminary results of independent interest.

Lemma 2.1.6. *Let E be a bounded BV set, and let $K_1, \ldots, K_p$ be non-overlapping cubes of diameters greater than or equal to $d > 0$. Then*

$$
\frac{1}{2m} \sum_{j=1}^{p} \|E \cap K_j\| < \frac{\sqrt{m}}{d} |E| + \|E\|.
$$

PROOF. Suppose $m = 1$. Then E is $\mathcal{L}^1$-equivalent to the union of disjoint cells $C_1, \ldots, C_q$, and $\|E\| = 2q$. Since the family

$$
\{ C_s \cap K_1, \ldots, C_s \cap K_p \}
$$

contains fewer than $(|C_s|/d) + 2$ one-dimensional cells,

$$\sum_{j=1}^{p} \|E \cap K_j\| \leq \sum_{j=1}^{p} \sum_{s=1}^{q} \|C_s \cap K_j\| = \sum_{s=1}^{q} \sum_{j=1}^{p} \|C_s \cap K_j\|$$

$$< 2 \sum_{s=1}^{q} \left(\frac{|C_s|}{d} + 2 \right) = 2 \left(\frac{|E|}{d} + \|E\| \right).$$

Suppose $m \geq 2$. Applying Fubini's theorem, inequality (1.9.2), and Theorem 1.9.1 to the previous inequality, we obtain

$$\sum_{j=1}^{p} \|E \cap K_j\| \leq \sum_{j=1}^{p} \sum_{i=1}^{m} \|E \cap K_j\|_i =$$

$$\sum_{i=1}^{m} \sum_{j=1}^{p} \int_{\mathbb{R}^{m-1}} \|(E \cap K_j)^y\| \, d\mathcal{H}^{m-1}(y) =$$

$$\sum_{i=1}^{m} \int_{\mathbb{R}^{m-1}} \left(\sum_{j=1}^{p} \|E^y \cap K_j^y\| \right) d\mathcal{H}^{m-1}(y) <$$

$$2 \sum_{i=1}^{m} \left(\frac{\sqrt{m}}{d} \int_{\mathbb{R}^{m-1}} |E^y| \, d\mathcal{H}^{m-1}(y) + \int_{\mathbb{R}^{m-1}} \|E^y\| \, d\mathcal{H}^{m-1}(y) \right) \leq$$

$$2 \left(\frac{m\sqrt{m}}{d} |E| + \sum_{i=1}^{m} \|E\|_i \right) \leq 2m \left(\frac{\sqrt{m}}{d} |E| + \|E\| \right). \qquad \square$$

Proposition 2.1.7. *For an additive function F on $\mathcal{BV}$ the following conditions are equivalent.*

(1) *If $\{B_i\}$ is a bounded sequence of BV sets with positive perimeters and $\lim d(B_i) = 0$, then*

$$\lim \frac{F(B_i)}{\|B_i\|} = 0 .$$

(2) *If $\{B_i\}$ is a bounded sequence of BV sets with positive perimeters and $\lim |B_i| = 0$, then*

$$\lim \frac{F(B_i)}{\|B_i\|} = 0 .$$

(3) *Given $\varepsilon > 0$, there is a $\theta > 0$ such that*

$$|F(B)| < \theta |B| + \varepsilon \|B\|$$

for every BV set $B \subset B(1/\varepsilon)$.

PROOF. Implication (2) $\Rightarrow$ (1) is trivial. Assuming (1) holds, choose an $\varepsilon > 0$ and find a $\delta > 0$ so that $|F(B)| < (\varepsilon/2m)\|B\|$ for each BV set $B \subset B(1/\varepsilon)$ with $d(B) \leq \delta$. Let $\theta := \varepsilon\sqrt{m}/\delta$ and select a BV set $E \subset B(1/\varepsilon)$. Covering the set E by nonoverlapping cubes $K_1, \ldots, K_q$ of the same diameters equal to δ and applying Lemma 2.1.6, we obtain

$$|F(E)| \leq \sum_{j=1}^{q}|F(E \cap K_j)| \leq \frac{\varepsilon}{2m}\sum_{j=1}^{q}\|E \cap K_j\|$$

$$< \varepsilon\left(\frac{\sqrt{m}}{\delta}|E| + \|E\|\right) = \theta|E| + \varepsilon\|E\|,$$

which establishes (3). It follows from (3) that

$$\frac{|F(B)|}{\|B\|} \leq \theta\frac{|B|}{\|B\|} + \varepsilon$$

for each BV set $B \subset B(1/\varepsilon)$ with $\|B\| > 0$. Thus to prove (3) $\Rightarrow$ (2), it suffices to show that $|B_i|/\|B_i\|$ approaches zero for each sequence $\{B_i\}$ of bounded BV sets with positive perimeters and $\lim|B_i| = 0$. But this is clear for $m = 1$, and it follows from the isoperimetric inequality when $m \geq 2$. $\square$

Remark 2.1.8. Let $F := \int v \cdot \nu \, d\mathcal{H}^{m-1}$ be the flux of $v \in C(\mathbb{R}^m; \mathbb{R}^m)$. Since v is uniformly continuous in $B[1/\varepsilon]$, given $\varepsilon > 0$, there is a $\delta > 0$ such that $|v(y) - v(x)| < \varepsilon$ for all $x, y \in B[1/\varepsilon]$ with $|y - x| < \delta$. Select a BV set $B \subset B(1/\varepsilon)$ with $d(B) < \delta$ and choose an $x \in B$. Then the Gauss-Green formula yields

$$|F(B)| = \left|\int_{\partial_* B}\left[v(y) - v(x)\right] \cdot \nu_B \, d\mathcal{H}^{m-1}(y)\right|$$

$$\leq \int_{\partial_* B}|v(y) - v(x)| \, d\mathcal{H}^{m-1}(y) < \varepsilon\|B\|.$$

Thus the flux of a continuous vector field is not merely a charge: it is a charge that satisfies the conditions of Proposition 2.1.7.

Let E be a locally BV set, and let $v : \mathrm{cl}_* E \to \mathbb{R}^m$ be a Borel measurable vector field that is bounded in every bounded subset of E. The additive function

$$F : A \mapsto \int_{\partial_*(A \cap E)} v \cdot \nu_{A \cap E} \, d\mathcal{H}^{m-1} : \mathcal{BV} \to \mathbb{R}$$

is called the *flux* of v, denoted by $\int v \cdot \nu \, d\mathcal{H}^{m-1}$. This terminology and notation extend that introduced in Example 2.1.4. Clearly $F = F \, \llcorner \, E$, and if $B \subset E$ is a locally BV set and G is the flux of the restricted vector field $v \upharpoonright \mathrm{cl}_* B$, then $G = F \mathbin{\llcorner} B$.

Definition 2.1.9. Let E be a locally BV set. A Borel measurable vector field $v : \mathrm{cl}_* E \to \mathbb{R}^m$ that is bounded in every bounded subset of E is called *charging* whenever its flux is a charge, necessarily in E.

If $B \subset E$ is a locally BV set and $v : \mathrm{cl}_* E \to \mathbb{R}^m$ is a charging vector field, then so is the restriction $v \upharpoonright \mathrm{cl}_* B$. By Example 2.1.4, each $w \in C(\mathbb{R}^m; \mathbb{R}^m)$ is charging, and changing w on an $\mathcal{H}^{m-1}$-negligible set gives a trivial example of a discontinuous charging vector field. A charging vector field whose flux differs from the flux of any continuous vector field is constructed in the following example of Buczolich.

Example 2.1.10. Assume $m \geq 2$, and for $\delta > 0$ define a function φ_δ in $C^\infty(\mathbb{R})$ by the formula

$$\varphi_\delta(t) := \begin{cases} \exp\left(\frac{t^2}{t^2 - \delta^2}\right) & \text{if } |t| < \delta, \\ 0 & \text{if } |t| \geq \delta. \end{cases}$$

The map $u_\delta : x \mapsto \left[\varphi_\delta(|x|), \ldots, \varphi_\delta(|x|)\right]$ belongs to $C^\infty(\mathbb{R}^m; \mathbb{R}^m)$, vanishes outside $B(\delta)$, and

$$|u_\delta|_\infty = |u_\delta(0)| = \sqrt{m}.$$

By a direct calculation,

$$\operatorname{div} u_\delta(x) = -2\delta^2 \frac{\varphi_\delta(|x|)}{\left(|x|^2 - \delta^2\right)^2} \sum_{i=1}^{m} \xi_i$$

for each $x = (\xi_1, \ldots, \xi_m)$ in $\mathbb{R}^m$. Since the function

$$t \mapsto \frac{\varphi_\delta(t)}{(t^2 - \delta^2)^2}$$

attains its maximum at $t = 0$,

$$\int_{B(\delta)} \left|\operatorname{div} u_\delta(x)\right| dx \leq \frac{2m}{\delta} |B(\delta)| = 2 \|B(\delta)\|. \qquad (*)$$

Enumerating a dense countable subset of $\mathbb{R}^m$, it is easy to construct inductively sequences $\{z_k\}$ in $\mathbb{R}^m$ and $\{\varepsilon_k\}$ in $\mathbb{R}$ so that the following conditions are satisfied:

   (i) $0 < \varepsilon_1 \leq 1/2$ and $0 < \varepsilon_{k+1} \leq \varepsilon_k/2$ for $k = 1, 2, \ldots$;
   (ii) for $k = 1, 2, \ldots$, the closed sets $B_k = B[z_k, \varepsilon_k]$ are disjoint;
   (iii) the open set $U = \bigcup_{k=1}^{\infty} B(z_k, \varepsilon_k)$ is dense in $\mathbb{R}^m$.

Note $0 < |U| < 2|B_1|$ and $\|U\| \leq \sum_{k=1}^{\infty} \|B_k\| < \infty$.

Claim 1. If $Z = \{z_1, z_2, \dots\}$ then $\mathbb{R}^m - U$ is contained in $\operatorname{cl} Z$.

Proof. Suppose there are $x \in \mathbb{R}^m - U$ and $\varepsilon > 0$ with $Z \cap B(x, \varepsilon) = \emptyset$. Since $\lim \varepsilon_k = 0$, we can find an integer $p \geq 1$ so that $B(x, \varepsilon/2) \cap B_k = \emptyset$ for each $k > p$. Now $x \notin U$ implies $B(x, \varepsilon/2) \not\subset \bigcup_{k=1}^{p} B_k$. Hence there are $y \in B(x, \varepsilon/2)$ and $\delta > 0$ with

$$B(y, \delta) \subset B(x, \varepsilon/2) - \bigcup_{k=1}^{p} B_k = B(x, \varepsilon/2) - \bigcup_{k=1}^{\infty} B_k \subset \mathbb{R}^m - U,$$

which contradicts condition (iii).

Let $0 < \delta_k \leq \varepsilon_k$ and $v_k(x) = u_{\delta_k}(x - z_k)$ for each $x \in \mathbb{R}^2$ and $k = 1, 2, \dots$. The vector field

$$v := \sum_{k=1}^{\infty} v_k$$

is well defined, $v(x) = 0$ for all x in $\mathbb{R}^m - U$, and $v(x) = v_k(x)$ whenever $x \in B_k$. Clearly v is bounded and Borel measurable, and Claim 1 implies that it is discontinuous at every point of $\mathbb{R}^m - U$.

Claim 2. The vector field v is charging.

Proof. Choose a bounded BV set A, and observe

$$\sum_{k=1}^{\infty} \int_{\partial_* A} |v_k \cdot \nu_A| \, d\mathcal{H}^{m-1} \leq \sum_{k=1}^{\infty} \int_{B_k \cap \partial_* A} |v_k| \, d\mathcal{H}^{m-1}$$

$$\leq \sqrt{m} \sum_{k=1}^{\infty} \mathcal{H}^{m-1}(B_k \cap \partial_* A)$$

$$= \sqrt{m} \, \mathcal{H}^{m-1} \left(\partial_* A \cap \bigcup_{k=1}^{\infty} B_k \right)$$

$$\leq \sqrt{m} \, \|A\| < \infty.$$

By Lebesgue's dominated convergence theorem,

$$\int_{\partial_* A} v \cdot \nu_A \, d\mathcal{H}^{m-1} = \sum_{k=1}^{\infty} \int_{\partial_* A} v_k \cdot \nu_A \, d\mathcal{H}^{m-1}.$$

In view of inequality $(*)$,

$$\sum_{k=1}^{\infty} \int_{\mathbb{R}^m} |\operatorname{div} v_k| \, d\mathcal{L}^m = \sum_{k=1}^{\infty} \int_{B_k} |\operatorname{div} v_k| \, d\mathcal{L}^m \leq 2 \sum_{k=1}^{\infty} \|B_k\| < \infty.$$

Thus the measurable function $g := \sum_{k=1}^{\infty} \operatorname{div} v_k$ belongs to $L^1(\mathbb{R}^m)$, and Lebesgue's dominated convergence theorem together with the Gauss-Green formula imply

$$\int_{\partial_* A} v \cdot \nu_A \, d\mathcal{H}^{m-1} = \sum_{k=1}^{\infty} \int_{\partial_* A} v_k \cdot \nu_A \, d\mathcal{H}^{m-1}$$

$$= \sum_{k=1}^{\infty} \int_A \operatorname{div} v_k \, \mathcal{L}^m = \int_A g \, d\mathcal{L}^m$$

for each bounded BV set A. It follows from Example 2.1.3 that the flux $\int v \cdot \nu \, d\mathcal{H}^{m-1}$ of v is a charge.

Claim 3. The charge $F := \int v \cdot \nu \, d\mathcal{H}^{m-1}$ is not the flux of a continuous vector field.

Proof. For $\delta > 0$, let $A(\delta)$ be the set of all points $x = (\xi_1, \ldots, \xi_m)$ in $B[\delta/2] - B(\delta/4)$ such that $\xi_i \geq 0$ for $i = 1, \ldots, m$. Observe that

$$\operatorname{div} u_\delta(x) < -\frac{1}{2\delta} \exp(-1) < 0$$

for every $x = (\xi_1, \ldots, \xi_m)$ in $A(\delta)$, and that $\|A(\delta)\| = \beta \delta^{m-1}$ where $\beta := \beta(m)$ is a positive constant. Thus

$$\left| \int_{A(\delta)} \operatorname{div} u_\delta(x) \, dx \right| > \gamma \|A(\delta)\|$$

where $\gamma := [\alpha(m)/(2\beta)] \exp(-1)$. Letting $A_k := z_k + A(\delta_k)$, the Gauss-Green formula yields

$$|F(A_k)| = \left| \int_{\partial_* A_k} v_k \cdot \nu_{A_k} \, d\mathcal{H}^{m-1} \right| = \left| \int_{A_k} \operatorname{div} v_k(x) \, dx \right|$$

$$= \left| \int_{A(\delta_k)} \operatorname{div} u_{\delta_k}(x) \, dx \right| > \gamma \|A(\delta_k)\| = \gamma \|A_k\|$$

for $k = 1, 2, \ldots$. As $\lim d(A_k) = 0$, the claim follows from Remark 2.1.8.

In contrast to Example 2.1.10, a bounded vector field with a single discontinuity need not be charging. This is clear in dimension one, and the next example shows it is also true in higher dimensions.

Example 2.1.11. Assume $m \geq 2$. For each x in $\mathbb{R}^m - \{0\}$, let

$$v(x) := \frac{1}{\sqrt{m}} \left(\cos \frac{\pi}{|x|^{m-1}}, \cdots, \cos \frac{\pi}{|x|^{m-1}} \right),$$

and let $v(0) := 0$. Since v is bounded and Borel measurable, we can define its flux $F := \int v \cdot \nu \, d\mathcal{H}^{m-1}$. For $k = 1, 2, \ldots$, let

$$A_k := \left\{ x \in \mathbb{R}^m : (2k+1)^{-\frac{1}{m-1}} \leq |x| \leq (2k)^{-\frac{1}{m-1}} \right\},$$

and observe

$$F(A_k) = m\,\alpha(m) \left(\frac{1}{2k+1} + \frac{1}{2k} \right) = \|A_k\|.$$

As $\sum_{k=1}^{\infty} \|A_k\| = \infty$, there are positive integers $p_i < q_i < p_{i+1}$ so that for $i = 1, 2, \ldots$, the sets $B_i := \bigcup_{k=p_i}^{q_i} A_k$ satisfy the inequality

$$m\,\alpha(m) \leq \|B_i\| \leq 2m\,\alpha(m).$$

Since $\lim |B_i| = 0$ and $F(B_i) = \|B_i\|$, the flux F is not a charge by Proposition 2.1.2.

The following example of Buczolich shows that the orthogonal projection $w = (f_1, 0, \ldots, 0)$ of a charging vector field $v = (f_1, \ldots, f_m)$ need not be charging.

Example 2.1.12. Assume $m = 2$. For $r > 0$, use [65, Lemma 10.4.1] to find a function φ_r in $C^\infty(\mathbb{R}; [0,1])$ so that

$$\varphi_r(t) := \begin{cases} 1 & \text{if } |t| \leq \frac{r}{2}, \\ 0 & \text{if } |t| \geq r, \end{cases}$$

and for $x = (\xi, \eta)$ in $\mathbb{R}^2$, let

$$u_r(x) := \begin{cases} \frac{\varphi_r(|x|)}{|x|}(-\eta, \xi) & \text{if } |x| > 0, \\ 0 & \text{if } |x| = 0. \end{cases}$$

Since $\operatorname{div} u_r(x) = 0$ for each x in $\mathbb{R}^2 - \{0\}$, the Gauss-Green formula implies

$$\int_{\partial_* A} u_r \cdot \nu_A \, d\mathcal{H}^{m-1} = \int_{\partial_*[A \cap B(\varepsilon)]} u_r \cdot \nu_{A \cap B(\varepsilon)} \, d\mathcal{H}^{m-1}$$

$$\leq \|A \cap B(\varepsilon)\|$$

for each bounded BV set A and each $\varepsilon > 0$. Letting $\varepsilon \to 0$ and using Proposition 1.9.2, we see that $\int_{\partial_* A} u_r \cdot \nu_A \, d\mathcal{H}^{m-1} = 0$ for every bounded BV set A.

For $n = 1, 2, \ldots$, divide the strip $[0,1] \times [2^{-n}, 2^{-n+1}]$ into non-overlapping squares $K_{n,1} \ldots, K_{n,2^n}$ of equal size. If $z_{n,i} = (\xi_{n,i}, \eta_{n,i})$

is the center of $K_{n,i}$, let $v_{n,i}(x) = u_{2^{-n-1}}(x - z_{n,i})$. By our earlier argument, $\int_{\partial_* A} v_{n,i} \cdot \nu_A \, d\mathcal{H}^{m-1} = 0$ for each bounded BV set A. Let

$$v := \sum_{n=1}^{\infty} \sum_{i=1}^{2^n} v_{n,i} \,,$$

and denote by f and g the first and second coordinate of v, respectively.

Claim 1. The vector field $v = (f, g)$ is charging.

Proof. The vector field v is Borel measurable and $|v|_\infty \leq 1$. We prove the claim by showing the flux of v is identically zero. To this purpose, choose a bounded BV set A, and observe that for $n = 1, 2, \ldots$ and $i = 1, \ldots, 2^n$, the balls $B_{n,i} := B(z_{n,i}, 2^{-n-1})$ are disjoint. Since $|v_{n,i}|_\infty \leq 1$, we obtain

$$\sum_{n=1}^{\infty} \sum_{i=1}^{2^n} \int_{\partial_* A} |v_{n,i} \cdot \nu_A| \, d\mathcal{H}^{m-1} \leq \sum_{n=1}^{\infty} \sum_{i=1}^{2^n} \int_{\partial_* A} |v_{n,i}| \, d\mathcal{H}^{m-1}$$

$$\leq \sum_{n=1}^{\infty} \sum_{i=1}^{2^n} \mathcal{H}^{m-1}(\partial_* A \cap B_{n,i})$$

$$\leq \mathcal{H}^{m-1}(\partial_* A) = \|A\| < \infty \,,$$

and by Lebesgue's dominated convergence theorem,

$$\int_{\partial_* A} v \cdot \nu_A \, d\mathcal{H}^{m-1} = \sum_{n=1}^{\infty} \sum_{i=1}^{2^n} \int_{\partial_* A} v_{n,i} \cdot \nu_A \, d\mathcal{H}^{m-1} = 0 \,.$$

Claim 2. The vector field $w := (f, 0)$ is not charging.

Proof. For $n = 1, 2, \ldots$ and $i = 1, \ldots, 2^n$, let

$$K_{n,i}^+ := [\xi_{n,i}, \xi_{n,i} + 2^{-n-1}] \times [\eta_{n,i}, \eta_{n,i} + 2^{-n-1}] \,,$$

and $B_n := \bigcup_{i=1}^{2^n} K_{n,i}^+$. Given an integer $n \geq 1$, observe $\|B_n\| = 2$ and

$$\int_{\partial B_n} w \cdot \nu_{B_n} \, d\mathcal{H}^1 = \sum_{i=1}^{2^n} \int_0^{2^{-n-1}} \varphi_{2^{-n-1}}(t) \, dt > \frac{1}{4} \,.$$

As $\lim |B_n| = 0$, the claim follows from Proposition 2.1.2.

Question 2.1.13. Is there a charge that is not the sum of the indefinite Lebesgue integral of a locally integrable function and the flux of a continuous vector field? Is every charge the flux of a charging vector field?

2.2. Spaces of charges

For a charge F and $n = 1, 2, \ldots$, let

$$\|F\|_n := \sup_{B \in \mathcal{BV}_n} |F(B)| \tag{2.2.1}$$

and observe $\|F\|_n \leq \|F\|_{n+1}$. Since each $(\mathcal{BV}_n, \mathcal{T})$ is a compact space and $\mathcal{BV} = \bigcup_{n=1}^{\infty} \mathcal{BV}_n$, it is easy to verify $\{\| \cdot \|_n : n = 1, 2, \ldots\}$ is a separating family of seminorms in CH [74, Definition 1.33]. This family defines a locally convex topology $\mathcal{S}$ in CH induced by an invariant metric

$$\rho(F, G) := \sum_{n=1}^{\infty} 2^{-n} \frac{\|F - G\|_n}{1 + \|F - G\|_n} \tag{2.2.2}$$

[74, Remark 1.38, (c)]. A sequence $\{F_i\}$ of charges $\mathcal{S}$-converges to a charge F if and only if $\{F_i\}$ converges to F uniformly on each $\mathcal{BV}_n$. It follows $(CH, \mathcal{S})$ is a complete space, and $CH(E)$ is a closed subset of $(CH, \mathcal{S})$ for each locally BV set E.

Observation 2.2.1. *If $A \in \mathcal{BV}_n$, then $\| \cdot \|_n$ is a norm in $CH(A)$.*

PROOF. Select a charge F in A with $\|F\|_n = 0$. Choose a figure C and find nonoverlapping cells $C_1, \ldots, C_p$ so that $C = \bigcup_{i=1}^{p} C_i$. Since $\|A \cap C_i\| \leq n$ by Corollary 1.9.4, we obtain

$$F(C) = \sum_{i=1}^{p} F(C_i) = \sum_{i=1}^{p} F(A \cap C_i) = 0.$$

As $\mathcal{F}$ is dense in $(\mathcal{BV}, \mathcal{T})$, the observation follows. $\qquad\qquad\qquad\square$

Lemma 2.2.2. *Let $r > 0$, and let $B \subset B(r)$ be a BV set. There are numbers $-r = t_0 < \cdots < t_p = r$ such that $p \leq \max\{1, 2\|B\|\}$ and*

$$\left\| B \cap \left([t_{i-1}, t_i) \times \mathbb{R}^{m-1} \right) \right\| < 1 + 4d(B)^{m-1}$$

for $i = 1, \ldots, p$.

PROOF. Let $\beta := 2d(B)^{m-1}$. If $\|B\| \leq 2\beta$, the lemma holds with $p = 1$. Thus assume $2\beta < \|B\|$, and for every $t \in [-r, r]$ let

$$B_t^- := B \cap \left([-r, t) \times \mathbb{R}^{m-1} \right) \quad \text{and} \quad B_t^+ := B \cap \left([t, r) \times \mathbb{R}^{m-1} \right).$$

It follows from Theorems 1.9.3 and 1.7.1, (1), respectively, that the function $t \mapsto \|B_t^-\|$ defined on the cell $[-r, r]$ is increasing and left continuous. Now $\|B_{-r}^-\| = 0$, $\|B_r^-\| = \|B\|$, and it is easy to deduce from Proposition 1.8.5 that

$$\lim_{s \to t+} \|B_s^-\| \leq \|B_t^-\| + \beta \quad \text{and} \quad \|B_t^-\| + \|B_t^+\| \leq \|B\| + \beta$$

for each $t \in [-r, r)$. We infer there is an $s \in (-r, r)$ such that

$$\frac{1}{2}\|B\| \le \|B_s^-\| \le \frac{1}{2}\|B\| + \beta,$$

and hence

$$\|B_s^+\| \le \|B\| - \|B_s^-\| + \beta \le \frac{1}{2}\|B\| + \beta.$$

Choose the least integer $n \ge 0$ for which $\|B\| \le 2^n$, and observe

$$2^n \le \max\{1, 2\|B\|\}.$$

If $\|B_s^-\| \le 2\beta$ do nothing; otherwise, repeat the previous argument with B replaced by B_s^-. Apply the same alternative to B_s^+. Inductively, we obtain numbers $-r = t_0 < \cdots < t_p = r$ such that $p \le 2^n$ and

$$\left\|B \cap \left([t_{i-1}, t_i) \times \mathbb{R}^{m-1}\right)\right\| \le 2^{-n}\|B\| + \beta \sum_{k=0}^{n-1} 2^{-k} < 1 + 2\beta$$

for $i = 1, \ldots, p$. $\qquad\square$

Corollary 2.2.3. *Let B be a BV subset of $B[k]$ where k is an integer with $k \ge 1 + 4d(B)^{m-1}$, and assume $\|B\| \ge k$. If F is a charge, then*

$$\big|F(B)\big| \le 2\|B\| \cdot \|F\|_k.$$

PROOF. By Lemma 2.2.2, the set B is the union of disjoint BV sets $B_1, \ldots, B_p$ such that $p \le 2\|B\|$ and $\|B_i\| < k$ for $i = 1, \ldots, p$. Thus

$$\big|F(B)\big| \le \sum_{i=1}^p \big|F(B_i)\big| \le p\|F\|_k \le 2\|B\| \cdot \|F\|_k,$$

since each B_i belongs to $\mathcal{BV}_k$. $\qquad\square$

Proposition 2.2.4. *Let A be a BV subset of $B[k]$ where k is an integer with $k \ge 1 + 4d(A)^{m-1}$. If F is a charge in A, then*

$$\|F\|_n \le 2(\|A\| + n)\|F\|_k$$

for each integer $n \ge k$. The seminorm $\|\cdot\|_k$ is a norm in $CH(A)$ that induces the subspace topology in $CH(A) \subset (CH, \mathcal{S})$. In particular, $(CH(A), \|\cdot\|_k)$ is a Banach space.

PROOF. Select an integer $n \ge k$, and a choose set $B \in \mathcal{BV}_n$. Observe

$$\big|F(B)\big| = \big|F(A \cap B)\big| \le \|F\|_k \le 2(\|A\| + n)\|F\|_k$$

if $\|A \cap B\| \leq k$, and by Corollary 2.2.3,

$$|F(B)| = |F(A \cap B)| \leq 2\|A \cap B\| \cdot \|F\|_k$$
$$\leq 2(\|A\| + \|B\|)\|F\|_k \leq 2(\|A\| + n)\|F\|_k$$

if $\|A \cap B\| > k$. Thus $\|F\|_n \leq 2(\|A\| + n)\|F\|_k$, and hence $\|F\|_k = 0$ implies $\|F\|_i = 0$ for $i = 1, 2, \ldots$. It follows $\{ \|\cdot\|_i : i = k, k+1, \ldots \}$ is a family of equivalent norms in $CH(A)$. $\qquad\square$

The next lemma facilitates an important characterization of charges presented in Proposition 2.2.6 below.

Lemma 2.2.5. *If F is a charge, then*

$$\lim \frac{F(B_i)}{\|B_i\|} = 0$$

for each bounded sequence $\{B_i\}$ of BV sets with $\lim \|B_i\| = \infty$.

PROOF. Choose $r > 0$ and a positive

$$\varepsilon < \frac{1}{\max\{r, 1 + 4(2r)^{m-1}\}}.$$

Use Proposition 2.1.2, (i) to find an $\eta > 0$ so that $|F(D)| < \varepsilon$ for each BV set $D \subset B(1/\varepsilon)$ with $\|D\| < 1/\varepsilon$ and $|D| < \eta$. Select an integer $q > (2r)^m/\eta$ and for $j = 1, \ldots, q$, let

$$C_j := \left[-r + (j-1)\frac{2r}{q}, -r + j\frac{2r}{q} \right) \times [-r, r]^{m-1}.$$

Now choose a BV set $B \subset B(r)$ with $\|B\| \geq q$. By Lemma 2.2.2, there are numbers $-r = t_0 < \cdots < t_p = r$ such that $p \leq 2\|B\|$ and the sets

$$B_i := B \cap \left([t_{i-1}, t_i) \times \mathbb{R}^{m-1} \right), \quad i = 1, \ldots, p,$$

have perimeters less than $1/\varepsilon$. The collection

$$\mathfrak{D} := \{ B_i \cap C_j : i = 1, \ldots, p; \ j = 1, \ldots, q \}$$

contains at most $p + q - 1$ nonempty sets and $B = \bigcup \mathfrak{D}$. We have

$$|B_i \cap C_j| \leq |C_j| < \eta \quad \text{and} \quad \|B_i \cap C_j\| \leq \|B_i\| < \frac{1}{\varepsilon},$$

since C_j are convex sets (Corollary 1.9.4). Thus

$$|F(B)| \leq \sum_{D \in \mathfrak{D}} |F(D)| < \varepsilon(p + q - 1) < 3\varepsilon\|B\|,$$

and the lemma follows. $\qquad\square$

Proposition 2.2.6. *An additive function F on $\mathcal{BV}$ is a charge if and only if given $\varepsilon > 0$, there is a $\theta > 0$ such that*

$$\big|F(B)\big| < \theta|B| + \varepsilon\big(\|B\| + 1\big)$$

for each BV set $B \subset B(1/\varepsilon)$.

PROOF. As the converse is obvious, assume F is a charge, and choose an $\varepsilon > 0$ and a BV set $B \subset B(1/\varepsilon)$. Using Lemma 2.2.5, find an integer $k \geq 1/\varepsilon$ so that $\|B\| \geq k$ implies $\big|F(B)\big| < \varepsilon\|B\|$. By continuity, there is an $\eta > 0$ such that $\big|F(B)\big| < \varepsilon$ whenever $\|B\| < k$ and $|B| < \eta$. If $\|B\| < k$ and $|B| \geq \eta$, then

$$\big|F(B)\big| \leq \|F\|_k \leq \|F\|_k \frac{|B|}{\eta}\,.$$

Thus selecting $\theta > \|F\|_k/\eta$, the desired inequality follows from the above alternatives. $\qquad\qquad\qquad\qquad\qquad\qquad\qquad\qquad\qquad\qquad\square$

2.3. Derivates

The *regularity* of a bounded BV set E is the number

$$r(E) := \begin{cases} \frac{|E|}{d(E)\|E\|} & \text{if } |E| > 0, \\ 0 & \text{if } |E| = 0. \end{cases}$$

The number $r(E)$ is well defined, since $|E| > 0$ implies $d(E)\|E\| > 0$; this is clear for $m = 1$, and it follows from the isoperimetric inequality for $m \geq 2$. The relationship between the regularity and shape of a bounded BV set E is given by the inequality

$$r(E)^m \leq \frac{1}{m^m \alpha(m)} s(E)\,. \tag{2.3.1}$$

Indeed, if $m = 1$ then $r(E) \leq s(E)/\alpha(1)$, since $\|E\| \geq 2 = \alpha(1)$ whenever $|E| > 0$. If $m \geq 2$ and $|E| > 0$, then

$$r(E)^m = \frac{|E|^{m-1}}{\|E\|^m} \cdot \frac{|E|}{d(E)^m} \leq \frac{1}{m^m \alpha(m)} s(E)$$

by the isoperimetric inequality. In view of inequality (2.3.1), regularity provides more information than shape: it controls *both* the shape and the perimeter of a bounded BV set.

The regularity of a ball is $1/(2m)$. Inequality (2.3.1) and the isodiametric inequality (1.4.1) imply

$$r(E) \leq \frac{1}{2m} \tag{2.3.2}$$

for every bounded BV set E.

Let F be a function defined on the family $\mathcal{BV}$, and let $x \in \mathbb{R}^m$. Given $\eta \geq 0$, let

$$\underline{D}_\eta F(x) := \sup_{\delta > 0} \inf_E \frac{F(E)}{|E|} \quad \text{and} \quad \overline{D}_\eta F(x) := \inf_{\delta > 0} \sup_E \frac{F(E)}{|E|}$$

where E is a BV set such that $d\bigl(E \cup \{x\}\bigr) < \delta$ and $r\bigl(E \cup \{x\}\bigr) > \eta$. The extended real numbers

$$\underline{D}F(x) := \inf_{\eta > 0} \underline{D}_\eta F(x) \quad \text{and} \quad \overline{D}F(x) := \sup_{\eta > 0} \overline{D}_\eta F(x)$$

are called, respectively, the *lower* and *upper derivate* of F at x. We say F is *derivable* at x whenever

$$\underline{D}F(x) = \overline{D}F(x) \neq \pm\infty,$$

and call this common value the *derivate* of F at x, denoted by $DF(x)$. Throughout this book, writing $DF(x) = c$ implies that F is derivable at x. We say F is *almost derivable* at x if $\overline{D}_\eta |F|(x) < \infty$ for each $\eta > 0$. Clearly, F is almost derivable at x whenever $\overline{D}|F|(x) < \infty$.

On $\mathbb{R}^m$, we define the extended real-valued functions $\underline{D}_\eta F$, $\overline{D}_\eta F$, $\underline{D}F$, and $\overline{D}F$ in the obvious way. Observe

$$\overline{D}F = -\underline{D}(-F) \quad \text{and} \quad \overline{D}_\eta F = -\underline{D}_\eta(-F) \tag{2.3.3}$$

for each $\eta \geq 0$. If $0 < \eta \leq \theta < 1/(2m)$, then

$$\underline{D}F \leq \underline{D}_\eta F \leq \underline{D}_\theta F \leq \overline{D}_\theta F \leq \overline{D}_\eta F \leq \overline{D}F. \tag{2.3.4}$$

On the other hand, if $\eta \geq 1/(2m)$, then $\underline{D}_\eta F = -\overline{D}_\eta F = \infty$ in view of inequality (2.3.2).

Let F be a function defined on $\mathcal{BV}$, and let $x \in \mathbb{R}^m$. If $DF(x)$ exists, then so does the well-known *ordinary derivate* of F at x [75, Chapter 4, Section 2]. If $m = 1$, the converse is true by Claim 2 of Example 2.3.2 below, although the *ordinary lower derivate* of F may differ from $\underline{D}F$ even when F is a charge. The latter fact can be established indirectly: assuming the ordinary lower derivate of each charge F is the same as $\underline{D}F$, one shows that each *Perron integrable* function [75, Chapter 6, Section 6] is $\mathcal{F}$-*integrable* according to [65, Definition 12.2.1]; this contradicts [65, Example 12.3.5]. For $m \geq 2$, the converse is false in a strong way. Indeed, it follows from [15] there are a measurable set E of positive measure and a charge F such that the ordinary derivate of F exists at each $x \in E$ and $DF(x)$ exists at no $x \in E$.

In general, we shall not be concerned with the special behavior of $\underline{D}_0 F$ and $\overline{D}_0 F$. Only for illustration, we present the following proposition.

Proposition 2.3.1. *If F is a function defined on $\mathcal{BV}$, then the extended real-valued functions $\underline{D}_0 F$ and $\overline{D}_0 F$ are, respectively, lower and upper semicontinuous.*

PROOF. Choose an $x \in \mathbb{R}^m$ and suppose $\underline{D}_0 F(x) > c$. There is a $\delta > 0$ such that $F(E)/|E| > c$ for each BV set $E \subset B(x, \delta)$ with $|E| > 0$. Let $y \in B(x, \delta)$, and find $\sigma > 0$ so that $B(y, \sigma) \subset B(x, \delta)$. Observe $E \subset B(x, \delta)$ whenever $d(E \cup \{y\}) < \sigma$, and conclude $\underline{D}_0 F(y) \geq c$. Thus $\underline{D}_0 F$ is lower semicontinuous at x, and the proposition follows from equality (2.3.3). $\qquad \square$

Example 2.3.2. Let $x \in \mathbb{R}^m$, and let $F := \int v \cdot \nu \, d\mathcal{H}^{m-1}$ be the flux of a locally bounded Borel measurable vector field $v : \mathbb{R}^m \to \mathbb{R}^m$.

Claim 1. If v is Lipschitz at x, then F is almost derivable at x.

Proof. Find positive numbers c and δ so that

$$|v(y) - v(x)| \leq c|y - x|$$

for each $y \in B(x, \delta)$. Select a positive $\eta < 1/(2m)$ and a BV set A with $d(A \cup \{x\}) < \delta$ and $r(A \cup \{x\}) > \eta$. Then

$$
\begin{aligned}
|F(A)| &= \left| \int_{\partial_* A} [v(y) - v(x)] \cdot \nu_A(y) \, d\mathcal{H}^{m-1}(y) \right| \\
&\leq \int_{\partial_* A} |v(y) - v(x)| \, d\mathcal{H}^{m-1}(y) \\
&\leq c \int_{\partial_* A} |y - x| \, d\mathcal{H}^{m-1}(y) \\
&\leq c \, d(A \cup \{x\}) \|A\| \leq \frac{c}{\eta} |A|
\end{aligned}
$$

by the Gauss-Green formula. We conclude $\overline{D}_\eta |F|(x) < \infty$.

Claim 2. If v is differentiable at x, then F is derivable at x and

$$DF(x) = \operatorname{div} v(x).$$

Proof. Choose positive numbers ε and $\eta < 1/(2m)$. Let

$$w(y) := v(x) + [Dv(x)](y - x)$$

for all $y \in \mathbb{R}^m$ (see Section 1.6). Then $\operatorname{div} w(y) = \operatorname{div} v(x)$ for every $y \in \mathbb{R}^m$, and there is a $\delta > 0$ such that

$$|v(y) - w(y)| < \varepsilon \eta |y - x|$$

for each $y \in B(x,\delta)$. Let A be a BV set with $d(A \cup \{x\}) < \delta$ and $r(A \cup \{x\}) > \eta$. Then

$$\operatorname{div} v(x)|A| = \int_A \operatorname{div} w(y)\, dy = \int_{\partial_* A} w \cdot \nu_A \, d\mathcal{H}^{m-1}$$

by the Gauss-Green formula. Consequently

$$\begin{aligned}
\left|\operatorname{div} v(x)|A| - F(A)\right| &= \left|\int_{\partial_* A} (w - v) \cdot \nu_A \, d\mathcal{H}^{m-1}\right| \\
&\leq \int_{\partial_* A} |w(y) - v(y)|\, d\mathcal{H}^{m-1}(y) \\
&< \varepsilon\eta \int_{\partial_* A} |y - x|\, d\mathcal{H}^{m-1}(y) \\
&\leq \varepsilon\eta\, d(A \cup \{x\}) \|A\| < \varepsilon|A|,
\end{aligned}$$

and the claim follows.

Appreciably stronger results than Claims 1 and 2 of Example 2.3.2 are established in Proposition 2.5.7 below.

In dimension one, it is clear that the sufficient conditions of Example 2.3.2 are also necessary. Not so in higher dimensions: following Buczolich, we prove that for a suitable choice of the numbers δ_k, the flux of the discontinuous vector field v constructed in Example 2.1.10 is derivable at each $x \in \mathbb{R}^m$.

Example 2.3.3. Assume $m \geq 2$, and define u_δ, z_k, ε_k, v_k, and U as in Example 2.1.10. Let v be the vector field of Example 2.1.10 corresponding to $\delta_k := \varepsilon_k^3$, $k = 1, 2, \ldots$, and let $F := \int v \cdot \nu \, d\mathcal{H}^{m-1}$.

According to Claim 2 of Example 2.3.2, the charge F is derivable at each $x \in U$. Select an x in $\mathbb{R}^m - U$, an $\eta > 0$, and a bounded BV set A with $r(A \cup \{x\}) > \eta$. Let $V_k := B(z_k, \delta_k)$ and $V := \bigcup_{k=1}^{\infty} V_k$. If $V \cap \operatorname{cl} A = \emptyset$, then $F(A) = 0$. Otherwise, there is the least integer $p \geq 1$ such that $\operatorname{cl} A$ meets V_p. As $x \notin B(z_p, \varepsilon_p)$, we have $d(A \cup \{x\}) > \varepsilon_p - \delta_p$, and in view of inequality (2.3.1),

$$\begin{aligned}
|A| = s(A \cup \{x\})d(A \cup \{x\})^m &\geq m^m \alpha(m)\eta^m d(A \cup \{x\})^m \\
&> \alpha(m)(m\eta)^m(\varepsilon_p - \delta_p)^m = \alpha(m)(m\eta)^m(1 - \varepsilon_p^2)^m \varepsilon_p^m \\
&> \alpha(m)\left(\frac{m\eta}{2}\right)^m \varepsilon_p^m.
\end{aligned}$$

By the Gauss-Green formula and inequality $(*)$ of Example 2.1.10,

$$
|F(A)| \leq \sum_{k=1}^{\infty} \left| \int_{\partial_* A} v_k \cdot \nu_A \, d\mathcal{H}^{m-1} \right| = \sum_{k=1}^{\infty} \left| \int_A \operatorname{div} v_k \, d\mathcal{L}^m \right|
$$

$$
\leq \sum_{k=p}^{\infty} \int_{V_k} |\operatorname{div} v_k| \, d\mathcal{L}^m \leq 2 \sum_{k=p}^{\infty} \|V_k\|
$$

$$
= 2m\,\alpha(m) \sum_{k=p}^{\infty} \delta_k^{m-1} < 4m\,\alpha(m)\varepsilon_p^{3(m-1)},
$$

and hence

$$
\left| \frac{F(A)}{|A|} \right| < 4m \left(\frac{2}{m\eta} \right)^m \varepsilon_p^{2m-3}.
$$

As the diameter of $A \cup \{x\}$ is getting smaller while $\operatorname{cl} A$ keeps intersecting V, the integer p approaches infinity. Since $m \geq 2$, the last inequality implies $DF(x) = 0$.

Lemma 2.3.4. *Let F be a function defined on $\mathcal{BV}$, let $x \in \mathbb{R}^m$, and let $\eta \geq 0$. If F is $\mathcal{T}$-continuous, then*

$$
\underline{D}_\eta F(x) = \sup_{\delta > 0} \inf_B \frac{F(B)}{|B|} \quad \text{and} \quad \overline{D}_\eta F(x) = \inf_{\delta > 0} \sup_B \frac{F(B)}{|B|}
$$

where B is an essentially clopen BV set such that

$$
B = \operatorname{cl}_* B, \quad x \in B, \quad d(B) < \delta, \quad \text{and} \quad r(B) > \eta. \qquad (*)
$$

PROOF. We prove only the first equality; the proof of the second one is analogous. Clearly

$$
\underline{D}_\eta F(x) \leq b := \sup_{\delta > 0} \inf_B \frac{F(B)}{|B|}
$$

where B is an essentially clopen BV set satisfying condition $(*)$. Proceeding toward a contradiction, suppose $\underline{D}_\eta F(x) < b$. There are $\delta > 0$, and a BV set E such that $d(E \cup \{x\}) < \delta$, $r(E \cup \{x\}) > \eta$, and

$$
\frac{F(E)}{|E|} < c := \inf_B \frac{F(B)}{|B|}
$$

where B is an essentially clopen BV set satisfying condition $(*)$. Select a positive $t < 1$ and $\alpha, \beta > 1$ such that

$$
d(E \cup \{x\}) < \frac{\delta}{\alpha} \quad \text{and} \quad r(E \cup \{x\}) > \eta \frac{\alpha\beta}{t}.
$$

Next find an open set $U \subset \mathbb{R}^m$ so that

$$\mathrm{cl}\left(E \cup \{x\}\right) \subset U \quad \text{and} \quad d(U) < \alpha\,d\!\left(E \cup \{x\}\right).$$

It follows from De Giorgi's theorem that there is an essentially clopen BV set $C \subset U$ with

$$C = \mathrm{cl}_* C, \quad |C| > t|E|, \quad \|C\| < \beta\|E\|, \quad \text{and} \quad \frac{F(C)}{|C|} < c.$$

If $x \in C$, then we have a contradiction, since

$$r\!\left(C \cup \{x\}\right) > \frac{t}{\alpha\beta}\, r\!\left(E \cup \{x\}\right) > \eta.$$

If $x \notin C$, choose an $\varepsilon > 0$ so that C does not meet $B[x,\varepsilon]$, and observe that a contradiction follows by considering the essentially clopen BV set $B := C \cup B[x,\varepsilon]$ for a sufficiently small ε. $\qquad\square$

Proposition 2.3.5. *Let F be a function defined on $\mathcal{BV}$, and let $\eta \geq 0$. If F is $\mathfrak{T}$-continuous, then the extended real-valued functions $\underline{D}_\eta F$ and $\overline{D}_\eta F$ are measurable, and so are $\underline{D}F$ and $\overline{D}F$.*

PROOF. In view of equalities (2.3.3) and the equality

$$\underline{D}F = \lim_{k \to \infty} \underline{D}_{1/k}F,$$

it suffices to prove the measurability of $\underline{D}_\eta F$. To this end, select a $c \in \mathbb{R}$ and let $E := \left\{x \in \mathbb{R}^m : \underline{D}_\eta F(x) < c\right\}$. Lemma 2.3.4 implies $x \in E$ if and only if there is an integer $i \geq 1$ such that for each integer $j \geq 1$ we can find an essentially clopen BV set B such that

$$B = \mathrm{cl}_* B, \quad x \in B, \quad d(B) < 1/j, \quad r(B) > \eta \quad \text{and} \quad \frac{F(B)}{|B|} < c - \frac{1}{i}.$$

Thus if $E_{i,j}$ is the union of all essentially clopen BV sets B satisfying the above condition, then

$$E := \bigcup_{i=1}^{\infty} \bigcap_{j=1}^{\infty} E_{i,j}.$$

It follows from Proposition 1.5.10 that E is a measurable set. $\qquad\square$

Given $\eta \geq 0$, we say that a sequence $\{B_i\}$ of bounded BV sets *η-converges* to an $x \in \mathbb{R}^m$ whenever

$$\lim d\!\left(B_i \cup \{x\}\right) = 0 \quad \text{and} \quad \liminf r\!\left(B_i \cup \{x\}\right) > \eta.$$

Clearly, if $\{B_i\}$ is an η-converging sequence in $\mathcal{BV}$, then it is bounded and all but finitely many sets B_i have positive measure. Thus if F

is a function defined on $\mathcal{BV}$, then $\liminf$ and $\limsup$ of the sequence $\{F(B_i)/|B_i|\}$ are defined. Instead of saying $\{B_i\}$ 0-converges to x, we usually say $\{B_i\}$ *tends* to x. The next observation is often useful. Its straightforward proof is left to the reader.

Observation 2.3.6. *Let F be a function defined on the family $\mathcal{BV}$, and let $x \in \mathbb{R}^m$. For $\theta > \eta > 0$, we have*

$$\underline{D}_\eta F(x) \le \inf_{\{B_i\}} \liminf \frac{F(B_i)}{|B_i|} \le \underline{D}_\theta F(x),$$

$$\overline{D}_\theta F(x) \le \sup_{\{B_i\}} \limsup \frac{F(B_i)}{|B_i|} \le \overline{D}_\eta F(x)$$

where $\{B_i\}$ is a sequence in $\mathcal{BV}$ that η-converges to x. In particular,

$$\underline{D}F(x) = \inf_{\{B_i\}} \liminf \frac{F(B_i)}{|B_i|} \quad \text{and} \quad \overline{D}F(x) = \sup_{\{B_i\}} \limsup \frac{F(B_i)}{|B_i|}$$

where $\{B_i\}$ is a sequence in $\mathcal{BV}$ tending to x. Furthermore, F is derivable at x if and only if a finite limit

$$\lim \frac{F(B_i)}{|B_i|}$$

exists for each sequence $\{B_i\}$ in $\mathcal{BV}$ tending to x; in which case all these limits have the same value equal to $DF(x)$.

Let F be a function defined on the family $\mathcal{F}_d$ of dyadic figures, and let $x \in \mathbb{R}^m$. Given $\eta \ge 0$, let

$$\underline{D}_\eta^d F(x) := \sup_{\delta > 0} \inf_C \frac{F(C)}{|C|} \quad \text{and} \quad \overline{D}_\eta^d F(x) := \inf_{\delta > 0} \sup_C \frac{F(C)}{|C|}$$

where $C \in \mathcal{F}_d$ is such that $x \in C$, $d(C) < \delta$, and $r(C) > \eta$. Using Lemmas 2.3.4 and 1.10.2, the reader can easily establish the following observation, whose proof is similar to that of Lemma 2.3.4.

Observation 2.3.7. *Let F be a $\mathcal{T}$-continuous function defined on $\mathcal{BV}$, and let $x \in \mathbb{R}^m$. If β is the constant from Lemma 1.10.2, then*

$$\underline{D}_\eta F(x) \le \underline{D}_\eta^d F(x) \le \underline{D}_{\beta\eta} F(x),$$

$$\overline{D}_{\beta\eta} F(x) \le \overline{D}_\eta^d F(x) \le \overline{D}_\eta F(x)$$

for each $\eta \ge 0$. In particular,

$$\underline{D}F(x) = \inf_{\eta > 0} \underline{D}_\eta^d F(x) \quad \text{and} \quad \overline{D}F(x) = \sup_{\eta > 0} \overline{D}_\eta^d F(x).$$

2.4. Derivability

We justify the term *almost derivable*, introduced in Section 2.3, by showing that a charge which is almost derivable at each point of a set $E \subset \mathbb{R}^m$, is derivable at almost all points of E (Theorem 2.4.3 below). This important result was established by Buczolich [18, Theorem 3.3]. Its applications are given in Chapter 3.

Let F be a function defined on the family of all dyadic cubes, and let $x \in \mathbb{R}^m$. Given $\eta \geq 0$, let

$$\underline{F}_\eta(x) := \sup_{\delta > 0} \inf_C \frac{F(C)}{|C|} \quad \text{and} \quad \overline{F}_\eta(x) := \inf_{\delta > 0} \sup_C \frac{F(C)}{|C|}$$

where C is a dyadic cube such that $d(C \cup \{x\}) < \delta$ and $r(C \cup \{x\}) > \eta$. Since the regularity of a cube is $1/(2m\sqrt{m})$, we have

$$\underline{D}_\eta F(x) \leq \underline{F}_\eta(x) \leq \underline{F}_\theta(x) \leq \overline{F}_\theta(x) \leq \overline{F}_\eta(x) \leq \overline{D}_\eta F(x) \qquad (2.4.1)$$

whenever $0 < \eta < \theta < 1/(2m\sqrt{m})$. If

$$\inf_{\eta > 0} \underline{F}_\eta(x) = \sup_{\eta > 0} \overline{F}_\eta(x) \neq \pm\infty,$$

we denote this common value by $F'(x)$.

Proposition 2.4.1. *Let $m = 1$, and denote by F the flux of a continuous vector field $v : \mathbb{R} \to \mathbb{R}$. Then F is derivable at $x \in \mathbb{R}$ if and only if $F'(x)$ exists, in which case v is differentiable at x and*

$$DF(x) = F'(x) = v'(x).$$

PROOF. Choose an $x \in \mathbb{R}$, and recall from Remark 2.1.5 that the charge F is determined by the relation

$$F([a, b]) = v(b) - v(a)$$

for each cell $[a, b]$. Since the dyadic rationals are dense in $\mathbb{R}$, the existence of $F'(x)$ implies that v is differentiable at x and $F'(x) = v'(x)$. Now $DF(x) = v'(x)$ by Claim 2 of Example 2.3.2. Conversely, if $DF(x)$ exists, then so does $F'(x)$ by inequality (2.4.1). $\qquad \square$

The next lemma and following theorem are versions of the classical *Ward theorem* [75, Chapter 4, Theorem 11.15].

Lemma 2.4.2. *Let F be an additive function on the family $\mathcal{F}_d$, and let $E \subset \mathbb{R}^m$ be such that for each $x \in E$ and each $\eta > 0$, either $\overline{F}_\eta(x) < \infty$ lanebreakor $\underline{F}_\eta(x) > -\infty$. Then $F'(x)$ exists for almost all $x \in E$.*

PROOF. Given $x \in E$, either $\overline{F}_{1/k}(x) < \infty$ or $\underline{F}_{1/k}(x) > -\infty$ for infinitely many integers $k \geq 1$. It follows E is the union of the sets

$$E_+ := \{x \in E : \overline{F}_\eta(x) < \infty \text{ for all } \eta > 0\},$$
$$E_- := \{x \in E : \underline{F}_\eta(x) > -\infty \text{ for all } \eta > 0\}.$$

In view of symmetry, it suffices to show that $F'(x)$ exists for almost all $x \in E_-$. Proceeding toward a contradiction, suppose the set

$$P := \Big\{x \in E_- : \inf_{\eta > 0} \underline{F}_\eta(x) < \sup_{\eta > 0} \overline{F}_\eta(x)\Big\}$$

has positive measure. Given $x \in P$, there is an $\eta_x > 0$ such that $-\infty < \underline{F}_\eta(x) < \overline{F}_\eta(x)$ for each positive $\eta \leq \eta_x$; in particular $\underline{F}_\eta(x)$ is a real number. Thus

$$P = \bigcup_{n=1}^{\infty} \Big\{x \in P : \eta_x > \frac{1}{n} \text{ and } \overline{F}_{\eta_x}(x) - \underline{F}_{\eta_x}(x) > \frac{1}{n}\Big\},$$

and there are a positive integer n and a set $Q \subset P$ of positive measure such that $\overline{F}_\eta(x) - \underline{F}_\eta(x) > 1/n$ for each $x \in Q$ and each positive $\eta \leq 1/n$. Fix a positive $\eta < 1/(4m\sqrt{m})$ so that $\overline{F}_\eta(x) - \underline{F}_\eta(x) > \eta$ for each $x \in Q$. Select an $\varepsilon > 0$, and for every integer k, let

$$Q_k := \{x \in Q : k\varepsilon < \underline{F}_\eta(x) \leq (k+1)\varepsilon\}.$$

As $Q = \bigcup_{k=-\infty}^{\infty} Q_k$, there is an integer p with $|Q_p| > 0$. Now letting $G := F - p\varepsilon \mathcal{L}^m$, a simple calculation reveals that

$$0 < \underline{G}_\eta(x) < 2\varepsilon \quad \text{and} \quad \overline{G}_\eta(x) > \eta \qquad (*)$$

for each $x \in Q_p$. The remainder of the proof depends only on inequalities $(*)$ and the properties of dyadic cubes stated in Section 1.1.

Claim 1. There are $R \subset Q_p$ and $\delta > 0$ such that $|R| > 0$, and $G(B) > 0$ for each dyadic cube B with $d(B) < \delta$ whose mother B^* meets R.

Proof. Given $x \in Q_p$, find a $\delta_x > 0$ so that $G(B) > 0$ for each dyadic cube B with $d(B \cup \{x\}) < \delta_x$ and $r(B \cup \{x\}) > \eta$. Since

$$Q_p = \bigcup_{i=1}^{\infty} \{x \in Q_p : \delta_x > 1/i\},$$

there is an integer $k \geq 1$ for which $R := \{x \in Q_p : \delta_x > 1/k\}$ has positive measure. Let $\delta := 1/(2k)$, and let B be a dyadic cube with

$d(B) < \delta$ and $R \cap B^* \neq \emptyset$. If $x \in R \cap B^*$, then

$$d(B \cup \{x\}) \leq d(B^*) < 2\delta < \delta_x,$$
$$r(B \cup \{x\}) \geq \frac{1}{2}r(B) = \frac{1}{4m\sqrt{m}} > \eta.$$

We conclude $G(B) > 0$, and Claim 1 is established.

Claim 2. There is a dyadic cube K such that

$$|R \cap K| > (1 - \varepsilon)|K|, \quad d(K) < \delta, \quad \text{and} \quad G(K) < 2\varepsilon|K|.$$

Proof. By the density theorem, R contains a density point z. Use Lemma 1.5.6 to find a positive $\sigma < \delta$ such that

$$|R \cap C| > (1 - \varepsilon)|C|$$

for each dyadic cube C with $d(C \cup \{z\}) < \sigma$ and $s(C \cup \{z\}) > \eta^m$. As $\underline{G}_\eta(z) < 2\varepsilon$, there is a dyadic cube K such that $d(K \cup \{z\}) < \sigma$, $r(K \cup \{z\}) > \eta$, and $G(K) < 2\varepsilon|K|$. From

$$\eta < r(K \cup \{z\}) = \frac{1}{2m}s(K \cup \{z\})^{\frac{1}{m}} < s(K \cup \{z\})^{\frac{1}{m}}$$

we obtain $|R \cap K| > (1 - \varepsilon)|K|$, which proofs Claim 2.

Now let $S := R \cap \operatorname{int} K$, and note that for each $x \in S$ and each $\rho > 0$, we can find a dyadic cube $C \subset K$ so that

$$d(C \cup \{x\}) < \rho, \quad r(C \cup \{x\}) > \eta, \quad \text{and} \quad G(C) > \eta|C|.$$

Thus if $\mathcal{C}$ is the collection of all dyadic cubes $C \subset K$ with $G(C) > \eta|C|$, then the family $\{C \cup \{x\} : C \in \mathcal{C} \text{ and } x \in S\}$ is a Vitali cover of S. Since $|S| \leq |K| < \infty$, by Vitali's theorem, there are disjoint sets $C_1, \ldots, C_s$ in $\mathcal{C}$ such that

$$\left| S - \bigcup_{j=1}^{s} C_j \right| < |S| - (1 - \varepsilon)|K|.$$

With no loss of generality we may assume that each C_j meets S. The previous inequality implies

$$\sum_{j=1}^{s} |C_j| > (1 - \varepsilon)|K|.$$

There is a finite family $\mathcal{D}$ of dyadic cubes such that K is the non-overlapping union of the figures $\bigcup_{j=1}^{s} C_j$ and $A := \bigcup \mathcal{D}$. For each $D \in \mathcal{D}$ find the largest (with respect to inclusion) dyadic cube D' with $D \subset D' \subset A$. Let $D_1, \ldots, D_r$ be nonoverlapping cubes from the family

$\{D' : D \in \mathfrak{D}\}$ whose union is A. Then K is the union of nonoverlapping dyadic cubes

$$D_1, \ldots, D_r, C_1, \ldots, C_s$$

where each $D_i^\star$ overlaps, and hence contains, some C_j. In particular, every $D_i^\star$ meets S, and Claim 1 implies $G(D_i) > 0$ for $i = 1, \ldots, r$. From this and Claim 2, we obtain

$$2\varepsilon|K| > G(K) = \sum_{i=1}^{r} G(D_i) + \sum_{j=1}^{s} G(C_j)$$

$$> \eta \sum_{j=1}^{s} |C_j| > \eta(1 - \varepsilon)|K|.$$

Consequently $2\varepsilon > \eta(1 - \varepsilon)$, contrary to the arbitrariness of ε. □

Theorem 2.4.3. *Let F be a charge, and let $E \subset \mathbb{R}^m$ be such that F is almost derivable at all $x \in E$. Then F is derivable at almost all $x \in E$.*

PROOF. Suppose $m \geq 2$; for $m = 1$, the theorem follows immediately from Proposition 2.4.1 and Lemma 2.4.2.

Claim 1. It suffices to prove the theorem for a measurable set E.

Proof. Let $E_{\max}$ be the set of all $x \in \mathbb{R}^m$ at which F is almost derivable. Then $E \subset E_{\max}$, and inequality (2.3.4) yields

$$E_{\max} = \bigcap_{k=1}^{\infty} \{x \in \mathbb{R}^m : \underline{D}_{1/k}|F|(x) < \infty\}.$$

By Proposition 2.3.5, the set $E_{\max}$ is measurable.

As there is nothing to prove if $|E| = 0$, assume E is a measurable set of positive measure. In view of inequality (2.4.1) and Lemma 2.4.2, we may further assume $F'(x)$ exists for all $x \in E$. Given $x \in E$, note

$$F'(x) = \sup_{\rho > 0} \inf_{C} \frac{F(C)}{|C|}$$

where C is a dyadic cube with $x \in C$ and $d(C) < \rho$. A proof completely analogous to that of Proposition 2.3.5 shows the function $x \mapsto F'(x)$ defined on E is measurable. This and Proposition 2.3.5 guarantee the measurability of sets involved in this proof. In particular, the set

$$E_1 := \{x \in E : \underline{D}F(x) < \overline{D}F(x)\}$$

is measurable. Proceeding toward a contradiction, suppose $|E_1| > 0$.

Claim 2. For all sufficiently small $\eta > 0$, the set

$$E(\eta) := \big\{x \in E_1 : \underline{D}_\eta F(x) < \overline{D}_\eta F(x)\big\}$$

is measurable and has positive measure.

Proof. Given $x \in E_1$, find an $\eta_x > 0$ so that $\underline{D}_\eta F(x) < \overline{D}_\eta F(x)$ for all positive $\eta < \eta_x$. The set $E' := \{x \in E_1 : \eta_x > \eta'\}$, not necessarily measurable, has positive measure for an $\eta' > 0$, and we see $E' \subset E(\eta)$ whenever $0 < \eta \leq \eta'$.

For the rest of the proof, fix an $\eta > 0$ so that $|E(\eta)| > 0$ and

$$\eta < \frac{1}{2(1+4m)\sqrt{m}} < \frac{1}{2m\sqrt{m}}.$$

Since $\overline{D}_\eta F \leq \overline{D}_\eta |F|$ and $-\underline{D}_\eta F = \overline{D}_\eta(-F) \leq \overline{D}_\eta |F|$, we have

$$-\infty < \underline{D}_\eta F(x) \leq F'(x) \leq \overline{D}_\eta F(x) < \infty$$

for each $x \in E(\eta)$. By symmetry, we can assume the measurable set

$$E_2 := \big\{x \in E(\eta) : \underline{D}_\eta F(x) \leq F'(x) < \overline{D}_\eta F(x)\big\}$$

has positive measure. Using a standard technique, find finite numbers $M > 0$, $c > 0$, and t so that the set E_3 of all $x \in E_2$ with

$$t - M < \underline{D}_\eta F(x) \leq F'(x) < t < t + c < \overline{D}_\eta F(x) < t + M$$

has positive measure. Letting $G := F - t\mathcal{L}^m$, it is easy to see that

$$E_3 := \big\{x \in E_2 : -M < \underline{D}_\eta G(x) \leq G'(x) < 0 < c < \overline{D}_\eta G(x) < M\big\}$$

is a measurable set of positive measure. For each $x \in E_3$ select the *largest* $\delta_x > 0$ so that the following holds:

- $G(C) < 0$ for each dyadic cube C such that $d\big(C \cup \{x\}\big) < \delta_x$ and $r\big(C \cup \{x\}\big) > \eta$;
- $-M < G(A)/|A| < M$ for each figure A with $x \in A$, $d(A) < \delta_x$, and $r(A) > \eta$.

As δ_x is maximal, it is easy to verify the function $x \mapsto \delta_x$ is upper semicontinuous, and hence measurable. Find a positive $\delta < 1$ so that the measurable set $P := \{x \in E_3 : \delta_x > \delta\}$ has positive measure, and observe that the following conditions are satisfied.

(i) $G(C) < 0$ for each dyadic cube C such that $P \cap C \neq \emptyset$ and $d(C) < \delta$.

(ii) $-M < G(A)/|A| < M$ for each figure A such that $P \cap A \neq \emptyset$, $d(A) < \delta$, and $r(A) > \eta$.

The remainder of the proof, which is quite technical, rests only on conditions (i), (ii), and the fact that $\overline{D}_\eta G(x) > c$ for each $x \in P$.

To simplify computations, we must introduced various constants. If $\beta > 1$ is the constant from Lemma 1.10.2, let $\tau := \eta/\beta$, $\tau' := \tau^{\frac{1}{m-1}}$,

$$\kappa := \frac{c\,\alpha(m)}{2} \left(\frac{m\tau}{6}\right)^m, \quad \gamma := \frac{\kappa}{2^{m+2}M}, \quad \theta := 3^{m-1}\left(\frac{\kappa}{4M}\right)^{\frac{m-1}{m}},$$

and $\varepsilon =: \gamma/p$ where $p > \max\{\gamma, (\tau\theta)^{-m}\}$ is an integer. While initially these choices appear completely mysterious, their meaning will transpire as the proof progresses.

According to the density theorem, there is a $z \in P$ which is a density point of P. By Lemma 1.5.6, we can find a positive

$$\sigma < \frac{\tau'\delta}{2\sqrt{m}} < \frac{\delta}{2\sqrt{m}}$$

so that $|S \cap P| > (1 - \varepsilon)|S|$ for each measurable set S with

$$d(S \cup \{z\}) < \sigma \quad \text{and} \quad s(S \cup \{z\}) > (3\sqrt{m})^{-m}.$$

Since $\overline{D}_\eta G(z) > c$, Observation 2.3.7 implies there is a dyadic figure A such that

$$z \in A, \quad d(A) < \frac{\sigma}{6\sqrt{m}}, \quad r(A) > \tau, \quad G(A) > c|A|.$$

Let $d := d(A)$, and find an integer k so that $2^{-k-1} < d \leq 2^{-k}$; note $k \geq 0$, since $d < 1$. Clearly z is contained in a dyadic cube J_1 of diameter $2^{-k}\sqrt{m}$, and we denote by $J_1, \ldots, J_{3^m}$ all dyadic cubes adjacent to J_1. Since $d(J_i \cup \{z\}) < 4d\sqrt{m} < \sigma$ and $s(J_i \cup \{z\}) \geq (2\sqrt{m})^{-m}$, we have

$$|J_i \cap P| > (1 - \varepsilon)|J_i| > 0 \quad \text{for} \quad i = 1, \ldots, 3^m.$$

In particular, each J_i meets P. Let $J := \bigcup_{i=1}^{3^m} J_i$, and observe

$$A \subset J, \quad z \in J, \quad d(J) < 6d\sqrt{m} < \sigma, \quad s(J) = (\sqrt{m})^{-m}.$$

As P is a measurable set,

$$|J - P| = |J| - |J \cap P| < \varepsilon|J|. \tag{1}$$

Since $r(A) > \tau$, the isoperimetric inequality yields

$$|A| > \alpha(m)(m\tau d)^m > \alpha(m)\left(\frac{m\tau}{6}\right)^m |J| = \frac{2\kappa}{c}|J|. \tag{2}$$

Let $\mathcal{C}$ be a finite collection of nonoverlapping dyadic cubes whose union is A. Subdividing the cubes of $\mathcal{C}$ into smaller dyadic cubes, we may assume each $C \in \mathcal{C}$ is contained in some J_i. Let

$$\mathcal{D} := \{C \in \mathcal{C} : C \cap P = \emptyset\} \quad \text{and} \quad D_0 := \bigcup \mathcal{D}.$$

Condition (i) implies $G(C) < 0$ for each $C \in \mathcal{C} - \mathcal{D}$, and hence

$$c|A| < G(A) = \sum_{C \in \mathcal{C}} G(C) < \sum_{D \in \mathcal{D}} G(D) = G(D_0). \qquad (3)$$

Each $D \in \mathcal{D}$ is contained in a unique J_i, and we denote by D' the largest (with respect to inclusion) dyadic subcube of J_i that contains D and still does not meet P. Let $B_0 := \bigcup\{D' : D \in \mathcal{D}\}$, and find a nonoverlapping subfamily $\mathcal{B}$ of $\{D' : D \in \mathcal{D}\}$ whose union is B_0. Keep in mind $d(B) < 2^{-k}\sqrt{m}$ for each $B \in \mathcal{B}$.

Claim 3. $D_0 = A \cap B_0$.

Proof. Choose cubes $B \in \mathcal{B}$ and $C \in \mathcal{C} - \mathcal{D}$. The cube C does not contain B, since B contains a $D \in \mathcal{D}$. As C meets P and B does not, the cube B does not contain C. Consequently B and C do not overlap, and neither do B_0 and $\bigcup(\mathcal{C} - \mathcal{D})$. The claim follows.

Inequalities (3) and (2) together with Claim 3 imply

$$G(A \cap B_0) = G(D_0) > c|A| > 2\kappa|J|. \qquad (4)$$

Denote by $\mathcal{K}$ the collection of all $K \in \mathcal{B}$ for which $G(A \cap K) \geq (\kappa/\varepsilon)|K|$. From $B_0 \subset J - P$ and inequality (1), we obtain

$$\sum_{B \in \mathcal{B} - \mathcal{K}} G(A \cap B) < \frac{\kappa}{\varepsilon} \sum_{B \in \mathcal{B} - \mathcal{K}} |B| \leq \frac{\kappa}{\varepsilon}|B_0| < \kappa|J|$$

and so by inequality (4),

$$\sum_{K \in \mathcal{K}} G(A \cap K) = G(A \cap B_0) - \sum_{B \in \mathcal{B} - \mathcal{K}} G(A \cap B)$$
$$> \kappa|J| = \kappa(3 \cdot 2^{-k})^m \geq \kappa(3d)^m > 0. \qquad (5)$$

In particular, the family $\mathcal{K}$ is not empty.

Let $K \in \mathcal{K}$ and $h_K := d(K)/\sqrt{m}$. As the choice of σ implies

$$2h_K \leq 2^{-k+1} < 4d < \sigma < \frac{\sigma}{\tau'} < \frac{\delta}{2\sqrt{m}},$$

we can choose a number l_K so that $2h_K \leq l_K \leq \sigma/\tau'$; more specific selections of l_K will be employed later. By the maximality of K, the mother K^* of K meets P. Thus there are a point x_K in $P \cap (K^* - K)$ and a cube $Q_K \subset K[x_K, l_K]$ such that

$$K \cap Q_K = \emptyset, \quad x_K \in Q_K, \quad d(Q_K) = l_K \sqrt{m}.$$

If $L \subset K$ is a figure, let $L' := (A \cap L) \cup Q_K$. Since $P \cap L' \neq \emptyset$ and $d(L') < 2l_K \sqrt{m} < \delta$, condition (ii) yields the following implication:

$$r(L') > \eta \implies -M < \frac{G(L')}{|L'|} < M. \tag{6}$$

Letting $L = \emptyset$, we see $G(Q_K)/|Q_K| > -M$. If $L \subset K$ is a cell, then

$$\begin{aligned}
\|L'\| &\leq \|A \cap L\| + \|Q_K\| \\
&\leq \|L\| + \mathcal{H}^{m-1}(\partial A \cap \operatorname{int} L) + \|Q_K\| \\
&\leq 2\|Q_K\| + \mathcal{H}^{m-1}(\partial A \cap \operatorname{int} L) \\
&\leq 4ml_K^{m-1} + \mathcal{H}^{m-1}(\partial A \cap \operatorname{int} K)
\end{aligned} \tag{7}$$

by Proposition 1.8.5. Finally, we shall need the inequality

$$\|A\| = \frac{|A|}{d\, r(A)} \leq \frac{d^{m-1}}{\tau}. \tag{8}$$

Claim 4. $G(A \cap K)/(4M) < (\sigma/\tau')^m$ for all $K \in \mathcal{K}$.

Proof. Suppose there is a $K \in \mathcal{K}$ which does not satisfy the claim, and let $l_K := \sigma/\tau'$. If $K' := (A \cap K) \cup Q_K$, then

$$|K'| \leq |K| + |Q_K| = h_K^m + l_K^m \leq 2l_K^m = 2\left(\frac{\sigma}{\tau'}\right)^m \leq \frac{G(A \cap K)}{2M}$$

and hence

$$\frac{G(K')}{|K'|} = \frac{G(A \cap K)}{|K'|} + \frac{G(Q_K)}{|Q_K|} \cdot \frac{|Q_K|}{|K'|} > 2M - M\frac{|Q_K|}{|K'|} > M.$$

In view of implication (6), the last inequality yields $r(K') \leq \eta$. Thus

$$\|K'\| \geq \frac{|K'|}{\eta\, d(K')} > \frac{|Q_K|}{\eta(2l_K\sqrt{m})} = \frac{l_K^{m-1}}{2\eta\sqrt{m}}.$$

By inequality (7) for $L = K$ and the choice of η,

$$\mathcal{H}^{m-1}(\partial A \cap \operatorname{int} K) > \frac{l_K^{m-1}}{2\eta\sqrt{m}} - 4ml_K^{m-1} > l_K^{m-1}.$$

Since $\|A\| \geq \mathcal{H}^{m-1}(\partial A \cap \operatorname{int} K)$, we obtain

$$\|A\| > l_K^{m-1} = \frac{\sigma^{m-1}}{\tau} > \frac{d^{m-1}}{\tau},$$

which contradicts inequality (8). The claim is established.

Given a cube $K := [a, b]^m$ in $\mathcal{K}$ and $a \le t < t' \le b$, let

$$K[t, t'] := [t, t'] \times [a, b]^{m-1}.$$

As G is a charge, the function $t \mapsto G\big(A \cap K[a, t]\big)$ is continuous in $[a, b]$. It follows there are numbers $a = t_0 < \cdots < t_p = b$ such that letting $K_i := K[t_{i-1}, t_i]$, we obtain $G(A \cap K_i) = G(A \cap K)/p$ for $i = 1, \ldots, p$. Our choice of ε and γ yields

$$G(A \cap K_i) = \frac{1}{p} G(A \cap K) \ge \frac{\kappa}{p\varepsilon} |K| = \frac{\kappa}{\gamma} h_K^m = 4M(2h_K)^m.$$

From this and Claim 4 we infer

$$2h_K \le \left[\frac{G(A \cap K)}{4pM} \right]^{\frac{1}{m}} \le \left[\frac{G(A \cap K)}{4M} \right]^{\frac{1}{m}} < \frac{\sigma}{\tau'}.$$

Let $l_K := \left[G(A \cap K)/(4pM) \right]^{\frac{1}{m}}$ and $K_i' := (A \cap K_i) \cup Q_K$ for $i = 1, \ldots, p$. Observe

$$|K_i'| \le |K| + |Q_K| = h_K^m + l_K^m < 2l_K^m = \frac{G(A \cap K_i)}{2M}$$

and hence

$$\frac{G(K_i')}{|K_i'|} = \frac{G(A \cap K_i)}{|K_i'|} + \frac{G(Q_K)}{|Q_K|} \cdot \frac{|Q_K|}{|K_i'|} > 2M - M\frac{|Q_K|}{|K_i'|} > M.$$

Implication (6) and the last inequality yield $r(K_i') \le \eta$. Proceeding as in the proof of Claim 4, we show that $\mathcal{H}^{m-1}(\partial A \cap \operatorname{int} K_i) > l_K^{m-1}$ for $i = 1, \ldots, p$. Thus

$$\mathcal{H}^{m-1}(\partial A \cap \operatorname{int} K) \ge \sum_{i=1}^{p} \mathcal{H}^{m-1}(\partial A \cap \operatorname{int} K_i)$$
$$> p l_K^{m-1} = p^{\frac{1}{m}} (4M)^{\frac{1-m}{m}} G(A \cap K)^{\frac{m-1}{m}}$$

for every $K \in \mathcal{K}$. These inequalities, Jensen's inequality [73, Theorem 3.3], inequality (5), and the choice of p yield

$$\|A\| \ge \sum_{K \in \mathcal{K}} \mathcal{H}^{m-1}(\partial A \cap \operatorname{int} K) > p^{\frac{1}{m}} (4M)^{\frac{1-m}{m}} \sum_{K \in \mathcal{K}} G(A \cap K)^{\frac{m-1}{m}}$$

$$\ge p^{\frac{1}{m}} (4M)^{\frac{1-m}{m}} \left[\sum_{K \in \mathcal{K}} G(A \cap K) \right]^{\frac{m-1}{m}} > p^{\frac{1}{m}} \theta d^{m-1} > \frac{d^{m-1}}{\tau}.$$

This again contradicts inequality (8). $\qquad\qquad\qquad\qquad\qquad\qquad\square$

2.5. Reduced charges

Observation 2.5.1. *Let $s \geq 0$. If a set $E \subset \mathbb{R}^m$ is $\mathcal{H}^s$-measurable and the measure $\mathcal{H}^s \, \llcorner \, E$ is locally finite, then*

$$\lim_{r \to 0+} \frac{\mathcal{H}^s\left(E \cap B[x,r]\right)}{r^s} = 0$$

for $\mathcal{H}^s$-almost all $x \in \mathbb{R}^m - E$.

PROOF. Since $\mathcal{H}^s\left[E \cap B(t)\right] < \infty$ for each $t > 0$, Theorem 1.5.11 yields

$$\lim_{r \to 0+} \frac{\mathcal{H}^s\left(E \cap B[x,r]\right)}{r^s} = \lim_{r \to 0+} \frac{\mathcal{H}^s\left(E \cap B(n) \cap B[x,r]\right)}{r^s} = 0$$

for $\mathcal{H}^s$-almost all $x \in B(n) - E$ and $n = 1, 2, \ldots$. $\square$

For a locally BV set E, the set of all $x \in \text{int}_* E$ for which

$$\lim_{r \to 0+} \frac{\mathcal{H}^{m-1}\left(B[x,r] \cap \partial_* E\right)}{r^{m-1}} = 0$$

is called the *critical interior* of E, denoted by $\text{int}_c E$. Since $\partial_* E$ is a Borel set by Proposition 1.5.1, and since $\mathcal{H}^{m-1} \, \llcorner \, \partial_* E$ is locally finite measure by Theorem 1.8.2, (3), Observation 2.5.1 implies

$$\mathcal{H}^{m-1}(\text{int}_* E - \text{int}_c E) = 0. \tag{2.5.1}$$

Lemma 2.5.2. *Let E be a locally BV set, $x \in \text{int}_c E$, and $\eta \geq 0$. If a sequence $\{C_i\}$ of bounded BV sets η-converges to x, then*

$$\lim \frac{|E \cap C_i|}{|C_i|} = 1$$

and the sequence $\{E \cap C_i\}$ also η-converges to x.

PROOF. Observe $\{C_i\}$ θ-converges to x for a $\theta > \eta$. Inequality (2.3.1) implies $\liminf s\left(C_i \cup \{x\}\right) \geq \alpha(m)(m\theta)^m$ for $i = 1, 2, \ldots$, and hence

$$\lim \frac{|E \cap C_i|}{|C_i|} = 1$$

according to Lemma 1.5.6. If $d_i := d\left(C_i \cup \{x\}\right)$ and $B_i := B[x, d_i]$, then $C_i \subset B_i$ and by Proposition 1.8.5,

$$\|E \cap C_i\| \leq \|C_i\| + \mathcal{H}^{m-1}(B_i \cap \partial_* E).$$

Applying inequality (2.3.1) again yields

$$\frac{d_i\|E\cap C_i\|}{|C_i|} \leq \frac{d_i\|C_i\|}{|C_i|} + \frac{1}{s(C_i\cup\{x\})}\cdot\frac{\mathcal{H}^{m-1}(B_i\cap\partial_*E)}{d_i^{m-1}}$$

$$\leq \frac{1}{\theta} + \frac{1}{\alpha(m)(\theta m)^m}\cdot\frac{\mathcal{H}^{m-1}(B_i\cap\partial_*E)}{d_i^{m-1}},$$

and since $x\in\mathrm{int}_c E$,

$$\limsup\frac{d_i\|E\cap C_i\|}{|C_i|} \leq \frac{1}{\theta}.$$

We conclude

$$\liminf r\big[(E\cap C_i)\cup\{x\}\big] = \liminf\left(\frac{|E\cap C_i|}{|C_i|}\cdot\frac{|C_i|}{d_i\|E\cap C_i\|}\right) \geq \theta > \eta.$$

$\square$

Proposition 2.5.3. *Let E be a locally BV set, $x\in\mathrm{int}_c E$, and let F be a function defined on $\mathcal{BV}$. For $\theta > \eta > 0$, we have*

$$\underline{D}_\eta(F\,\mathsf{L}\,E)(x) \leq \inf_{\{E_i\}}\liminf\frac{F(E_i)}{|E_i|} \leq \underline{D}_\theta(F\,\mathsf{L}\,E)(x),$$

$$\overline{D}_\theta(F\,\mathsf{L}\,E)(x) \leq \sup_{\{E_i\}}\limsup\frac{F(E_i)}{|E_i|} \leq \overline{D}_\eta(F\,\mathsf{L}\,E)(x)$$

where $\{E_i\}$ is a sequence in $\mathcal{BV}(E)$ that η-converges to x. Additional conclusions analogous to those stated in Observation 2.3.6 are also true.

PROOF. Clearly

$$b := \inf_{\{B_i\}}\liminf\frac{(F\,\mathsf{L}\,E)(B_i)}{|B_i|} \leq c := \inf_{\{E_i\}}\liminf\frac{F(E_i)}{|E_i|}$$

where $\{B_i\}$ and $\{E_i\}$ are sequence in $\mathcal{BV}$ and $\mathcal{BV}(E)$, respectively, that η-converge to x. Assuming $b < c$, there is a sequence $\{C_i\}$ in $\mathcal{BV}$ that η-converges to x and for which

$$c > \lim\frac{(F\,\mathsf{L}\,E)(C_i)}{|C_i|} = \lim\left(\frac{F(E\cap C_i)}{|E\cap C_i|}\cdot\frac{|E\cap C_i|}{|C_i|}\right)$$

$$\geq \liminf\frac{F(E\cap C_i)}{|E\cap C_i|}$$

by Lemma 2.5.2. In addition, Lemma 2.5.2 implies $\{E\cap C_i\}$ η-converges to x, a contradiction. Thus $b = c$, and the proposition follows from Observation 2.3.6 and symmetry. $\square$

Corollary 2.5.4. *Let F be a function defined on $\mathcal{BV}$, let E be a locally BV set, and let $x \in \text{int}_c E$. Then*

$$\underline{D}F(x) \leq \underline{D}(F \llcorner E)(x) \leq \overline{D}(F \llcorner E)(x) \leq \overline{D}F(x).$$

In particular, $F \llcorner E$ is derivable at x whenever F is, in which case

$$D(F \llcorner E)(x) = DF(x).$$

Remark 2.5.5. Let E be a locally BV set, and let $F := F \llcorner E$ be a function defined on $\mathcal{BV}$. While it is clear that $DF(x) = 0$ for each x in $\mathbb{R}^m - \text{cl}\, E$, little can be said about the derivability of F at points of the set $\text{cl}\, E - E$. We shall see in Remark 6.1.2, (1) that there is a charge F in a bounded BV set A such that the set N of all points $x \in \text{cl}\, A - A$ for which $\overline{D}|F|(x) = \infty$ is rather large: the measure $\mathcal{H}^{m-1} \llcorner N$ is not σ-finite.

Let $E \subset \mathbb{R}^m$ and $x \in E \cap \text{int}_* E$. A map $\phi : E \to \mathbb{R}^n$ is called *differentiable* at x *relative* to E if there is a linear map $L : \mathbb{R}^m \to \mathbb{R}^n$ satisfying the following condition: given $\varepsilon > 0$, we can find a $\delta > 0$ so that

$$\big|\phi(y) - \phi(x) - L(y - x)\big| < \varepsilon|y - x|$$

for each $y \in E \cap B(x, \delta)$. The linear map L, which is unique by Observation 2.5.6 below, is called the *differential* of ϕ at x *relative* to E, denoted by $D_E\phi(x)$. Clearly, the concepts of differentiability and relative differentiability coincide at the interior points of E.

Observation 2.5.6. *If a linear map L satisfying the above condition exists, then it is unique.*

PROOF. Seeking a contradiction, assume there are distinct linear maps L_1 and L_2 satisfying the following condition: given $\varepsilon > 0$, we can find a $\delta > 0$ so that

$$\big|\phi(y) - \phi(x) - L_i(y - x)\big| < \varepsilon|y - x|$$

for each $y \in E \cap B(x, \delta)$ and $i = 1, 2$. By our assumption, the linear map $L := L_1 - L_2$ has positive norm $\|L\|$. Let $\varepsilon := \|L\|/4$, find a corresponding $\delta > 0$, and consider the open sets

$$S := \Big\{y \in \mathbb{R}^m : \big|L(y)\big| > 2\varepsilon|y|\Big\} \quad \text{and} \quad T := x + S.$$

If $S = \emptyset$, then $\|L\| \leq 2\varepsilon = \|L\|/2$. This contradiction shows $|S| > 0$. As $rS = S$ for each $r > 0$, it is easy to see $0 \in \text{cl}_* S$. Hence $x \in \text{cl}_* T$, and Corollary 1.5.7 yields $x \in \mathrm{cl}_*(T \cap E)$. In particular, the intersection $T \cap E \cap B(x, \delta)$ is not empty. Since

$$
\begin{aligned}
2\varepsilon|y - x| < \left|L(y - x)\right| &= \left|L_1(y - x) - L_2(y - x)\right| \\
&\leq \left|L_1(y - x) - \phi(y) + \phi(x)\right| + \left|\phi(y) - \phi(x) - L_2(y - x)\right| \\
&< 2\varepsilon|y - x|
\end{aligned}
$$

for each y in $T \cap E \cap B(x, \delta)$, we have a contradiction. □

Let E be a measurable set, let $x \in \mathrm{int}_* E$, and let $v = (v_1, \dots, v_m)$ be a vector field defined on $\mathrm{cl}_* E$ that is differentiable at x relative to $\mathrm{cl}_* E$; note $\mathrm{int}_* E = \mathrm{int}_*(\mathrm{cl}_* E)$ by Corollary 1.5.3. The *divergence* of v at x *relative* to $\mathrm{cl}_* E$ is the number

$$
\mathrm{div}_* v(x) := \mathrm{tr}\, D_{\mathrm{cl}_* E} v(x)
$$

where $\mathrm{tr}\, D_{\mathrm{cl}_* E} v(x)$ denotes the usual *trace* of the linear transformation $D_{\mathrm{cl}_* E} v(x) : \mathbb{R}^m \to \mathbb{R}^m$ [31, Chapter 6, Section 8]. Observe $\mathrm{div}_* v(x)$ may be defined even when the partial derivatives $(\partial v_i / \partial x_i)(x)$ have no meaning; however,

$$
\mathrm{div}_* v(x) = \mathrm{div}\, v(x)
$$

whenever $\mathrm{div}\, v(x)$ is defined, in particular, whenever $x \in \mathrm{int}\,(\mathrm{cl}_* E)$.

Proposition 2.5.7. *Let E be a locally BV set, let $x \in \mathrm{int}_c E$, and let $F := \int v \cdot \nu \, d\mathcal{H}^{m-1}$ be the flux of a bounded Borel measurable vector field $v : \mathrm{cl}_* E \to \mathbb{R}^m$.*

(i) *If v is Lipschitz at x, then F is almost derivable at x.*

(ii) *If v is differentiable at x relative to $\mathrm{cl}_* E$, then F is derivable at x and $DF(x) = \mathrm{div}_* v(x)$.*

PROOF. The proofs of both claims are similar; they resemble the proofs of Claims 1 and 2 of Example 2.3.2. Let v be Lipschitz at x. There are positive numbers c and δ such that

$$
\left|v(y) - v(x)\right| \leq c|y - x|
$$

for each y in $B(x, \delta) \cap \mathrm{cl}_* E$. Given a positive $\theta < 1/(2m)$, choose a positive $\eta < \theta$ and a sequence $\{E_i\}$ in $\mathcal{BV}(E)$ that η-converges to x. If i is sufficiently large, then $E_i \subset B(x, \delta)$ and

$$|F(E_i)| = \left| \int_{\partial_* E_i} [v(y) - v(x)] \cdot \nu_{E_i}(y) \, d\mathcal{H}^{m-1}(y) \right|$$

$$\leq c \int_{\partial_* E} |y - x| \, d\mathcal{H}^{m-1}(y)$$

$$\leq cd\big(E_i \cup \{x\}\big) \|E_i\| \leq \frac{2c}{\eta} |E_i|.$$

We conclude $\limsup \big[|F(E_i)|/|E_i|\big] \leq 2c/\eta$. Since $F = F \, \mathsf{L} \, E$, an application of Proposition 2.5.3 shows $\overline{D}_\theta |F|(x) \leq 2c/\eta < \infty$.

Let v be differentiable at x relative to $\mathrm{cl}_* E$, and let $\varepsilon > 0$. Select a sequence $\{E_i\}$ in $\mathcal{BV}(E)$ that tends to x, and find an $\eta > 0$ so that $\liminf r\big(E_i \cup \{x\}\big) > \eta$. For each $y \in \mathbb{R}^m$, let

$$w(y) := v(x) + \big[D_{\mathrm{cl}_* E} v(x)\big](y - x).$$

Then $\mathrm{div}\, w(y) = \mathrm{div}_* v(x)$ for all $y \in \mathbb{R}^m$, and there is a $\delta > 0$ with

$$\big|v(y) - w(y)\big| < \varepsilon \eta |y - x|$$

for every y in $B(x, \delta) \cap \mathrm{cl}_* E$. The Gauss-Green formula yields

$$\mathrm{div}_* v(x)|E_i| = \int_{E_i} \mathrm{div}\, w(y) \, dy = \int_{\partial_* E_i} w \cdot \nu_{E_i} \, d\mathcal{H}^{m-1}$$

for $i = 1, 2, \dots$. If i is sufficiently large, then $E_i \subset B(x, \delta)$ and

$$\big|\mathrm{div}_* v(x)|E_i| - F(E_i)\big| = \left| \int_{\partial_* E_i} (w - v) \cdot \nu_{E_i} \, d\mathcal{H}^{m-1} \right|$$

$$< \varepsilon \eta \int_{\partial_* E_i} |y - x| \, d\mathcal{H}^{m-1}(y)$$

$$\leq \varepsilon \eta \, d\big(E_i \cup \{x\}\big) \|E_i\| < \varepsilon |E_i|.$$

From the arbitrariness of ε, we conclude $\lim \big[F(E_i)/|E_i|\big] = \mathrm{div}_* v(x)$. Another application of Proposition 2.5.3 completes the proof. $\qquad \square$

The results of this section suggest that the following definition may be useful.

Definition 2.5.8. Let E be a locally BV set, and let $F := \int v \cdot \nu \, d\mathcal{H}^{m-1}$ be the flux of a bounded Borel measurable vector field $v : \mathrm{cl}_* E \to \mathbb{R}^m$. If F is derivable at $x \in \mathrm{int}_c E$, we call the number

$$\mathfrak{div}\, v(x) := DF(x)$$

the *mean divergence* of v at x.

Remark 2.5.9. We list a few facts indicating why $\eth\mathfrak{iv}\, v(x)$ is a natural generalization of $\mathrm{div}_* v(x)$, and a fortiori of $\mathrm{div}\, v(x)$.

(i) According to Proposition 2.5.7, (ii),

$$\eth\mathfrak{iv}\, v(x) = \mathrm{div}_* v(x)$$

whenever $x \in \mathrm{int}_c E$ and v is differentiable at x relative to $\mathrm{cl}_* E$.

(ii) Assume that v is a charging vector field, and that F is almost derivable at almost all $x \in \mathrm{int}_c E$. Then F is derivable at almost all $x \in \mathrm{int}_c E$ by Theorem 2.4.3, and hence the function

$$\eth\mathfrak{iv}\, v : x \mapsto \eth\mathfrak{iv}\, v(x)$$

is defined almost everywhere in E. If F is almost derivable at *all* points $x \in \mathrm{int}_c E$, then Theorem 3.6.13 below implies that the Gauss-Green formula

$$F = \int \eth\mathfrak{iv}\, v(x)\, dx$$

holds whenever $\eth\mathfrak{iv}\, v$ belongs to $L^1_\ell(E)$.

(iii) In view of Example 2.1.4 and Proposition 2.5.7, (i), the assumptions of paragraph (ii) are satisfied for each $v \in C(\mathrm{cl}\, E; \mathbb{R}^m)$ that is Lipschitz at almost all $x \in \mathrm{int}_c E$, or at all $x \in \mathrm{int}_c E$.

(iv) Assume $v \in C(\mathrm{cl}\, E; \mathbb{R}^m)$ is Lipschitz at almost all $x \in \mathrm{int}_* E$. In view of Whitney's theorem, the vector field v can be extended to a vector field $w \in C(\mathbb{R}^m; \mathbb{R}^m)$ that is still Lipschitz at almost all $x \in \mathrm{int}_* E$. By Stepanoff's theorem, w is differentiable at almost all $x \in \mathrm{int}_* E$. If w is differentiable at $x \in \mathrm{int}_* E$, then v is differentiable at x relative to $\mathrm{cl}_* E$ and

$$\mathrm{div}\, w(x) = \mathrm{div}_* v(x)\,.$$

In particular, if w is differentiable at $x \in \mathrm{int}_* E$, then $\mathrm{div}\, w(x)$ is determined by v and does not depend on the extension w. Up to a negligible set, this fact follows also from Lemma 1.6.3.

(v) We show in Corollary 5.2.7 below that under mild hypothesis, $\eth\mathfrak{iv}\, v$ is the distributional divergence of v.

Concepts similar to mean divergence were introduced previously. The reader may compare the Gauss-Green theorem alluded to in item (ii) with those proved in [76, Theorems 3 and 7] and [3, Theorem 5.5].

2.6. Partitions

A *partition* is a collection (possibly empty)

$$P := \{(A_1, x_1), \ldots, (A_p, x_p)\}$$

where $A_1, \ldots, A_p$ are disjoint bounded BV sets, and $x_1, \ldots, x_p$ are points of $\mathbb{R}^m$, not necessarily distinct. The bounded BV set

$$[P] := \bigcup_{i=1}^{p} A_i$$

is called the *body* of P. Given an $\eta \geq 0$ and a nonnegative function δ defined on a set $E \subset \mathbb{R}^m$, we say P is

- η-*regular* if $r(A_i \cup \{x_i\}) > \eta$ for $i = 1, \ldots, p$;
- δ-*fine* if $x_i \in E$ and $d(A_i \cup \{x_i\}) < \delta(x_i)$ for $i = 1, \ldots, p$.

A partition P is 0-regular whenever each A_i has positive measure. If P is a δ-fine partition, then every x_i belongs to $\{\delta > 0\} = E - \{\delta = 0\}$. We introduce two additional concepts:

- a set $T \subset \mathbb{R}^m$ is called *thin* whenever the measure $\mathcal{H}^{m-1} \llcorner T$ is σ-finite;
- a nonnegative function δ defined on a set $E \subset \mathbb{R}^m$ is called a *gage* on E whenever $\{\delta = 0\}$ is a thin set.

Variants of δ-fine partitions for a positive function δ were introduced independently by Henstock [30], Kurzweil [42], and McShane [53]. Still independently, Mawhin [50, 49] and Pfeffer [61, 62] applied various forms of regularity restrictions to δ-fine partitions. Finally, Pfeffer [64, 63] replaced positive functions by gages.

Lemma 2.6.1 (Cousin). *Let σ be a positive function defined on $K[r]$ where $r \geq 1$ is an integer. There are nonoverlapping dyadic cubes K_i and points $x_i \in K_i$ such that $d(K_i) < \sigma(x_i)$ for $i = 1, \ldots, p$, and*

$$K[r] = \bigcup_{i=1}^{p} K_i.$$

PROOF. Call a cube $L \subset K[r]$ *nice* if there are nonoverlapping dyadic cubes L_i and points $x_i \in L_i$ such that $d(L_i) < \sigma(x_i)$ for $i = 1, \ldots, p$, and $L = \bigcup_{i=1}^{p} L_i$; otherwise call L *faulty*. Observe a cube $L \subset K[r]$ is nice whenever it is the union of nonoverlapping nice cubes.

Proceeding toward a contradiction, suppose $L := K[r]$ is faulty. As L is the union of nonoverlapping dyadic cubes $L^1, \ldots, L^{(2r)^m}$ whose diameters equal 1, among $L^1, \ldots, L^{(2r)^m}$ there is a faulty cube, denoted by C_1. As C_1 is the union of nonoverlapping dyadic cubes $C_1^1, \ldots, C_1^{2^m}$ whose diameters equal $1/2$, among $C_1^1, \ldots, C_1^{2^m}$ there is a faulty cube, denoted by C_2. Construct inductively a decreasing sequence $\{C_n\}$ of faulty dyadic cubes such that $d(C_n) = 1/2^n$ for $n = 1, 2, \ldots$. Now

$$\bigcap_{n=1}^{\infty} C_n = \{x\}$$

for an $x \in K[r]$. As $\sigma(x) > 0$, we have $d(C_k) < \sigma(x)$ for a sufficiently large integer k. This implies C_k is nice, a contradiction. $\qquad \square$

Remark 2.6.2. The collection $\{(K_1, x_1), \ldots, (K_p, x_p)\}$ of Cousin's lemma is not a partition, since the cubes $K_1, \ldots, K_p$ are not disjoint. However, removing suitable $(m-1)$-dimensional faces of the cubes K_i, we obtain a σ-fine partition

$$P := \{(B_1, x_1), \ldots, (B_p, x_p)\}$$

such that $K[r] = [P]$ and $\operatorname{cl} B_i = \operatorname{cl}_* B_i = K_i$ for $i = 1, \ldots, p$.

Proposition 2.6.3. *Let $A \in \mathcal{BV}$, and let δ be a positive function defined on $\operatorname{cl} A$. There is a δ-fine partition Q with $[Q] = A$.*

PROOF. Find an integer $r \geq 1$ so that $A \subset K[r]$, and define a positive function σ on $K[r]$ by the formula:

$$\sigma(x) := \begin{cases} \delta(x) & \text{if } x \in \operatorname{cl} A, \\ \operatorname{dist}(x, A) & \text{if } x \in K[r] - \operatorname{cl} A. \end{cases}$$

In view of Remark 2.6.2, there is a σ-fine partition P with $[P] = K[r]$. From the choice of σ, we see that $A \cap B = \emptyset$ whenever $(B, x) \in P$ and $x \notin \operatorname{cl} A$. It follows

$$Q := \{(A \cap B, x) : (B, x) \in P \text{ and } x \in \operatorname{cl} A\}$$

is the desired partition. $\qquad \square$

The next lemma, which generalizes the classical result of Besicovitch [5], was proved by Howard [34].

Lemma 2.6.4. *Let $F_1, \ldots, F_k$ be charges, let*

$$F := \max\{|F_1|, \ldots, |F_k|\},$$

and let σ be a gage on $K[r]$ where $r \geq 1$ is an integer. Given $\varepsilon > 0$, there are nonoverlapping dyadic cubes $K_i \subset K[r]$ and points $x_i \in K_i$ such that $d(K_i) < \sigma(x_i)$ for $i = 1, \ldots, q$, and

$$F\left(K[r] - \bigcup_{i=1}^{q} K_i\right) < \varepsilon.$$

PROOF. Let $K := K[r]$, and observe the function F is $\mathfrak{T}$-continuous and *subadditive* in the following sense:

$$F(A \cup B) \le F(A) + F(B) \qquad (*)$$

whenever $A, B \in \mathcal{BV}$ do not overlap. The set $T := \{\sigma = 0\}$ is the union of sets T_n such that $\mathcal{H}^{m-1}(T_n) < c_n < \infty$ for $n = 1, 2, \ldots$. Choose an $\varepsilon > 0$, and let $\kappa > 1$ be the constant from Lemma 1.3.2. For an integer $n \ge 1$, let

$$\varepsilon_n := \min\left\{ \frac{(\sqrt{m})^{m-1}}{2m\kappa c_n}, \frac{\varepsilon}{2^n} \right\}.$$

There is an $\eta_n > 0$ so that $F(B) < \varepsilon_n$ for each $B \in \mathcal{BV}(K)$ with $\|B\| < 1/\varepsilon_n$ and $|B| < \eta_n$. Use Lemma 1.3.2 to find a family $\mathcal{C}_n$ of dyadic cubes of diameters less than $\varepsilon_n \eta_n$ such that

$$T_n \subset \operatorname{int} \bigcup \mathcal{C}_n \quad \text{and} \quad \sum_{C \in \mathcal{C}_n} d(C)^{m-1} < \kappa c_n.$$

If $\mathcal{E}$ is a finite subfamily of $\mathcal{C}_n$ and $E := \bigcup \mathcal{E}$, then $F(E) < \varepsilon_n$; indeed,

$$\|E\| \le \sum_{C \in \mathcal{E}} \|C\| \le \frac{2m}{(\sqrt{m})^{m-1}} \sum_{C \in \mathcal{E}} d(C)^{m-1} < \frac{2m\kappa c_n}{(\sqrt{m})^{m-1}} \le \frac{1}{\varepsilon_n},$$

$$|E| \le \sum_{C \in \mathcal{E}} |C| < \sum_{C \in \mathcal{E}} d(C)\|C\| < \varepsilon_n \eta_n \sum_{C \in \mathcal{E}} \|C\| < \eta_n.$$

Let $\mathcal{C}$ be a nonoverlapping subfamily of $\mathcal{C}_0 := \bigcup_{n=1}^{\infty} \mathcal{C}_n$ whose union is the same as that of $\mathcal{C}_0$, and define a positive function σ_+ on K by the formula:

$$\sigma_+(x) := \begin{cases} \min\{d(C) : C \in \mathcal{C} \text{ and } x \in C\} & \text{if } x \in K \cap T, \\ \sigma(x) & \text{if } x \in K - T. \end{cases}$$

By Cousin's lemma, there are nonoverlapping dyadic cubes K_i with $K = \bigcup_{i=1}^{p} K_i$ and $x_i \in K_i$ such that $d(K_i) < \sigma_+(x_i)$ for $i = 1, \ldots, p$. Since

$$T = \bigcup_{n=1}^{\infty} T_n \subset \bigcup_{n=1}^{\infty} \left(\operatorname{int} \bigcup \mathcal{C}_n\right) \subset \operatorname{int} \bigcup_{n=1}^{\infty} \left(\bigcup \mathcal{C}_n\right) = \operatorname{int} \bigcup \mathcal{C},$$

the definition of σ_+ implies that for each $x_i \in T$, there is a cube $C \in \mathcal{C}$ with $K_i \subset C \subset K$.

Let $\mathcal{D}$ be the collection of all $C \in \mathcal{C}$ that are contained in K and contain some of the cubes $K_1, \ldots, K_p$. Since $\mathcal{C}$ is a nonoverlapping family, the collection $\mathcal{D}$ is finite and $D := \bigcup \mathcal{D}$ is a figure. After a suitable reordering, there is an integer q with $0 \le q \le p$ such that the cubes $K_{q+1}, \ldots, K_p$ are contained in D, and none of the cubes $K_1, \ldots, K_q$ is contained in D. In particular, $x_i \notin T$ whenever $i \le q$. The definition of σ_+ implies $d(K_i) < \sigma(x_i)$ for $i = 1, \ldots, p$. Let

$$A := \bigcup_{i=1}^{q} K_i \quad \text{and} \quad B := \bigcup_{i=q+1}^{p} K_i.$$

Assume K_i with $i \le q$ overlaps $C \in \mathfrak{D}$. Since $K_i \not\subset C$, we have $C \subset K_i$. However, C contains a cube K_j with $j > q$, a contradiction; since K_i and K_j do not overlap. Consequently the figures A and D do not overlap. As $B \subset D$ and $A \cup B = K$, we conclude $B = D$. For $n = 1, 2, \ldots$, let

$$\mathfrak{D}_n := \mathfrak{D} \cap \mathfrak{C}_n - \bigcup_{j=1}^{n-1} \mathfrak{D}_j \quad \text{and} \quad D_n := \bigcup \mathfrak{D}_n.$$

Since $\{\mathfrak{D}_n\}$ is a disjoint collection of finite families of nonoverlapping dyadic cubes, $\{D_n\}$ is a nonoverlapping collection of figures. As the union of the families $\mathfrak{D}_n$ is the finite family $\mathfrak{D}$, there is an integer $s \ge 1$ such that $\mathfrak{D}_n = \emptyset$ for each $n > s$. Thus $D = \bigcup_{n=1}^{s} D_n$, and inequality $(*)$ implies

$$F(K - A) = F(D) \le \sum_{n=1}^{s} F(D_n) < \sum_{n=1}^{s} \varepsilon_n \le \sum_{n=1}^{s} \frac{\varepsilon}{2^n} < \varepsilon. \qquad \square$$

Corollary 2.6.5. *Let* $F_1, \ldots, F_k$ *be charges, let*

$$F := \max\{|F_1|, \ldots, |F_k|\},$$

and let σ *be a gage on* $K[r]$ *where* $r \ge 1$ *is an integer. Given* $\varepsilon > 0$, *there is a* σ-*fine partition* $Q := \{(B_1, x_1), \ldots, (B_q, x_q)\}$ *such that* $[Q] \subset K[r]$,

$$F\big(K[r] - [Q]\big) < \varepsilon$$

and each B_i *is simultaneously a cube and a dyadic figure.*

PROOF. Let $K := K[r]$, and let $K_1, \ldots, K_q$ and $x_1, \ldots, x_q$ be as in Lemma 2.6.4. By the $\mathfrak{T}$-continuity of F, in the interior of each dyadic cube K_i, we can find a cube B_i that is a dyadic figure and for which the measure $|K_i - B_i|$ is so small that

$$\sum_{i=1}^{q} F(K_i - B_i) < \varepsilon - F\left(K - \bigcup_{i=1}^{q} K_i \right);$$

note $\|K_i - B_i\| \leq 2\|K_i\|$. Now $Q := \{(B_1, x_1), \ldots, (B_q, x_q)\}$ is the desired partition, since

$$F(K - [Q]) \leq F\left(K - \bigcup_{i=1}^{q} K_i\right) + F\left(\bigcup_{i=1}^{q} [K_i - B_i]\right)$$

$$\leq F\left(K - \bigcup_{i=1}^{q} K_i\right) + \sum_{i=1}^{q} F(K_i - B_i) < \varepsilon. \qquad \square$$

Lemma 2.6.6. *Let E be a locally BV set. Given $\theta > \eta > 0$, there is a gage Δ on $\mathrm{cl}_* E$ with the following property: if $\{(B_1, x_1), \ldots, (B_q, x_q)\}$ is a θ-regular Δ-fine partition, then*

$$\{(E \cap B_1, x_1), \ldots, (E \cap B_q, x_q)\}$$

is an η-regular Δ-fine partition.

PROOF. It follows from Lemma 2.5.2 that given $x \in \mathrm{int}_c E$, we can find a $\Delta_x > 0$ so that $r[(E \cap B) \cup \{x\}] > \eta$ for each BV set B with

$$d(B \cup \{x\}) < \Delta_x \quad \text{and} \quad r(B \cup \{x\}) > \theta.$$

Since $T := \mathrm{cl}_* E - \mathrm{int}_c E$ is a thin set, letting $\Delta_x = 0$ for each $x \in T$, the function $\Delta : x \mapsto \Delta_x$ is the desired gage on $\mathrm{cl}_* E$. $\qquad \square$

Theorem 2.6.7. *Let $F_1, \ldots, F_k$ be charges, let*

$$F := \max\{|F_1|, \ldots, |F_k|\},$$

and let A be a bounded BV set. Suppose $\eta < 1/(2m\sqrt{m})$ and ε are positive numbers. For each gage δ on $\mathrm{cl}_* A$, there is an η-regular δ-fine partition P such that $[P] \subset A$ and $F(A - [P]) < \varepsilon$.

PROOF. Note $1/(2m\sqrt{m})$ is the regularity of a cube, and choose a θ with $\eta < \theta < 1/(2m\sqrt{m})$. In view of Proposition 1.10.5 and the $\mathcal{T}$-continuity of F, there is an essentially closed set $C \subset A$ with $F(A - C) < \varepsilon/2$. Let

$$G := F \llcorner C = \max\{|F_1 \llcorner C|, \ldots, |F_k \llcorner C|\},$$

and choose an integer $r \geq 1$ so that $K := K[r]$ contains C. If Δ is a gage on $\mathrm{cl}_* C$ associated with η and θ according to Lemma 2.6.6, then

$$\sigma(x) := \begin{cases} \min\{\delta(x), \Delta(x)\} & \text{if } x \in \mathrm{cl}_* C, \\ \mathrm{dist}\,(x, \mathrm{cl}_* C) & \text{if } x \in K - \mathrm{cl}_* C, \end{cases}$$

defines a gage σ on K. By Corollary 2.6.5, there is an θ-regular σ-fine partition $Q := \{(B_1, x_1), \ldots, (B_q, x_q)\}$ such that $G(K - [Q]) < \varepsilon/2$ and $[Q] \subset K$. After a suitable reordering, we may assume

$$\{x_1, \ldots, x_q\} \cap \mathrm{cl}_* C = \{x_1, \ldots, x_p\}$$

for an integer p with $0 \leq p \leq q$. If $A_i := B_i \cap C$, then our choice of σ shows $|A_i| = 0$ for $i = p + 1, \ldots, q$, and Lemma 2.6.6 implies the collection $P := \{(A_1, x_1), \ldots, (A_p, x_p)\}$ is an η-regular δ-fine partition. Since $[P] \subset C \subset A$ and since the sets $C - [P]$ and $C - [Q]$ are equivalent,

$$F(A - [P]) \leq F(A - C) + F(C - [P])$$
$$< \frac{\varepsilon}{2} + F(C - [Q]) = \frac{\varepsilon}{2} + G(K - [Q]) < \varepsilon$$

and the theorem is proved. $\qquad\qquad\qquad\qquad\qquad\qquad\qquad\qquad\square$

Lemma 2.6.8. *Let N be a negligible set, and let $\varepsilon > 0$. There is an absolutely continuous Radon measure $\mu < \varepsilon$ such that*

$$\lim \frac{\mu(B_k)}{|B_k|} = \infty$$

whenever $\{B_k\}$ is a sequence of subsets of $\mathbb{R}^m$ with $|B_k| > 0$ for all k and $\lim d(B_k \cup \{x\}) = 0$ for an $x \in N$. In particular, the restriction $G := \mu \restriction \mathbf{BV}$ is a charge and $\underline{D}G(x) = \infty$ for each $x \in N$.

PROOF. Find a decreasing sequence $\{U_i\}$ of open sets containing N so that $|U_i| < \varepsilon 2^{-i}$ for $i = 1, 2, \ldots$. Observe

$$\mu := \sum_{i=1}^{\infty} (\mathcal{L}^m \mathsf{L} \, U_i)$$

is an absolutely continuous Radon measure, and $\mu(\mathbb{R}^m) < \varepsilon$. Choose an $x \in N$ and a sequence $\{B_k\}$ with $\lim d(B_k \cup \{x\}) = 0$. If $i \geq 1$ is an integer, then $B_k \subset U_i$ for all sufficiently large k. Thus $\mu(B_k) \geq i|B_k|$ for all sufficiently large k, and the lemma follows. $\qquad\qquad\square$

Theorem 2.6.9. *Let N and T be sets that are, respectively, negligible and thin, and let $0 < \eta < 1/(2m\sqrt{m})$. Suppose F is a charge in a locally BV set E which satisfies the following conditions:*

- $\underline{D}_\eta F(x) \geq 0$ *for each $x \in E - N$,*
- $\underline{D}_\eta F(x) > -\infty$ *for each $x \in \mathrm{int}_c E - T$.*

Then F is nonnegative, in particular $\underline{D}F \geq 0$.

PROOF. In view of Corollary 1.5.3, Theorem 1.8.2, and equality (2.5.1), making N and T larger, we may assume

$$\mathrm{cl}_* E - E \subset N \quad \text{and} \quad \mathrm{cl}_* E - \mathrm{int}_c E \subset T.$$

Choose an $\varepsilon > 0$, and select a charge G associated with ε and the negligible set N according to Lemma 2.6.8. Letting $H := F + G$, observe $\underline{D}_\eta H(x) \geq 0$ whenever $x \in \mathrm{cl}_* E - T$. Given $x \in \mathrm{cl}_* E - T$, find a $\delta_x > 0$ so that $H(C) > -\varepsilon |C|$ for each BV set C with $r(C \cup \{x\}) > \eta$ and $d(C \cup \{x\}) < \delta_x$. If $A \in \mathcal{BV}(E)$, then the formula

$$\delta(x) := \begin{cases} \delta_x & \text{if } x \in \mathrm{cl}_* A - T, \\ 0 & \text{if } x \in \mathrm{cl}_* A \cap T, \end{cases}$$

defines a gage δ on $\mathrm{cl}_* A$. By Theorem 2.6.7, there is an η-regular δ-fine partition $P := \{(A_1, x_1), \ldots, (A_p, x_p)\}$ such that $[P] \subset A$ and

$$-\varepsilon < H\big(A - [P]\big) = H(A) - \sum_{i=1}^p H(A_i)$$

$$\leq H(A) + \varepsilon \sum_{i=1}^p |A_i| < F(A) + \varepsilon\big(1 + |A|\big).$$

The theorem follows from the arbitrariness of ε. $\qquad\square$

A partition $P := \{(A_1, x_1), \ldots, (A_p, x_p)\}$ is called *dyadic* if each A_i is a dyadic figure containing x_i. For bounded BV sets that are essentially clopen, we obtain a result analogous to Theorem 2.6.7 where partitions are replaced by dyadic partitions. Unlike in ordinary partitions, the points $x_1, \ldots, x_p$ in a dyadic partition P are distinct. To show this apparent difference is inessential, we need the following lemma.

Lemma 2.6.10. *Let δ be a nonnegative function on a set $E \subset \mathbb{R}^m$. There is a countable set $C \subset E$ such that the function σ defined on E by the formula*

$$\sigma(x) := \begin{cases} \delta(x) & \text{if } x \in E - C, \\ 0 & \text{if } x \in C, \end{cases}$$

has no strict local maximum.

PROOF. For $r > 0$, denote by C_r the set of all $x \in E$ for which

$$E_{x,r} := \big\{y \in E \cap B(x, r) : \delta(y) \geq \delta(x)\big\}$$

is a countable set. Choose a $z \in \mathbb{R}^m$, and let $D := C_r \cap B(z, r/2)$.

Claim. The set D is countable.

Proof. Given $x \in D$, let $D_x := \big\{y \in D : \delta(y) \geq \delta(x)\big\}$ and note $y \in B(x, r)$ and $\delta(y) \geq \delta(x)$ for each $y \in D_x$. Thus $D_x \subset E_{x,r}$, in particular, D_x is a countable set. Now denote by d the infimum of δ in D. If $d = \delta(x_0)$ for an $x_0 \in D$, then $D = D_{x_0}$

is countable. Otherwise, find a sequence $\{x_n\}$ in D with $\lim \delta(x_n) = d$, and observe $D = \bigcup_{n=1}^{\infty} D_{x_n}$ is again countable.

The claim implies C_r is a locally countable set. As $\mathbb{R}^m$ is a Lindelöf space [21, Corollary 4.1.16], the set C_r is countable, and so is the set $C := \bigcup_{k=1}^{\infty} C_{1/k}$. Let σ be a function on E defined as in the statement of the lemma. Clearly, σ attains the absolute minimum at each $x \in C$. If $x \in E - C$, then each $E_{x,1/k}$ is an uncountable set. As C is countable, there are points y_k in $E_{x,1/k} - (C \cup \{x\})$, and

$$\sigma(y_k) = \delta(y_k) \geq \delta(x) = \sigma(x)$$

for $k = 1, 2, \ldots$. Since the sequence $\{y_k\}$ converges to x, the function σ does not have a strict local maximum at x. $\qquad \square$

Although we shall not need this, we note that Lemma 2.6.10 holds (with the same proof) when E is any separable metrizable space.

Proposition 2.6.11. *Let* $F_1, \ldots, F_k$ *be charges, let* $F := \max\{|F_1|, \ldots, |F_k|\}$, *and let* A *be a bounded essentially clopen BV set. Suppose* $\eta < 1/(2m\sqrt{m})$ *and* ε *are positive numbers. Then for each gage* δ *on* $\mathrm{cl}_* A$, *there is an* η-regular δ-fine dyadic partition P such that $[P] \subset \mathrm{int}_* A$ and $F(\mathrm{int}_* A - [P]) < \varepsilon$.

PROOF. Choose an integer $r \geq 1$ so that $K := K[r]$ contains A, and let

$$G := F \, \llcorner \, \mathrm{int}_* A = \max\{|F_1 \, \llcorner \, \mathrm{int}_* A|, \ldots, |F_k \, \llcorner \, \mathrm{int}_* A|\}.$$

In view of Lemma 2.6.10, we may assume δ has no strict local maximum. As A is an essentially clopen BV set, the formula

$$\sigma(x) := \begin{cases} \min\{\delta(x), \mathrm{dist}\,(x, \partial_* A)\} & \text{if } x \in \mathrm{cl}_* A, \\ \mathrm{dist}\,(x, \partial_* A) & \text{if } x \in K - \mathrm{cl}_* A, \end{cases}$$

defines a gage σ on K. By Corollary 2.6.5, there is an η-regular σ-fine partition $Q := \{(B_1, x_1), \ldots, (B_q, x_q)\}$ such that each B_i is a dyadic figure, $[Q] \subset K$, and $G(K - [Q]) < \varepsilon$. After a suitable reordering there is an integer p with $0 \leq p \leq q$ and

$$\{x_1, \ldots, x_q\} \cap \mathrm{int}_* A = \{x_1, \ldots, x_x\}.$$

It follows from the definition of σ that $P' := \{(B_1, x_1), \ldots, (B_p, x_p)\}$ is an η-regular δ-fine partition with $[P'] \subset \mathrm{int}_* A$, and $B_i \cap \mathrm{int}_* A = \emptyset$ for $i = p+1, \ldots, q$. In particular,

$$F(\mathrm{int}_* A - [P']) = G(K - [Q]) < \varepsilon.$$

Since δ has no strict local maximum, there are distinct points $y_1, \ldots, y_p$ in $\mathrm{int}_* A$ such that $\{(B_1, y_1), \ldots, (B_p, y_p)\}$ is still an η-regular δ-fine partition. Now if C_i is a sufficiently small dyadic cube containing y_i, then

$$P = \{(B_1 \cup C_1, y_1), \ldots, (B_p \cup C_p, y_p)\}$$

is a dyadic partition with $[P] \subset \mathrm{int}_* A$ that is still η-regular δ-fine. Using the $\mathfrak{T}$-continuity of F, the cubes C_i can be selected so small that $F(\mathrm{int}_* A - [P]) < \varepsilon$. $\qquad \square$

3

Variations of charges

Following ideas of Thomson [84], we associate with each function F defined on the family $\mathcal{BV}$ certain measures, called the variational measures, or simply variations, of F. If F is a charge, we show that important properties of F, such as derivability, are determined by the behavior of these measures. In general, variational measures provide more information about charges than the classical variation.

3.1. Some classical concepts

Although the concepts and results of this section are well known, they are presented in a less familiar context of BV sets. Thus to avoid any ambiguities, we shall formulate precisely the definitions and give detailed proofs of all claims.

Let F be an *additive* function on $\mathcal{BV}$. For $A \in \mathcal{BV}$, we let

$$VF(A) := \sup_{\mathcal{A}} \sum_{i=1}^{p} |F(A_i)|$$

where $\mathcal{A} = \{A_1, \ldots, A_p\}$ is a disjoint subfamily of $\mathcal{BV}(A)$. The extended real-valued function

$$VF : A \mapsto VF(A)$$

defined on $\mathcal{BV}$ is called the *classical variation*, or simply the *variation*, of F. Clearly, $|F| \leq VF$ and the equality occurs whenever $F \geq 0$. Moreover,

$$V(F \, \mathsf{L} \, E) = (VF) \, \mathsf{L} \, E \tag{3.1.1}$$

for each locally BV set E.

90

Proposition 3.1.1. *Let* F *be an additive function on* $\mathcal{BV}$ *such that* $VF < \infty$. *Then* VF *is an additive function on* $\mathcal{BV}$, *which is a charge whenever* F *is a charge.*

PROOF. Choose disjoint bounded BV sets A and B. If

$$\mathcal{A} := \{A_1, \ldots, A_p\} \quad \text{and} \quad \mathcal{B} := \{B_1, \ldots, B_q\}$$

are disjoint subfamilies of $\mathcal{BV}(A)$ and $\mathcal{BV}(B)$, respectively, then $\mathcal{A} \cup \mathcal{B}$ is a disjoint subfamily of $\mathcal{BV}(A \cup B)$. Thus

$$\sum_{i=1}^{p} |F(A_i)| + \sum_{j=1}^{q} |F(B_j)| \le VF(A \cup B),$$

and the arbitrariness of $\mathcal{A}$ and $\mathcal{B}$ yields

$$VF(A) + VF(B) \le VF(A \cup B).$$

Assuming the previous inequality is strict, we can find a disjoint subfamily $\{C_1, \ldots, C_r\}$ of $\mathcal{BV}(A \cap B)$ such that

$$VF(A) + VF(B) < \sum_{i=1}^{r} |F(C_i)|$$

$$\le \sum_{i=1}^{r} |F(A \cap C_i)| + \sum_{i=1}^{r} |F(B \cap C_i)|$$

$$\le VF(A) + VF(B);$$

since $\{A \cap C_1, \ldots, A \cap C_r\}$ and $\{B \cap C_1, \ldots, B \cap C_r\}$ are, respectively, disjoint subfamilies of $\mathcal{BV}(A)$ and $\mathcal{BV}(B)$. This contradiction shows VF is additive.

Now assume F is a charge, and seeking a contradiction, suppose VF is not a charge. By Proposition 2.1.2, (ii), there is an $\varepsilon > 0$ and a sequence $\{A_i\}$ of BV subsets of $B_\varepsilon := B(1/\varepsilon)$ such that

$$\sup \|A_i\| < \infty, \quad \lim |A_i| = 0, \quad \text{and} \quad \inf VF(A_i) > \varepsilon.$$

Find a finite disjoint family $\mathcal{A}_1$ contained in $\mathcal{BV}(A_1)$, and hence in $\mathcal{BV}(B_\varepsilon)$, so that $\sum_{A \in \mathcal{A}_1} |F(A)| > \varepsilon$. Proceeding inductively, assume $\mathcal{A}_n \subset \mathcal{BV}(B_\varepsilon)$ is a finite disjoint family with $\sum_{A \in \mathcal{A}_n} |F(A)| > n\varepsilon$. Since $\{A - A_i\} \to A$ for each $A \in \mathcal{A}_n$, there is an A_k such that

$$\sum_{A \in \mathcal{A}_n} |F(A - A_k)| > n\varepsilon.$$

Now find a finite disjoint family $\mathcal{B} \subset \mathcal{BV}(A_k)$ with $\sum_{B \in \mathcal{B}} |F(B)| > \varepsilon$, and observe

$$\mathcal{A}_{n+1} := \{A - A_k : A \in \mathcal{A}_n\} \cup \mathcal{B}$$

is a finite disjoint subfamily of $\mathcal{BV}(B_\varepsilon)$ for which

$$\sum_{A \in \mathcal{A}_{n+1}} |F(A)| > (n+1)\varepsilon.$$

It follows $VF(B_\varepsilon) = \infty$, a contradiction.   □

In the literature, the following corollary is known as *Jordan's decomposition theorem*.

Corollary 3.1.2. *Let F be a charge, and assume $VF < \infty$. There is a unique pair F_1, F_2 of nonnegative charges such that*

$$F = F_1 - F_2 \quad and \quad VF = F_1 + F_2.$$

PROOF. Given F and VF, the unique solution of the linear equations $F = F_1 - F_2$ and $VF = F_1 + F_2$ is the pair of charges

$$F_1 := \frac{VF + F}{2} \quad \text{and} \quad F_2 := \frac{VF - F}{2},$$

which are nonnegative, since $|F| \le VF$.   □

For functions defined on $\mathcal{BV}$, we introduce two more classical concepts: absolute continuity and singularity.

- An additive function F on $\mathcal{BV}$ is called *absolutely continuous* (abbreviated as AC) in a locally BV set E if given $\varepsilon > 0$, there is a $\delta > 0$ such that $|F(B)| < \varepsilon$ for each BV set $B \subset E \cap B(1/\varepsilon)$ with $|B| < \delta$.

- A function F defined on $\mathcal{BV}$ is called *singular* in a set $E \subset \mathbb{R}^m$ if $DF(x) = 0$ for almost all $x \in E$.

Remark 3.1.3. The following simple facts are noteworthy.

(1) An additive function on $\mathcal{BV}$ that is AC in $\mathbb{R}^m$ is a charge, but not every charge is AC in $\mathbb{R}^m$.

(2) Let F be an additive function on $\mathcal{BV}$, and let E be a locally BV set. If F is AC in E, then it is AC in each locally BV set $A \subset E$. Moreover, F is AC in E if and only if $F \llcorner E$ is AC in $\mathbb{R}^m$.

(3) Let F be a function defined on $\mathcal{BV}$, and let E be a locally BV set. If F is singular, then $F \llcorner E$ is singular in E by Corollary 2.5.4. If F is singular in E and E is essentially closed, then $F \llcorner E$ is singular in $\mathbb{R}^m$.

An additive function F on $\mathcal{BV}$ that is AC in $\mathbb{R}^m$ is called AC. Similarly, a function F defined on $\mathcal{BV}$ that is singular in $\mathbb{R}^m$ is called singular. The abbreviation AC will be used only for additive functions on $\mathcal{BV}$, in particular, for charges; when talking about measures, absolute continuity will not be abbreviated.

Proposition 3.1.4. *If F is an additive function on $\mathcal{BV}$, then the following statements are equivalent.*

(1) *F is AC.*

(2) *VF is AC.*

(3) *$F = F_1 - F_2$ where F_1 and F_2 are nonnegative AC functions.*

(4) *Given $\varepsilon > 0$, there is a $\theta > 0$ such that $\big|F(B)\big| < \theta|B| + \varepsilon$ for each BV set $B \subset B(1/\varepsilon)$.*

PROOF. Implications (2) $\Rightarrow$ (1), (3) $\Rightarrow$ (1), and (4) $\Rightarrow$ (1) are obvious. Choose an $\varepsilon > 0$, and let $B_\varepsilon := B(1/\varepsilon)$.

(1) $\Rightarrow$ (2): Find a $\delta > 0$ so that $\big|F(B)\big| < \varepsilon$ for each BV set $B \subset B_\varepsilon$ with $|B| < \delta$. Let B be such a set, and let $B_1, \ldots, B_p$ be a disjoint subfamily of $\mathcal{BV}(B)$. After a suitable reordering, we may assume $F(B_i) \geq 0$ for $i = 1, \ldots, k$ and $F(B_i) < 0$ for $i = k + 1, \ldots, p$ where $0 \leq k \leq p$ is an integer. Thus

$$\sum_{i=1}^{p} \big|F(B_i)\big| = F\left(\bigcup_{i=1}^{k} B_i\right) - F\left(\bigcup_{i=k+1}^{p} B_i\right) < 2\varepsilon,$$

and hence $VF(B) \leq 2\varepsilon$. Since $B(1/\varepsilon)$ is the union of finitely many BV sets whose measures are less than δ, we infer $VF(B_\varepsilon) < \infty$. It follows $VF < \infty$, and by Proposition 3.1.1, VF is an additive function on $\mathcal{BV}$. Since $VF(B) < 2\varepsilon$ for each BV set $B \subset B_\varepsilon$ with $|B| < \delta$, we see the variation VF is AC.

(2) $\Rightarrow$ (4): Find a $\delta > 0$ so that $VF(B) < \varepsilon$ for each BV set $B \subset B_\varepsilon$ with $|B| < \delta$. Let $\theta := VF(B_\varepsilon)/\delta$, and select a BV set $B \subset B_\varepsilon$. If $|B| < \delta$, then

$$\big|F(B)\big| \leq VF(B) < \varepsilon \leq \theta|B| + \varepsilon,$$

and if $|B| > \delta$, then

$$\big|F(B)\big| \leq VF(B) \leq VF(B_\varepsilon)\frac{|B|}{\delta} < \theta|B| + \varepsilon.$$

Finally, as (1) $\Rightarrow$ (2), the implication (1) $\Rightarrow$ (3) is established by letting $F_1 := VF$ and $F_2 := VF - F$. $\square$

To appreciate the strength of absolute continuity, the reader should compare the next proposition with Theorem 2.6.9.

Proposition 3.1.5. *Let F be an additive function on $\mathcal{BV}$, and let $0 < \eta < 1/(2m\sqrt{m})$. If F is AC in a locally BV set E and $\underline{D}_\eta F(x) \geq 0$ for almost all $x \in E$, then $F \llcorner E \geq 0$. In particular, if F is both AC in E and singular in E, then $F \llcorner E = 0$.*

PROOF. Choose an $A \in \mathcal{BV}(E)$, and an $\varepsilon > 0$ so that $A \subset B(1/\varepsilon)$. Since $G := F \llcorner E$ is AC, there is a $\delta > 0$ such that $\big|G(C)\big| < \varepsilon$ for every BV set $C \subset B(1/\varepsilon)$ with $|C| < \delta$. If $\eta < \theta < 1/(2m\sqrt{m})$, then Proposition 2.5.3 implies $\underline{D}_\theta G(x) \geq \underline{D}_\eta F(x)$ for each $x \in \text{int}_c E$. Thus we can find an open set U so that $|U| < \delta$ and $\underline{D}_\theta G(x) \geq 0$ for each $x \in \text{cl}_* A - U$. There is a positive function σ defined on $\text{cl}_* A$ satisfying the following conditions:

- if $x \in \text{cl}_* A - U$, then $G(C) > -\varepsilon|C|$ for each BV set C with
$$d\big(C \cup \{x\}\big) < \sigma(x) \quad \text{and} \quad r\big(C \cup \{x\}\big) > \theta;$$
- if $x \in U \cap \text{cl}_* A$, then $B\big[x, \sigma(x)\big] \subset U$.

It follows from Theorem 2.6.7 that there is an θ-regular σ-fine partition $P := \big\{(C_1, x_1), \ldots, (C_p, x_p)\big\}$ such that $[P] \subset A$ and
$$\Big|G(A - [P])\Big| < \varepsilon.$$

Since $\bigcup_{x_i \in U} C_i \subset U$, we obtain
$$F(A) = G(A) = G\big(A - [P]\big) + G\left(\bigcup_{x_i \in U} C_i\right) + \sum_{x_i \notin U} G(C_i)$$
$$> -2\varepsilon - \varepsilon \sum_{x_i \notin U} |C_i| \geq -\varepsilon\big(2 + |A|\big),$$

and $F(A) \geq 0$ by the arbitrariness of ε. $\square$

Lemma 3.1.6. *If μ is a locally finite Borel measure, then the following statements are equivalent.*

    (i) *$\mu(B) = 0$ for each negligible Borel set B.*

    (ii) *Given $\varepsilon > 0$, there is a $\delta > 0$ such that $\mu(E) < \varepsilon$ for each set $E \subset B(1/\varepsilon)$ with $|E| < \delta$.*

In particular, every absolutely continuous Radon measure μ satisfies condition (ii).

PROOF. If (ii) does not hold, then there is an $\varepsilon > 0$ and a sequence $\{E_i\}$ of subsets of $B_\varepsilon := B(1/\varepsilon)$ such that $|E_i| < 2^{-i}$ and $\mu(E_i) \geq \varepsilon$ for $i = 1, 2, \ldots$. Find Borel sets B_i so that $E_i \subset B_i \subset B_\varepsilon$ and $|B_i| = |E_i|$. Since $\varepsilon \leq \mu(E_i) \leq \mu(B_i) \leq \mu(B_\varepsilon) < \infty$, the Borel set

$$B := \bigcap_{j=1}^{\infty} \bigcup_{i=j}^{\infty} B_i$$

is negligible and $\mu(B) \geq \varepsilon$. Consequently (i) does not hold.

Conversely if (ii) holds, and E is a negligible set, then $\mu(E) < \varepsilon$ for any $\varepsilon > 0$. Thus $\mu(E) = 0$ and (i) holds. $\qquad \square$

Proposition 3.1.7. *Let E be a locally BV set, and let F be the indefinite integral of $f \in L_\ell^1(E)$. Then F is AC.*

PROOF. The zero extension $g := \overline{f}$ belongs to $L_{\mathrm{loc}}^1(\mathbb{R}^m)$, and so the Radon measure $\mu := \mathcal{L}^m \, \mathsf{L} \, |g|$ is absolutely continuous. Choose an $\varepsilon > 0$, and use Lemma 3.1.6 to find a $\delta > 0$ so that $\mu(B) < \varepsilon$ for each set $B \subset B(1/\varepsilon)$ with $|B| < \delta$. Since the charge F is the restriction of the signed measure $\mathcal{L}^m \, \mathsf{L} \, g$ to $\mathcal{BV}$, we have $|F(A)| \leq \mu(A)$ for each bounded BV set A. The proposition follows. $\qquad \square$

We close this section by establishing the fundamental result about derivability of the indefinite integral F of $f \in L_\ell^1(E)$ where E is a locally BV set. Unlike the classical proof, which rests on the derivability of additive functions with finite variation [75, Chapter 4, Theorem 6.3], our argument is based on a Riemann type approximation of F.

Proposition 3.1.8. *Let E be a locally BV set, and let F be the indefinite integral of $f \in L_\ell^1(E)$. Given $\varepsilon > 0$, there is a positive function δ on $B(1/\varepsilon)$ such that*

$$\sum_{i=1}^{p} \left| \overline{f}(x_i) |A_i| - F(A_i) \right| < \varepsilon$$

for each δ-fine partition $\{(A_1, x_1), \ldots, (A_p, x_p)\}$.

PROOF. Choose an $\varepsilon > 0$, and use the Vitali-Carathéodory theorem [73, Theorem 2.25] to find extended real-valued functions g and h defined on $B_\varepsilon := B(1/\varepsilon)$ so that g is upper semicontinuous, h is lower semicontinuous, $g \leq \overline{f} \leq h$, and

$$\int_{B_\varepsilon} \left[h(x) - g(x) \right] dx < \varepsilon .$$

There is a positive function δ defined on B_ε such that $B[x,\delta(x)] \subset B_\varepsilon$, and the inequalities

$$g(y) < \overline{f}(x) + \varepsilon \quad \text{and} \quad h(y) > \overline{f}(x) - \varepsilon$$

hold whenever $x \in B_\varepsilon$ and $y \in B[x,\delta(x)]$. If $\{(A_1,x_1),\ldots,(A_p,x_p)\}$ is a δ-fine partition, then

$$\int_{A_i} g(x)\,dx \le F(A_i) \le \int_{A_i} h(x)\,dx,$$

$$\int_{A_i} g(x)\,dx - \varepsilon|A_i| \le \overline{f}(x_i)|A_i| \le \int_{A_i} h(x)\,dx + \varepsilon|A_i|$$

for $i = 1,\ldots,p$. Consequently

$$\sum_{i=1}^p \left| \overline{f}(x_i)|A_i| - F(A_i) \right| \le \sum_{i=1}^p \left(\int_{A_i} [h(x) - g(x)]\,dx + \varepsilon|A_i| \right)$$

$$\le \int_{B_\varepsilon} [h(x) - g(x)]\,dx + \varepsilon|B_\varepsilon|$$

$$< \varepsilon(1 + |B_\varepsilon|). \qquad \square$$

Theorem 3.1.9. *Let E be a locally BV set, and let $f \in L^1_\ell(E)$. If $F := \int f\,d\mathcal{L}^m$ is the indefinite integral of f, then*

$$DF(x) = \overline{f}(x)$$

for almost all $x \in \mathbb{R}^m$.

PROOF. Denote by N the set of all $x \in \mathbb{R}^m$ at which either F is not derivable or $DF(x) \ne \overline{f}(x)$. Lemma 2.3.4 implies the following: given $x \in N$, there is a $\beta_x > 0$ such that for each $\gamma > 0$, we can find a closed BV set C with $x \in C$, $d(C) < \gamma$, $r(C) > \beta_x$, and

$$\left| \overline{f}(x)|C| - F(C) \right| \ge \beta_x|C|.$$

Fix an integer $k \ge 1$, let $N_k := \{x \in N \cap B(k) : \beta_x > 1/k\}$, and choose a positive $\varepsilon < 1$. By Proposition 3.1.8, there is a positive function δ on $B_\varepsilon := B(1/\varepsilon)$ such that

$$\sum_{i=1}^p \left| \overline{f}(x_i)|A_i| - F(A_i) \right| < \frac{\varepsilon}{k}$$

for each δ-fine partition $\{(A_1,x_1),\ldots,(A_p,x_p)\}$. Let $\mathcal{C}$ be the collection of all $(1/k)$-regular closed BV sets C such that $d(C) < \delta(x_C)$ for an $x_C \in C \cap N_k$, and

$$\left| \overline{f}(x_C)|C| - F(C) \right| \geq \frac{1}{k}|C| .$$

By inequality (2.3.1), the family $\mathcal{C}$ is a Vitali cover of N_k, and Vitali's theorem yields disjoint sets $C_1, C_2, \ldots$ in $\mathcal{C}$ with $\left| N_k - \bigcup_i C_i \right| = 0$. Since each $\left\{ (C_1, x_{C_1}), \ldots, (C_p, x_{C_p}) \right\}$ is a δ-fine partition,

$$|N_k| \leq \sum_i |C_i| \leq k \sum_i \left| \overline{f}(x_{C_i})|C_i| - F(C_i) \right| \leq \varepsilon .$$

From the arbitrariness of ε, we see that N_k is a negligible set, and so is the set $N = \bigcup_{k=1}^{\infty} N_k$. $\qquad\qquad\qquad\qquad\qquad\qquad\qquad\qquad\qquad\square$

3.2. The essential variation

A nonnegative function δ defined on a set $E \subset \mathbb{R}^m$ is called an *essential gage* whenever $\{\delta = 0\}$ is a negligible set. Clearly, each gage is an essential gage, but not vice versa.

Let F be a function defined on $\mathcal{BV}$, and let $\eta \geq 0$. For a nonnegative function δ defined on a set $A \subset \mathbb{R}^m$, let

$$F(\delta, \eta) := \sup_P \sum_{i=1}^{p} |F(A_i)|$$

where $P := \left\{ (A_1, x_1), \ldots, (A_p, x_p) \right\}$ is an η-regular δ-fine partition; as usual, we define $\sum_{i=1}^{p} |F(A_i)| := 0$ whenever $P = \emptyset$. Observe

$$F(\sigma, \theta) \leq F(\delta, \eta) \tag{3.2.1}$$

whenever $0 \leq \eta \leq \theta$ and σ is a nonnegative function defined on a set $B \subset A$ such that $\sigma(x) \leq \delta(x)$ for each $x \in B$. If $E \subset \mathbb{R}^m$, we let

$$V_{\mathrm{ess},\eta} F(E) := \inf_\delta F(\delta, \eta)$$

where δ is an essential gage on E, and

$$V_{\mathrm{ess}} F(E) := \sup_{\eta > 0} V_{\mathrm{ess},\eta} F(E) .$$

Of the extended real-valued functions

$$V_{\mathrm{ess},\eta} F : E \mapsto V_{\mathrm{ess},\eta} F(E) \quad \text{and} \quad V_{\mathrm{ess}} F : E \mapsto V_{\mathrm{ess}} F(E)$$

defined for all $E \subset \mathbb{R}^m$, the latter is called the *essential variation* of F. In a different setting, the concept of essential variation was introduced in [6].

Proposition 3.2.1. *Let F be a function defined on $\mathcal{BV}$, and select an $\eta \geq 0$. The extended real-valued functions $V_{\mathrm{ess},\eta}F$ and $V_{\mathrm{ess}}F$ are absolutely continuous Borel regular measures. Moreover, $V_{\mathrm{ess},\eta}F = 0$ whenever $\eta \geq 1/(2m)$.*

PROOF. If $E \subset \mathbb{R}^m$ and $|E| = 0$, then $\delta : x \mapsto 0$ is an essential gage on E. As only the empty partition is δ-fine, $V_{\mathrm{ess},\eta}F(E) = 0$. By inequality (2.3.2), only the empty partition is η-regular for $\eta \geq 1/(2m)$, and hence $V_{\mathrm{ess},\eta}F = 0$ whenever $\eta \geq 1/(2m)$.

Let $B \subset A \subset \mathbb{R}^m$, and let δ be an essential gage on A. Then $\sigma := \delta \restriction B$ is an essential gage on B, and

$$V_{\mathrm{ess},\eta}F(B) \leq F(\sigma, \eta) \leq F(\delta, \eta)$$

according to (3.2.1). The inequality $V_{\mathrm{ess},\eta}F(B) \leq V_{\mathrm{ess},\eta}F(A)$ follows from the arbitrariness of δ.

Let E be the union of a countable family $\{E_n\}$ of subsets of $\mathbb{R}^m$. Assume first the sets E_n are disjoint. If δ_n is an essential gage on E_n, define an essential gage δ on E by letting $\delta(x) := \delta_n(x)$ for $x \in E_n$. Given a δ-fine partition P, the partitions $P_n := \{(A, x) \in P : x \in E_n\}$ are δ_n-fine and $P = \bigcup_n P_n$. It follows

$$V_{\mathrm{ess},\eta}F(E) \leq F(\delta, \eta) \leq \sum_n F(\delta_n, \eta)$$

and, as δ_n are arbitrary, $V_{\mathrm{ess},\eta}F(E) \leq \sum_n V_{\mathrm{ess},\eta}F(E_n)$. If E_n are any sets, not necessarily disjoint, the previous results imply

$$V_{\mathrm{ess},\eta}F(E) \leq \sum_n V_{\mathrm{ess},\eta}F\left(E_n - \bigcup_{k<n} E_k\right) \leq \sum_n V_{\mathrm{ess},\eta}F(E_n).$$

Let $A, B \subset \mathbb{R}^m$ be such that $d := \mathrm{dist}(A, B) > 0$, and let $\delta < d/2$ be an essential gage on $A \cup B$. Then $\delta_A := \delta \restriction A$ and $\delta_B := \delta \restriction B$ are essential gages on A and B, respectively. If P_A and P_B are, respectively, δ_A-fine and δ_B-fine partitions, then $P_A \cup P_B$ is a δ-fine partition; indeed, the choice of δ implies $A' \cap B' = \emptyset$ whenever $(A', x) \in P_A$ and $(B', y) \in P_B$. Thus

$$V_{\mathrm{ess},\eta}F(A) + V_{\mathrm{ess},\eta}F(B) \leq F(\delta_A, \eta) + F(\delta_B, \eta) \leq F(\delta, \eta),$$

and by the arbitrariness of δ,

$$V_{\text{ess},\eta}F(A) + V_{\text{ess},\eta}F(B) \leq V_{\text{ess},\eta}F(A \cup B).$$

It follows $V_{\text{ess},\eta}F$ is a metric, and hence Borel, measure.

We show next the measure $V_{\text{ess},\eta}F$ is Borel regular. For this purpose, select an $E \subset \mathbb{R}^m$ with $V_{\text{ess},\eta}F(E) < \infty$, and choose an integer $n \geq 1$. There is an essential gage δ_n on E such that

$$F(\delta_n, \eta) < V_{\text{ess},\eta}F(E) + \frac{1}{n}.$$

For an integer $k \geq 1$, let $E_k := \{x \in E : \delta_n(x) > 1/k\}$. Further let $\rho(x) := 1/k$ for each $x \in E_k$, and $\sigma(x) := 1/k$ for each $x \in \text{cl } E_k$. Given an η-regular σ-fine partition $\{(B_1, y_1), \ldots, (B_p, y_p)\}$, there is an η-regular ρ-fine partition $\{(B_1, x_1), \ldots, (B_p, x_p)\}$ obtained by moving $y_i \in \text{cl } E_k$ to a nearby $x_i \in E_k$. This and inequality (3.2.1) yield

$$V_{\text{ess},\eta}F(\text{cl } E_k) \leq F(\sigma, \eta) = F(\rho, \eta) \leq F(\delta_n, \eta) < V_{\text{ess},\eta}F(E) + \frac{1}{n}.$$

The set $C_n := \bigcup_{k=1}^{\infty} \text{cl } E_k$ is Borel, and since $\{\text{cl } E_k\}$ is an increasing sequence, we obtain $V_{\text{ess},\eta}F(C_n) \leq V_{\text{ess},\eta}F(E) + 1/n$. As the null set $\{\delta_n = 0\}$ is contained in a negligible Borel set N_n, the set E is contained in a Borel set $B_n := C_n \cup N_n$. The absolute continuity of $V_{\text{ess},\eta}F$ yields

$$V_{\text{ess},\eta}F(B_n) = V_{\text{ess},\eta}F(C_n) \leq V_{\text{ess},\eta}F(E) + \frac{1}{n}.$$

The Borel set $B := \bigcap_{n=1}^{\infty} B_n$ still contains E and

$$V_{\text{ess},\eta}F(E) \leq V_{\text{ess},\eta}F(B) = \lim V_{\text{ess},\eta}F(B_n) \leq V_{\text{ess},\eta}F(E).$$

Since $\{V_{\text{ess},1/k}F\}$ is an increasing sequence of measures whose limit is $V_{\text{ess}}F$ the properties of $V_{\text{ess}}F$ follow readily from those of $V_{\text{ess},\eta}F$. For illustration, we show the measure $V_{\text{ess}}F$ is Borel regular. To this end, select an $E \subset \mathbb{R}^m$, and find Borel sets D_k so that

$$E \subset D_k \quad \text{and} \quad V_{\text{ess},1/k}F(E) = V_{\text{ess},1/k}F(D_k)$$

for $k = 1, 2, \ldots$. The Borel set $D := \bigcap_{k=1}^{\infty} D_k$ still contains E, and

$$\begin{aligned}
V_{\text{ess}}F(E) &\leq V_{\text{ess}}F(D) = \lim V_{\text{ess},1/k}F(D) \\
&\leq \lim V_{\text{ess},1/k}F(D_k) = \lim V_{\text{ess},1/k}F(E) \\
&= V_{\text{ess}}F(E).
\end{aligned}$$ $\qquad\square$

Proposition 3.2.2. *If F is a charge and $VF < \infty$, then*

$$V_{\mathrm{ess}}F(A) \leq VF(A)$$

for each bounded BV set A. In particular, $V_{\mathrm{ess}}F$ is locally finite.

PROOF. Working toward a contradiction, assume $VF(A) < V_{\mathrm{ess}}F(A)$ for a bounded BV set A, and select an $r > 0$ so that $A \subset B(r)$. According to Proposition 1.10.5, there is a sequence $\{B_k\}$ of essentially closed BV subsets of $B(r) - A$ such that $\{B_k\} \to B(r) - A$. It follows each $U_k := B(r) - \mathrm{cl}_* B_k$ is an open bounded BV sets containing $\mathrm{int}_* A$, and $\{U_k\} \to A$. According to Propositions 2.1.2 and 3.1.1, we have $\lim VF(U_k) = VF(A)$. Fix an integer $k \geq 1$ with

$$VF(U_k) < V_{\mathrm{ess}}F(A) = V_{\mathrm{ess}}F(\mathrm{int}_* A);$$

note the equality $V_{\mathrm{ess}}F(A) = V_{\mathrm{ess}}F(\mathrm{int}_* A)$ follows from the density theorem and absolute continuity of the measure $V_{\mathrm{ess}}F$. There is an $\eta > 0$ so that $VF(U_k) < F(\delta, \eta)$ for every essential gage δ on $\mathrm{int}_* A$. As $\delta : x \mapsto \mathrm{dist}\,(x, \partial U_k)$ is a positive gage on $\mathrm{int}_* A$, we can find an η-regular δ-fine partition $\{(A_1, x_1), \ldots, (A_p, x_p)\}$ for which

$$VF(U_k) < \sum_{i=1}^{p} |F(A_i)|.$$

This is a contradiction, since $\{A_1, \ldots, A_p\}$ is a disjoint subfamily of $\mathcal{BV}(U_k)$ by the choice of δ. $\square$

Whether Proposition 3.2.2 holds without the assumption $VF < \infty$ is unclear. However, the next example shows the inequality $V_{\mathrm{ess}}F \leq VF$ can be strict.

Example 3.2.3 (The devil's staircase). For each cell $J := [a, b]$, we define intervals $J_- := (-\infty, a)$, $J_+ := (b, \infty)$, and

$$J_0 := [a, a + h], \quad J_1 := (a + h, b - h), \quad J_2 := [b - h, b],$$

where $h := |J|/3$. Select a fixed one-dimensional cell B, and define open intervals B_-, B_+, B_1, and cells B_0, B_2 as above. Assuming a cell $B_{i_1 \cdots i_k}$ has been defined for a k-tuple $\{i_1, \ldots, i_k\}$ where $i_j = 0, 2$ for $j = 1, \ldots, k$, let $B_{i_1 \cdots i_k i} := (B_{i_1 \cdots i_k})_i$ for $i = 0, 1, 2$. Observe

$$C_B := \bigcap_{k=1}^{\infty} \bigcup_{i_k = 0, 2} B_{i_1 \cdots i_k} = B - \bigcup_{k=0}^{\infty} \bigcup_{i_k = 0, 2} B_{i_1 \cdots i_k 1}$$

is the *Cantor ternary set* in B. For each $t \in C_B$, there is a unique infinite sequence $\{i_1(t), i_2(t), \ldots\}$ such that $i_k(t) = 0, 2$ and $t \in B_{i_1(t) \cdots i_k(t)}$ for $k = 1, 2, \ldots$. Moreover, for each $t \in B - C_B$, there is a unique finite sequence $\{i_1(t), \ldots, i_{n(t)}(t)\}$ such that $n(t) \geq 0$, $i_k(t) = 0, 2$ for $k = 1, \ldots, n(t)$, and $t \in B_{i_1(t) \cdots i_k(t)1}$. The *devil's staircase* on B is a function $s_B : \mathbb{R} \to [0, 1]$ defined as follows:

$$
s_B(x) := \begin{cases} \sum_{k=1}^{\infty} i_k(t) 2^{-k} & \text{if } t \in C_B, \\ \sum_{k=1}^{n(t)} i_k(t) 2^{-k} + 2^{-n(t)-1} & \text{if } t \in B - C_B, \end{cases}
$$

$s(t) := 0$ if $t \in B_-$, and $s(t) := 1$ if $t \in B_+$. It is well known and easy to prove that s_B is a continuous increasing function which is constant on each connected component of $\mathbb{R} - C_B$ [29, Section 19, Exercise 3]. The charge S associated with s_B according to Remark 2.1.5 is a nonnegative charge in B, and $VS(B) = S(B) = 1$. On the other hand, since C_B is a closed negligible set, the function $\delta : x \to \text{dist}(x, C_B)$ is an essential gage on B. It follows $V_{\text{ess}} S(B) = 0$. As S is clearly singular, S is not AC according to Proposition 3.1.5.

In the setting of an abstract measure space with a derivation base, the following theorem was proved in [11, Theorem 1].

Theorem 3.2.4. *Let F be a $\mathfrak{T}$-continuous function defined on $\mathcal{BV}$, and let E be a measurable set. Then*

$$
V_{\text{ess}} F(E) = \int_E \overline{D} |F|(x) \, dx, \quad \text{and} \quad V_{\text{ess},\eta} F(E) = \int_E \overline{D}_\eta |F|(x) \, dx
$$

for each positive $\eta < 1/(2m)$.

PROOF. Since $V_{\text{ess}} F = V_{\text{ess}} |F|$ and $V_{\text{ess},\eta} F = V_{\text{ess},\eta} |F|$, we may assume $F \geq 0$. Select a positive $\eta < 1/(2m)$, and observe the nonnegative extended real-valued function $\overline{D}_\eta F$ is measurable by Proposition 2.3.5 (this and the application of Lemma 2.3.4 below are the only places where the $\mathfrak{T}$-continuity of F is used). Thus the integral $I := \int_E \overline{D}_\eta F(x) \, dx$ exists, possibly equal to ∞.

Claim 1. If $I < \infty$ or $V_{\text{ess},\eta} F(E) < \infty$, then

$$
E_\infty := \{x \in E : \overline{D}_\eta F(x) = \infty\}
$$

is a negligible set.

Proof. If $I < \infty$, the statement is clear. So assume $V_{\text{ess},\eta} F(E) < \infty$, and fix an integer $k \geq 1$. In view of Lemma 2.3.4, given $x \in E_\infty$ and

$\rho > 0$, there is a closed BV set C such that $x \in C$, $d(C) < \rho$, $r(C) > \eta$, and $F(C) > k|C|$. Find an essential gage δ on E_∞ so that

$$F(\delta, \eta) < V_{\text{ess},\eta} F(E_\infty) + 1 \leq V_{\text{ess},\eta} F(E) + 1 < \infty.$$

The family $\mathfrak{C}$ of all closed BV sets C such that $d(C) < \delta(x_C)$ for an $x_C \in C \cap E_\infty$, $r(C) > \eta$, and $F(C) > k|C|$ is a Vitali cover of $E_\infty - \{\delta = 0\}$. By Vitali's theorem, there are disjoint sets $C_1, C_2, \ldots$ in $\mathfrak{C}$ with $\left| E_\infty - \bigcup_i C_i \right| = 0$; for $\{\delta = 0\}$ is a negligible set. Since each $\{(C_1, x_{C_1}), \ldots, (C_p, x_{C_p})\}$ is an η-regular δ-fine partition, we obtain

$$|E_\infty| \leq \sum_i |C_i| < \frac{1}{k} \sum_i F(C_i) \leq \frac{1}{k} F(\delta, \eta) < \frac{1}{k} \left[V_{\text{ess},\eta} F(E) + 1 \right],$$

and the claim follows by letting $k \to \infty$.

In view of Claim 1, we may assume $|E_\infty| = 0$. As this implies

$$V_{\text{ess},\eta} F(E_\infty) = \int_{E_\infty} \overline{D}_\eta |F|(x)\, dx = 0,$$

we may further assume $E_\infty = \emptyset$. Now the measurable sets

$$E_n := \left\{ x \in E \cap B(n) : \overline{D}_\eta F(x) < n \right\}, \quad n = 1, 2, \ldots,$$

form an increasing sequence whose union is E, and so it suffices to prove

$$V_{\text{ess},\eta} F(E_n) = \int_{E_n} \overline{D}_\eta F(x)\, dx$$

for each integer $n \geq 1$.

Consequently, we may assume from the onset that $I < \infty$ and that $E \subset U$ where U is an open set of finite measure. The first assumption allows us to define a function G on $\mathcal{BV}$ by letting

$$G(C) := F(C) - \int_C \overline{D}_\eta F(x) \chi_E(x)\, dx$$

for each $C \in \mathcal{BV}$. Observe the set $N := \left\{ x \in E : \overline{D}_\eta G(x) \neq 0 \right\}$ is negligible according to Theorem 3.1.9.

Claim 2. $V_{\text{ess},\eta} F(E) \leq I$.

Proof. Choose an $\varepsilon > 0$, and given $x \in E - N$, select a $\delta_x > 0$ so that $B[x, \delta(x)] \subset U$ and $G(C) < \varepsilon|C|$ for every BV set C with $r(C \cup \{x\}) > \eta$ and $d(C \cup \{x\}) < \delta_x$. The formula

$$\delta(x) := \begin{cases} \delta_x & \text{if } x \in E - N, \\ 0 & \text{if } x \in N \end{cases}$$

defines an essential gage on E. If $\{(C_1, x_1), \ldots, (C_p, x_p)\}$ is an η-regular δ-fine partition, then

$$\sum_{i=1}^{p} F(C_i) = \sum_{i=1}^{p} \left[G(C_i) + \int_{C_i \cap E} \overline{D}_\eta F(x) \, dx \right]$$

$$< \varepsilon \sum_{i=1}^{p} |C_i| + I \leq \varepsilon |U| + I.$$

Thus $V_{\text{ess},\eta} F(E) \leq F(\delta, \eta) \leq \varepsilon |U| + I$, and the claim follows from the arbitrariness of ε.

Claim 3. $V_{\text{ess},\eta} F(E) = I$.

Proof. Proceeding toward a contradiction, assume $V_{\text{ess},\eta} F(E) < I$, and select an essential gage σ on E so that $F(\sigma, \eta) < I$. Making σ smaller, we may assume $N \subset \{\sigma = 0\}$ and $B[x, \sigma(x)] \subset U$ for each x in $E - \{\sigma = 0\}$. Choose an $\varepsilon > 0$, and let $\mathcal{B}$ be the family of all closed BV sets B such that $d(B) < \sigma(x_B)$ for an

$$x_B \in B \cap E, \quad r(B) > \eta, \quad \text{and} \quad G(B) > -\varepsilon |B|.$$

In view of Lemma 2.3.4, the family $\mathcal{B}$ is a Vitali cover of $E - \{\sigma = 0\}$. As above, Vitali's theorem implies there are disjoint sets $B_1, B_2, \ldots$ in $\mathcal{B}$ with $|E - \bigcup_i B_i| = 0$. Since each $\{(B_1, x_{B_1}), \ldots, (B_p, x_{B_p})\}$ is an η-regular σ-fine partition, we obtain

$$I \leq \sum_i \int_{B_i \cap E} \overline{D} F(x) \, dx = \sum_i \left[F(B_i) - G(B_i) \right]$$

$$< \sum_i F(B_i) + \varepsilon \sum_i |B_i| \leq F(\sigma, \eta) + \varepsilon |U|.$$

The arbitrariness of ε yields a contradiction.

As the increasing sequences $\{V_{\text{ess},1/k} F\}$ and $\{\overline{D}_{1/k} F\}$ converge to $V_{\text{ess}} F$ and $\overline{D} F$, respectively, the equality $V_{\text{ess}} F(E) = \int_E \overline{D} F(x) \, dx$ follows from Claim 3 by the monotone convergence theorem. $\square$

Corollary 3.2.5. *A $\mathfrak{T}$-continuous function F defined on $\mathcal{BV}$ is singular in a set $E \subset \mathbb{R}^m$ if and only if $V_{\text{ess}} F(E) = 0$.*

PROOF. Suppose $V_{\text{ess}} F(E) = 0$. As $V_{\text{ess}} F$ is a Borel regular measure, there is a measurable set B such that $E \subset B$ and

$$\int_B \overline{D} |F|(x) \, dx = V_{\text{ess}} F(B) = V_{\text{ess}} F(E) = 0.$$

It follows F is singular in B, and hence in E.

Conversely, let F be singular in a set E. Then up to a negligible set, E is contained in the measurable set $A := \{x \in \mathbb{R}^m : \overline{D}|F|(x) = 0\}$. As $V_{\text{ess}}F(A) = 0$ by Theorem 3.2.4, the corollary follows from the absolute continuity of $V_{\text{ess}}F$. □

Theorem 3.2.6. *For a charge F and a set $E \subset \mathbb{R}^m$, the following conditions are equivalent.*

(1) *The measure $(V_{\text{ess}}F) \, \mathsf{L} \, E$ is σ-finite.*

(2) *The measure $(V_{\text{ess},\eta}F) \, \mathsf{L} \, E$ is σ-finite for each $\eta > 0$.*

(3) *The charge F is derivable almost everywhere in E.*

PROOF. The implication (1) $\Rightarrow$ (2) is obvious, since $V_{\text{ess},\eta}F \leq V_{\text{ess}}F$ for each $\eta > 0$.

(2) $\Rightarrow$ (3): Let $k \geq 1$ be an integer. If $(V_{\text{ess},1/k}F) \, \mathsf{L} \, E$ is σ-finite, then $E \subset \bigcup_{i=1}^{\infty} E_i$ and $V_{\text{ess},1/k}F(E_i) < \infty$ for $i = 1, 2, \ldots$. As $V_{\text{ess},1/k}F$ is a Borel regular measure, we may assume each E_i is a measurable set. Using Theorem 3.2.4, find a negligible set $N_k \subset E$ so that $\overline{D}_{1/k}|F|(x)$ is finite for each $x \in E - N_k$. The set $N := \bigcup_{k=1}^{\infty} N_k$ is negligible, and it is easy to see that F is almost derivable at each $x \in E - N$. By Theorem 2.4.3, the charge F is derivable almost everywhere in E.

(3) $\Rightarrow$ (1): If F is derivable almost everywhere in E, then up to a negligible set, E is contained in the union of the sets

$$E_k := \{x \in B(1/k) : \overline{D}|F|(x) < k\}, \quad k = 1, 2, \ldots,$$

which are measurable by Proposition 2.3.5. Now

$$V_{\text{ess}}F(E_k) = \int_{E_k} \overline{D}|F|(x) \, dx \leq k \big| B(1/k) \big| < \infty$$

according to Theorem 3.2.4, and the σ-finiteness of $(V_{\text{ess}}F) \, \mathsf{L} \, E$ follows from the absolute continuity of $V_{\text{ess}}F$. □

Corollary 3.2.7. *Let F be a charge. If $VF < \infty$, then F is derivable almost everywhere and DF belongs to $L^1_{\text{loc}}(\mathbb{R}^m)$. If $F \geq 0$, then*

$$\int DF \, d\mathcal{L}^m \leq F.$$

PROOF. In view of Proposition 3.2.2, the first claim follows from Theorems 3.2.6 and 3.2.4. If $F \geq 0$, then $VF = F < \infty$, and

$$\int_B DF(x) \, dx = V_{\text{ess}}F(B) \leq VF(B) = F(B)$$

for each bounded BV set B by Theorem 3.2.4. □

Our next result is the *Lebesgue's decomposition theorem* for a charge whose classical variation is finite.

Theorem 3.2.8. *Let F be a charge with $VF < \infty$. There is a unique pair F_{AC}, F_S of charges such that F_{AC} is AC, F_S is singular, and*

$$F = F_{AC} + F_S.$$

Moreover, $F_{AC} = \int DF(x)\,dx$.

PROOF. Using Corollary 3.2.7, define charges

$$F_{AC} := \int DF(x)\,dx \quad \text{and} \quad F_S := F - F_{AC},$$

and observe they have the required properties by Proposition 3.1.7 and Theorem 3.1.9. Let G_{AC}, G_S be another pair of charges having the properties stipulated in the proposition. Then the charge

$$F_{AC} - G_{AC} = G_S - F_S$$

is both AC and singular. Proposition 3.1.5 implies $G_{AC} = F_{AC}$ and $G_S = F_S$. $\square$

The next corollary together with Proposition 3.1.7 give a complete characterization of AC functions.

Corollary 3.2.9. *Let F be an additive function on $\mathcal{BV}$. If F is AC, then it is derivable almost everywhere, DF belongs to $L^1_{\mathrm{loc}}(\mathbb{R}^m)$, and*

$$\int DF(x)\,dx = F.$$

PROOF. By Remark 3.1.3, (i) and Proposition 3.1.4, the function F is a charge and $VF < \infty$. The corollary follows from the uniqueness of Lebesgue's decomposition, since $F = F_{AC}$. $\square$

An immediate consequence of Corollary 3.2.9 and Definition 2.5.8 is the following version of the Gauss-Green theorem.

Corollary 3.2.10. *Let E be a locally BV set, and let $v : \mathrm{cl}_* E \to \mathbb{R}^m$ be a bounded Borel measurable vector field. Suppose the flux of v is AC. Then the mean divergence $\mathfrak{div}\, v(x)$ is defined for almost all $x \in E$, the function $\mathfrak{div}\, v : x \mapsto \mathfrak{div}\, v(x)$ belongs to $L^1_\ell(E)$, and*

$$\int_A \mathfrak{div}\, v\,d\mathcal{L}^m = \int_{\partial_* A} v \cdot \nu_A\,d\mathcal{H}^{m-1}$$

for each bounded BV set $A \subset E$.

Proposition 3.2.11. *If an additive function F on $\mathcal{BV}$ is AC, then*

$$VF(A) = V_{\text{ess}}F(A)$$

for each bounded BV set A.

PROOF. Select a bounded BV set A, and let $\{A_1, \ldots, A_p\}$ be a disjoint subfamily of $\mathcal{BV}(A)$. By Corollary 3.2.9 and Theorem 3.2.4,

$$\sum_{i=1}^{p} |F(A_i)| = \sum_{i=1}^{p} \left| \int_{A_i} DF(x)\, dx \right| \leq \sum_{i=1}^{p} \int_{A_i} |DF(x)|\, dx$$

$$\leq \int_{A} |DF(x)|\, dx = \int_{A} D|F|(x)\, dx = V_{\text{ess}}\, F(A).$$

Thus $VF(A) \leq V_{\text{ess}}F(A)$, and an application of Proposition 3.2.2 completes the proof. □

3.3. The integration problem

For readers with a rudimentary knowledge of categories [78, Chapter 1, Sections 1 and 2], we give a brief intuitive outline of our main goal. To avoid technicalities, the presentation is deliberately vague; in particular, the underlying category is not specified. A rigorous treatment is given in Section 3.7 below.

Let E be a locally BV set. A *primitive* of a function f defined on E is any charge F in E such that $DF(x) = f(x)$ for almost all $x \in E$. A function defined on E which has a primitive is called *potentially integrable* in E. According to Proposition 2.3.5, each potentially integrable function in E is measurable, and it follows from Luzin's theorem [75, Chapter 7, Theorem 2.3] that the converse is true in dimension one. In higher dimensions, it is unclear whether every measurable function defined on E is also potentially integrable in E. The next observation shows that a primitive of a potentially integrable function in E is not unique whenever int $E \neq \emptyset$.

Observation 3.3.1. *For each cell A, there is a nonnegative charge S_A in A such that S_A is singular and $S_A(A) = 1$.*

PROOF. Let $A := \prod_{i=1}^{m} [a_i, b_i]$, and let $s := s_{[a_1, b_1]}$ be the devil's staircase on $[a_1, b_1]$ defined in Example 3.2.3. Letting

$$v(x) := \frac{b_1 - a_1}{|A|} \left(s(\xi_1), 0, \ldots, 0 \right)$$

for each $x = (\xi_1, \ldots, \xi_m)$ in $\mathbb{R}^m$ defines a vector field $v \in C(\mathbb{R}^m; \mathbb{R}^m)$, and it is easy to see that the flux S_A of v is the desired charge. $\square$

Notation 3.3.2. Let E be a locally BV set. Throughout the remainder of the book we shall use the following families.

$CH_D(E)$: all *primitives* in E, i.e., all charges in E that are *derivable almost everywhere* in E.

$CH_S(E)$: all charges in E that are *singular* in E.

$CH_{AC}(E)$: all charges in E that are AC in E, or equivalently AC.

$CH_{AD}(E)$: all charges F in E that are *almost derivable* at each point $x \in \mathrm{cl}_* E - T_F$ where T_F a thin set.

$M(E)$: all equivalence classes of *measurable* functions defined on the set E.

$PI(E)$: all equivalence classes of *potentially integrable* functions in the set E.

An easy verification that each of these families is a linear spaces is left to the reader.

If E is a locally BV set, then there is a *short exact sequence* [78, Chapter 4, Section 5]

$$0 \longrightarrow CH_S(E) \xrightarrow{\subset} CH_D(E) \xrightarrow{D} PI(E) \longrightarrow 0 \qquad (3.3.1)$$

where the linear map $D : F \mapsto DF$ is called the *derivation*. Since each linear space has an algebraic basis [33, Chapter 1, Section 2], the derivation D has a right inverse. In other words, there is a linear map

$$I_E : PI(E) \to CH_D(E),$$

such that $D \circ I_E$ is the identity map of $PI(E)$. As I_E assigns to every potentially integrable function in E a unique primitive, we may call it a *universal integration* on E. Unfortunately, a universal integration on E has little practical value — its utility can be compared to that of a *Banach limit* [74, Chapter 3, Exercise 4]. Indeed, the map I_E is neither unique nor constructive and, above all, it is not a *natural transformation* of functors. In contrast, when the derivation D is properly interpreted, the short exact sequence (3.3.1) is functorial.

A more fruitful approach to integration relies on finding a linear space $X(E) \subset CH_D(E)$ such that $X(E) \cap CH_S(E) = \{0\}$. Then the derivation $D : X(E) \to M(E)$ is an injective map, and if the space $X(E)$ is functorial, the inverse map

$$I : D\big[X(E)\big] \to X(E)$$

is a natural transformation. The following examples illustrate the idea.

- It follows from Theorem 3.1.9 and Corollary 3.2.9 that the map
  $D : CH_{AC}(E) \to L^1_\ell(E)$ is bijective; its inverse is the map

$$\int : f \mapsto \int f \, d\mathcal{L}^m : L^1_\ell(E) \to CH_{AC}(E)$$

defined in Example 2.1.3.

- Theorem 2.4.3 shows $CH_{AD}(E)$ is a subspace of $CH_D(E)$, and
  it follows from Theorem 2.6.9 that $CH_{AD}(E) \cap CH_S(E) = \{0\}$.
  Hence the map $D : CH_{AD}(E) \to M(E)$ is injective, and we show
  in Section 5.5 below that its inverse is given by a Riemann type
  integral which extends the Lebesgue integral.

In general, neither of the spaces $CH_{AC}(E)$ and $CH_{AD}(E)$ contains the
other (Examples 3.3.3 and 3.3.4 below). However, in Section 3.6 below,
we shall exhibit a well behaved space $CH_*(E) \subset CH_D(E)$ such that

$$CH_{AC}(E) + CH_{AD}(E) \subset CH_*(E), \qquad (3.3.2)$$

$$CH_*(E) \cap CH_S(E) = \{0\}. \qquad (3.3.3)$$

A simple definition of the space $CH_*(E)$ rests on a property of Hausdorff
measures which is established in the next section.

Example 3.3.3. Define an $f \in C(\mathbb{R})$ by letting

$$f(t) := \begin{cases} t^2 \sin \frac{1}{t^2} & \text{if } t \neq 0, \\ 0 & \text{if } t = 0, \end{cases}$$

and observe

$$f'(t) = \begin{cases} 2t \sin \frac{1}{t^2} - \frac{2}{t} \cos \frac{1}{t^2} & \text{if } t \neq 0, \\ 0 & \text{if } t = 0. \end{cases}$$

Thus the charge F associated with f according to Remark 2.1.5 belongs to $CH_{AD}(\mathbb{R})$. On the other hand, since $VF([0,1]) = \infty$, Proposition 3.1.4 implies F is not AC.

Example 3.3.4. This example is the same as that presented in [75,
Chapter 6, Section 7]. Let $C := C_B$ be the Cantor ternary set in the
cell $B := [0,1]$ (see Example 3.2.3). Then $B - C = \bigcup_{i=1}^\infty (a_i, b_i)$ where
$h_i := b_i - a_i > 0$. Letting

$$k_i := \sum_{n=i}^\infty h_n \quad \text{and} \quad \theta_i := \sqrt{k_i} - \sqrt{k_{i+1}},$$

it is easy to verify $\sum_{i=1}^{\infty} \theta_i = 1$ and $\lim(\theta_i/h_i) = \infty$. Define a nonnegative function f on B by the formula

$$f(t) := \begin{cases} \dfrac{\theta_i}{\sqrt{(t-a_i)(b_i-t)}} & \text{if } t \in (a_i, b_i), \\ 0 & \text{if } t \in C, \end{cases}$$

and observe

$$\int_B f(t)\, dt = \sum_{i=1}^{\infty} \int_{a_i}^{b_i} f(t)\, dt = \pi \sum_{i=1}^{\infty} \theta_i = \pi.$$

Hence $f \in L^1(B)$, and the indefinite integral $F := \int f\, d\mathcal{L}^1$ of f is a charge in B which is AC.

Select a point x in $D := C - \bigcup_{i=1}^{\infty} \{a_i, b_i\}$, and note that for each $r > 0$ both intervals $(x, x + r)$ and $(x - r, x)$ contain infinitely many intervals (a_i, b_i). Since

$$\inf\{f(t) : t \in (a_i, b_i)\} = f\left(\tfrac{1}{2}[a_i + b_i]\right) = 2\frac{\theta_i}{h_i},$$

we see that given $k > 0$, there is an $r > 0$ such that $B_r := B(x, r)$ is contained in B and $f(t) > k$ for each t in $B_r - C$; the same is true for $x \in C - \{0, 1\}$, but we do not need this. Since C is a negligible set,

$$\lim_{r \to 0+} \frac{F(B_r)}{|B_r|} = \lim \frac{1}{|B_r|} \int_{B_r} f(t)\, dt = \infty$$

and we see that F is not almost derivable at x. Now $\mathcal{H}^0$ is the counting measure, and so the uncountable set D is not thin. Consequently F does not belong to $CH_{AD}(\mathbb{R})$.

3.4. An excursion to Hausdorff measures

The *essence* of a set $E \subset \mathbb{R}^m$ is the closed set

$$\operatorname{ess} E := \operatorname{supp}(\mathcal{L}^m \, \llcorner \, E).$$

Thus an $x \in \mathbb{R}^m$ belongs to $\operatorname{ess} E$ if and only if $|E \cap B(x, r)| > 0$ for each $r > 0$. Clearly, $\operatorname{ess} E$ depends only on the equivalence class of E, and $\operatorname{cl}_* E \subset \operatorname{ess} E \subset \operatorname{cl} E$. By the density theorem, $|E - \operatorname{ess} E| = 0$.

Note the set $\operatorname{ess} E - E$ may have positive measure even when E is a measurable set. Indeed, modifying the construction of the Cantor ternary set, it is easy to obtain a closed nowhere dense set C of positive measure contained in a cell B. Since $B - C$ is an open dense subset of B, we have $\operatorname{ess}(B - C) = B$.

We say a set $E \subset \mathbb{R}^m$ is *essential* if $E \neq \emptyset$ and $E = \operatorname{ess} E$. Each essential set has positive measure, and if a set E has positive measure, then $\operatorname{ess} E$ is an essential set.

The main result of this section is Theorem 3.4.1 below. It is due to Buczolich [19, Theorem 4.1], and solves a nontrivial problem of the measure-category type, cf. [58]. We shall apply it in Section 3.5 below.

Theorem 3.4.1. *Let $0 \leq s < m$, and let E be a measurable set of positive measure. There are a compact essential set $C \subset E$ and a G_δ set $D \subset C$ that satisfy the following conditions:*

(i) *D is dense in C, and $|D| = 0$;*
(ii) *given a set $Z \subset \mathbb{R}^m$ with $\mathcal{H}^s(Z) = 0$, we can find a compact essential set $C_Z \subset C - Z$ so that $D \cap C_Z$ is dense in C_Z.*

The proof of Theorem 3.4.1 is rather technical and proceeds through a series of lemmas. For the purpose of this proof, we introduce conventions, notation, and terminology which shall not be used outside the present section.

We fix a number s with $0 < s < m$, and let $t = 2^{s-m}$. For $n = 0, 1, \ldots$, denote by $\mathfrak{Q}(n)$ the family of all dyadic cubes of diameter $2^{-n}\sqrt{m}$, and let $\mathfrak{Q}_n := \bigcup_{k=n}^{\infty} \mathfrak{Q}(k)$. Note $\mathfrak{Q}_0$ is the family of all dyadic cubes of any diameter. The elements of $\mathfrak{Q}(n)$ are referred to as *n-cubes*. We leave the routine proof of the next observation to the reader.

Observation 3.4.2. *If $\{n_i\}$ is a strictly increasing sequence of nonnegative integers and $\mathfrak{Q} := \bigcup_{i=1}^{\infty} \mathfrak{Q}(n_i)$, then*

$$U = \bigcup \{Q \in \mathfrak{Q} : Q \subset U\}$$

for every open set $U \subset \mathbb{R}^m$.

For each integer $i \geq 1$, select an integer

$$p_i \geq 1 + \frac{is}{m-s}, \quad \text{and let} \quad \eta_i := 1 - \frac{1-t}{2^{m(p_i+i)}}.$$

A compact essential set $C \subset \mathbb{R}^m$ is called *amiable* if there are positive integers n_i such that for $i = 1, 2, \ldots$ the following conditions hold:

(i) $n_{i+1} > n_i + p_i + i$;

(ii) $|C \cap Q| \geq \eta_i |Q|$ whenever Q is an n_i-cube that meets C.

We establish Theorem 3.4.1 by showing that each measurable set of positive measure contains an amiable subset (Corollary 3.4.4 below), and that amiable sets have the desired properties.

Unfortunately, carrying out this simple program is a tedious process. We begin with a technical lemma, which facilitates the first part of our plan.

Lemma 3.4.3. *Let $K \subset \mathbb{R}^m$ be a compact set of positive measure, and let $k \geq 0$ be an integer. Suppose $Q_j \in \mathfrak{Q}_0$ and $0 < \nu_j < 1$ are such that*

$$|K \cap Q_j| > \nu_j |Q_j|$$

for $j = 1, \ldots, k$. If $0 < \nu < 1$, then for every sufficiently large integer $n \geq 1$ there is a nonempty compact set $K_n \subset K$ such that

- *$|K_n \cap Q_j| > \nu_j |Q_j|$ for $j = 1, \ldots, k$,*
- *$|K_n \cap Q| > \nu |Q|$ for each $Q \in \mathfrak{Q}(n)$ that meets K_n.*

PROOF. By Lemma 1.5.6, for every $x \in \operatorname{int}_* K$ there is an integer $n_x \geq 1$ such that $|K \cap Q| > \nu |Q|$ for each $Q \in \mathfrak{Q}_{n_x}$ with $x \in Q$. The sets

$$E_n := \{x \in \operatorname{int}_* K : n_x \leq n\}, \quad n = 1, 2, \ldots,$$

form an increasing sequence whose union is $\operatorname{int}_* K$. By Corollary 1.5.3,

$$\lim |E_n| = |K| \quad \text{and} \quad \lim |E_n \cap Q_j| = |K \cap Q_j|.$$

Thus there is an integer $p \geq 1$ such that

$$|E_n| > 0 \quad \text{and} \quad |E_n \cap Q_j| > \nu_j |Q_j|$$

for each $n \geq p$ and $j = 1, \ldots, k$. Fixing an integer $n \geq p$, observe

$$\mathcal{K} := \{Q \in \mathfrak{Q}(n) : E_n \cap Q \neq \emptyset\}$$

is a finite family that covers E_n, and $|K \cap Q| > \nu |Q|$ for every $Q \in \mathcal{K}$. In the interior of each $Q \in \mathcal{K}$ select a cube Q' and let $C := \bigcup \{Q' : Q \in \mathcal{K}\}$. Taking Q' only slightly smaller than Q, we can assure that the following claims hold:

(i) $|E_n \cap C| > 0$;

(ii) $|E_n \cap C \cap Q_j| > \nu_j |Q_j|$ for $j = 1, \ldots, k$;

(iii) $|K \cap Q'| > \nu |Q|$ for each $Q \in \mathcal{K}$.

Every n-cube which meets the compact set $K_n := K \cap C$ overlaps a cube from $\mathcal{K}$, and hence belongs to $\mathcal{K}$. Thus by claim (iii),

$$|K_n \cap Q| = |K \cap Q'| > \nu|Q|$$

for each $Q \in \mathfrak{Q}(n)$ that meets K_n. Since $E_n \cap C \subset K_n$, the lemma follows from claims (i) and (ii). □

Corollary 3.4.4. *Each measurable set E of positive measure contains an amiable subset.*

PROOF. Select a compact set $K \subset E$ with $|K| > 0$, and for $i = 1, 2, \ldots,$ construct integers $n_i \geq 1$ and nonempty compact sets K_i so that

- $n_{i+1} > n_i + p_i + i$,
- $K_{i+1} \subset K_i \subset K$,
- $|K_i \cap Q| > \eta_k|Q|$ whenever $1 \leq k \leq i$ is an integer and Q is an n_k-cube that meets K_i.

This can be done inductively. Indeed, applying Lemma 3.4.3 to the case $k = 0$ and $\nu = \eta_1$, we find K_1 and n_1. As only finitely many n_k-cubes with $k \leq i$ meet K_i, another application of Lemma 3.4.3 facilitates the inductive step. Now let $K_0 := \bigcap_{i=1}^{\infty} K_i$, and observe

$$|K_0 \cap Q| = \lim |K_i \cap Q| \geq \eta_k|Q|$$

for each n_k-cube Q that meets K_0. Since K_0 is nonempty, it follows $|K_0| > 0$, and we conclude $\operatorname{ess} K_0$ is an amiable subset of E. □

Throughout the remainder of this section, let C be an amiable set with the associated sequence $\{n_i\}$ of positive integers. We set

$$n_i' := n_i + p_i \quad \text{and} \quad n_i'' := n_i + p_i + i.$$

For each integer $i \geq 1$ and each n_i'-cube Q, select an n_i''-cube $Q^* \subset Q$, and let

$$D_i := \bigcup_{Q \in \mathfrak{Q}(n_i')} \operatorname{int} Q^* \quad \text{and} \quad D_C := \bigcap_{j=1}^{\infty} \bigcup_{i=j}^{\infty} D_i.$$

Since each $\bigcup_{i=j}^{\infty} D_i$ is a dense open subset of $\mathbb{R}^m$, by Baire's category theorem, D_C is a dense G_δ subset of $\mathbb{R}^m$. Select a $K \in \mathfrak{Q}(0)$, and observe it contains $2^{mn_i'}$ connected components of D_i. The measure of each of these components equals $2^{-mn_i''}$, and so

$$|K \cap D_C| = \lim_{j \to \infty} \sum_{i=j}^{\infty} |K \cap D_i| = \lim_{i \to \infty} \sum_{i=j}^{\infty} 2^{-mi} = 0.$$

Since $\mathbf{Q}(0)$ is a countable family and $\mathbb{R}^m = \bigcup \mathbf{Q}(0)$, the set D_C is negligible. In several steps, we show the negligible G_δ set $D = C \cap D_C$ satisfies conditions (i) and (ii) of Theorem 3.4.1. The above density proof serves as a model for proving condition (i). The proof of condition (ii) requires additional lemmas.

Observation 3.4.5. *If Q is an n_i-cube which meets C, then*

$$|C \cap L^*| \geq t|L^*|$$

for each n_i'-cube $L \subset Q$.

PROOF. As C is an amiable set,

$$|Q - C| = |Q| - |C \cap Q| \leq (1 - \eta_i)|Q|$$
$$= \frac{1-t}{2^{m(p_i+i)}}|Q| = (1-t)|L^*|$$

by the choice of η_i. The observation follows:

$$|C \cap L^*| = |L^*| - |L^* - C| \geq |L^*| - |Q - C| \geq t|L^*|. \qquad \square$$

Corollary 3.4.6. *The set $D := C \cap D_C$ is dense in C.*

PROOF. Select integers $i \geq j \geq 1$ and an n_i-cube Q that meets C. It follows from Observation 3.4.5 that Q meets $C \cap D_i$. Thus each $Q \in \bigcup_{i=j}^\infty \mathbf{Q}(n_i)$ which meets C meets also $C \cap \bigcup_{i=j}^\infty D_i$. In view of Observation 3.4.2, the set $C \cap \bigcup_{i=j}^\infty D_i$ is a relatively open and dense subset of C. An application of Baire's category theorem completes the argument. $\qquad \square$

If $\mathcal{A}$ is a family of sets and B is a set (not necessarily in $\mathcal{A}$), let

$$\mathcal{A}(B) := \{A \in \mathcal{A} : A \subset B\}.$$

Given $\varepsilon > 0$ and $Z \subset \mathbb{R}^m$, we say a nonoverlapping family $\mathcal{Z}$ of dyadic cubes is an ε-*cover* of Z whenever the following conditions hold:

(1) $Z \subset \operatorname{int} \bigcup \mathcal{Z}$;
(2) $\sum_{K \in \mathcal{Z}} d(K)^s < \varepsilon$;
(3) $\sum_{K \in \mathcal{Z}(Q)} d(K)^s < d(Q)^s$ for each dyadic cube Q not in $\mathcal{Z}$.

While the members of an ε-cover $\mathcal{Z}$ need not have small diameters, condition (3) suggests $\mathcal{Z}$ covers Z with certain efficiency.

Lemma 3.4.7. *If $Z \subset \mathbb{R}^m$ and $\mathcal{H}^s(Z) = 0$, then Z has an ε-cover for each $\varepsilon > 0$.*

PROOF. The existence of a nonoverlapping family $\mathcal{K}_0 \subset \mathcal{Q}_0$ that satisfies conditions (1) and (2) follows from Lemma 1.3.2. Proceeding inductively, for $n = 0, 1, \ldots$, we construct families $\mathcal{K}_n$ and $\mathcal{L}_n$ so that the following conditions hold: $\mathcal{L}_n$ consists of all $L \in \mathcal{Q}(n) - \mathcal{K}_n$ satisfying the inequality

$$d(L)^s \leq \sum_{K \in \mathcal{K}_n(L)} d(K)^s,$$

and once $\mathcal{L}_n$ has been constructed,

$$\mathcal{K}_{n+1} := \mathcal{K}_n - \bigcup_{L \in \mathcal{L}_n} \mathcal{K}_n(L).$$

Let $\mathcal{K} := \bigcap_{n=0}^{\infty} \mathcal{K}_n$ and $\mathcal{L} := \bigcup_{n=0}^{\infty} \mathcal{L}_n$. Observe $\mathcal{K} \cap \mathcal{L} = \emptyset$, and $\mathcal{Z} := \mathcal{K} \cup \mathcal{L}$ is a nonoverlapping family such that $\bigcup \mathcal{K}_0 \subset \bigcup \mathcal{Z}$; in particular, $Z \subset \operatorname{int} \bigcup \mathcal{Z}$. Furthermore,

$$\sum_{K \in \mathcal{Z}} d(K)^s = \sum_{K \in \mathcal{K}} d(K)^s + \sum_{n=0}^{\infty} \sum_{L \in \mathcal{L}_n} d(L)^s$$

$$\leq \sum_{K \in \mathcal{K}} d(K)^s + \sum_{n=0}^{\infty} \sum_{L \in \mathcal{L}_n} \sum_{K \in \mathcal{K}_n(L)} d(K)^s$$

$$= \sum_{K \in \mathcal{K}} d(K)^s + \sum_{K \in \mathcal{K}_0 - \mathcal{K}} d(K)^s = \sum_{K \in \mathcal{K}_0} d(K)^s < \varepsilon.$$

To show condition (3) is also satisfied, select a $Q \in \mathcal{Q}(n) - \mathcal{Z}$, and observe Q does not belong to $\mathcal{K}_n$; indeed, since

$$\mathcal{K}_n \cap \mathcal{Q}(n) = \mathcal{K}_k \cap \mathcal{Q}(n)$$

for each $k \geq n$, every n-cube in $\mathcal{K}_n$ belongs to $\mathcal{K}$, and hence to $\mathcal{Z}$. As $Q \notin \mathcal{L}_n$ and $Q \in \mathcal{Q}(n) - \mathcal{K}_n$, the lemma follows:

$$d(Q)^s > \sum_{K \in \mathcal{K}_n(Q)} d(K)^s$$

$$= \sum_{K \in \mathcal{K}(Q)} d(K)^s + \sum_{k=n}^{\infty} \sum_{L \in \mathcal{L}_k(Q)} \sum_{K \in \mathcal{K}_k(L)} d(K)^s$$

$$\geq \sum_{K \in \mathcal{K}(Q)} d(K)^s + \sum_{k=n}^{\infty} \sum_{L \in \mathcal{L}_k(Q)} d(L)^s$$

$$= \sum_{K \in \mathcal{K}(Q)} d(K)^s + \sum_{L \in \mathcal{L}(Q)} d(L)^s = \sum_{K \in \mathcal{Z}(Q)} d(K)^s. \qquad \square$$

Observation 3.4.8. *Let $\mathcal{K}$ be a nonoverlapping family of dyadic cubes, and let Q be a dyadic cube. If $\left| Q \cap \bigcup \mathcal{K} \right| \geq t|Q|$ then*

$$d(Q)^s \leq \sum_{K \in \mathcal{K}(Q)} d(K)^s .$$

PROOF. As there is nothing to prove otherwise, we may assume that all cubes in $\mathcal{K}(Q)$ are properly contained in Q. Thus $d(Q)/d(K) \geq 2$ for each $K \in \mathcal{K}(Q)$, and

$$
\begin{aligned}
d(Q)^s &= \frac{\left(\sqrt{m}\right)^m}{d(Q)^{m-s}} |Q| \leq \frac{\left(\sqrt{m}\right)^m}{d(Q)^{m-s}} \cdot \frac{1}{t} \left| Q \cap \bigcup \mathcal{K} \right| \\
&= 2^{m-s} \frac{\left(\sqrt{m}\right)^m}{d(Q)^{m-s}} \sum_{K \in \mathcal{K}(Q)} |K| \\
&= 2^{m-s} \sum_{K \in \mathcal{K}(Q)} \left[\frac{d(K)}{d(Q)} \right]^{m-s} d(K)^s \leq \sum_{K \in \mathcal{K}(Q)} d(K)^s . \qquad \square
\end{aligned}
$$

The following lemma shows that for an n_i-cube Q the conclusion of Observation 3.4.8 can be derived from a more convenient assumption. This is a crucial step, since it allows us to exploit ε-covers whose existence was established in Lemma 3.4.7.

Lemma 3.4.9. *Let $\mathcal{K}$ be a nonoverlapping family of dyadic cubes, and let Q be an n_i-cube such that $\left| L^* \cap \bigcup \mathcal{K} \right| \geq t|L^*|$ for each n_i'-cube L contained in Q. Then*

$$d(Q)^s \leq \sum_{K \in \mathcal{K}(Q)} d(K)^s .$$

PROOF. We may assume each $K \in \mathcal{K}(Q)$ is properly contained in Q. Let $L_1, \ldots, L_r$ be all n_i'-cubes contained in Q. Observe $r = 2^{mp_i}$ and $|L_j| = |Q|/r$. Moreover, our assumption and Observation 3.4.8 yield

$$d(L_j^*)^s \leq \sum_{K \in \mathcal{K}(L_j^*)} d(K)^s \leq \sum_{K \in \mathcal{K}(L_j)} d(K)^s \qquad (*)$$

for $j = 1, \ldots, r$. After a suitable reordering, there is an integer k with $0 \leq k \leq r$ such that L_j is contained in some $K_j \in \mathcal{K}(Q)$ when $j \leq k$, and L_j is contained in no $K \in \mathcal{K}(Q)$ when $j > k$. It follows that for

$j = k+1, \ldots, r$ each element of $\mathcal{K}(L_j)$ is a proper subcube of L_j. Now

$$\mathcal{K}(Q) = \{K_1, \ldots, K_k\} \cup \bigcup_{j=k+1}^{r} \mathcal{K}(L_j),$$

since $\mathcal{K}$ is a nonoverlapping family. As $d(L_j) \leq d(K_j) \leq d(Q)/2$ for $j = 1, \ldots, k$, we obtain

$$\frac{k}{r}|Q| = \sum_{j=1}^{k} |L_j| = \frac{1}{(\sqrt{m})^m} \sum_{j=1}^{k} d(L_j)^{m-s} d(L_j)^s$$

$$\leq \frac{1}{(\sqrt{m})^m} \left[\frac{d(Q)}{2}\right]^{m-s} \sum_{j=1}^{k} d(K_j)^s$$

$$= \frac{|Q|}{d(Q)^s 2^{m-s}} \sum_{j=1}^{k} d(K_j)^s .$$

The previous inequality implies

$$\sum_{j=1}^{k} d(K_j)^s \geq \frac{k}{r} d(Q)^s 2^{m-s} . \qquad (**)$$

Observe that due to our choice of p_i, we have the estimate

$$(m-s)p_i - si \geq m - s .$$

This inequality together with inequality $(*)$ facilitate the following calculation:

$$\sum_{j=k+1}^{r} \sum_{K \in \mathcal{K}(L_j)} d(K)^s \geq \sum_{j=k+1}^{r} d(L_j^*)^s$$

$$= (r-k)d(Q)^s 2^{-s(p_i+i)}$$

$$= (r-k)d(Q)^s 2^{-mp_i} 2^{(m-s)p_i - si}$$

$$\geq \frac{r-k}{r} d(Q)^s 2^{m-s} .$$

From the previous inequality and inequality $(**)$, it is easy to deduce the desired result:

$$\sum_{K \in \mathcal{K}(Q)} d(K)^s = \sum_{j=1}^{k} d(K_j)^s + \sum_{j=k+1}^{r} \sum_{K \in \mathcal{K}(L_j)} d(K)^s$$

$$\geq d(Q)^s 2^{m-s} > d(Q)^s . \qquad \square$$

Now we have established all preliminary results and are ready to proof the main theorem — still a nontrivial task.

PROOF OF THEOREM 3.4.1. Clearly, it suffices to prove the theorem when $m - 1 < s < m$. By Corollary 3.4.4, the set E contains an amiable set C. We assume C is the set of the paragraph preceding Observation 3.4.5, and use freely the notation introduced in that paragraph. The negligible G_δ set $D \subset C$ is dense in C by Corollary 3.4.6. Thus we only need to show that D satisfies condition (ii) of Theorem 3.4.1. To this end, choose a set $Z \subset \mathbb{R}^m$ with $\mathcal{H}^s(Z) = 0$, and a positive $\varepsilon < |C|$. Enlarging Z, we may assume Z contains the boundaries of all dyadic cubes. By Lemma 3.4.7, the set Z has an ε-cover $\mathcal{Z}$, and we let $U := \text{int} \bigcup \mathcal{Z}$ and $C_Z := C - U$. As $d(K) \le \sqrt{m}$ for each $K \in \mathcal{Z}$,

$$|U| \le \left| \bigcup \mathcal{Z} \right| = \sum_{K \in \mathcal{Z}} |K| = \sum_{K \in \mathcal{Z}} \left[\frac{d(K)}{\sqrt{m}} \right]^m$$

$$\le \sum_{K \in \mathcal{Z}} \left[\frac{d(K)}{\sqrt{m}} \right]^s \le \sum_{K \in \mathcal{Z}} d(K)^s < \varepsilon < |C|.$$

Thus $|C_Z| > 0$, and in view of Observation 3.4.2, the set C_Z is essential whenever $|Q \cap C_Z| > 0$ for each $Q \in \bigcup_{i=1}^\infty \mathbf{Q}(n_i)$ that meets C_Z. Proceeding toward a contradiction, suppose there is an n_i-cube Q such that Q meets C_Z and $|Q \cap C_Z| = 0$. As $\partial Q \subset Z \subset U$, the interior of Q meets C_Z and consequently $Q \notin \mathcal{Z}$. Since C is amiable,

$$\left| Q \cap \bigcup \mathcal{Z} \right| \ge |Q \cap U| \ge |Q \cap (C \cap U)|$$

$$= |Q \cap C| \ge \eta_i |Q| > t|Q|$$

and Observation 3.4.8 yields $d(Q)^s \le \sum_{K \in \mathcal{Z}(Q)} d(K)^s$. This contradicts the definition of ε-covers.

It remains to show $D \cap C_Z$ is dense in C_Z. To this end, select integers $i \ge j \ge 1$ and an n_i-cube Q that meets C_Z. As observed in the previous paragraph, Q does not belong to $\mathcal{Z}$. Aiming at a contradiction, suppose Q does not meet $C_Z \cap D_i$. If $L \subset Q$ is an n'_i-cube, then D_i contains $\text{int} L^*$. Hence $C_Z \cap \text{int} L^* = \emptyset$, and consequently $C \cap \text{int} L^* \subset U \cap \text{int} L^*$. Since $\partial L^* \subset Z \subset U$, we have $C \cap L^* \subset U \cap L^*$. By Observation 3.4.5,

$$\left| L^* \cap \bigcup \mathcal{Z} \right| \ge |L^* \cap U| \ge |L^* \cap C| \ge t|L^*|$$

and Lemma 3.4.9 implies $d(Q)^s \le \sum_{K \in \mathcal{Z}(Q)} d(K)^s$. As this contradicts the definition of ε-covers, we conclude that each $Q \in \bigcup_{i=j}^\infty \mathbf{Q}(n_i)$ which meets C_Z meets also $C_Z \cap \bigcup_{i=j}^{\infty} D_i$. It follows from Observation 3.4.2 that the set $C_Z \cap \bigcup_{i=j}^{\infty} D_i$ is relatively open and dense in C. An application of Baire's category theorem completes the proof. $\qquad\square$

3.5. The critical variation

Replacing essential gages by gages in the definition of essential variation given at the beginning of Section 3.2, we obtain another variational measure, which is more informative than the essential variation. Explicitly, given a function F defined on the family $\mathcal{BV}$, a set $E \subset \mathbb{R}^m$, and $\eta \geq 0$, we let

$$V_{*,\eta}F(E) := \inf_{\delta} F(\delta, \eta)$$

where δ is a gage on E, and then define

$$V_* F(E) := \sup_{\eta > 0} V_{*,\eta}F(E) .$$

Of the extended real-valued functions

$$V_{*,\eta}F : E \mapsto V_{*,\eta}F(E) \quad \text{and} \quad V_* F : E \mapsto V_* F(E)$$

defined for all $E \subset \mathbb{R}^m$, the latter is called the *critical variation* of F. The following inequalities are obvious:

$$V_{\mathrm{ess},\eta}F \leq V_{*,\eta}F \quad \text{and} \quad V_{\mathrm{ess}}F \leq V_* F . \tag{3.5.1}$$

Let F and G be functions defined on $\mathcal{BV}$, and let δ and σ be gages on a set $E \subset \mathbb{R}^m$. Then $\rho = \min\{\delta, \sigma\}$ is a gage on E, and (3.2.1) yields

$$V_{*,\eta}(F + G)(E) \leq (F + G)(\rho, \eta) \leq F(\rho, \eta) + G(\rho, \eta)$$
$$\leq F(\delta, \eta) + G(\sigma, \eta)$$

for each $\eta \geq 0$. It follows $V_{*,\eta}(F + G) \leq V_{*,\eta}F + V_{*,\eta}G$ for each $\eta \geq 0$, and consequently

$$V_*(F + G) \leq V_* F + V_* G . \tag{3.5.2}$$

As the Hausdorff measures are Borel regular, each thin set is contained in a Borel thin set. In view of this, the proof of the Proposition 3.5.1 below is completely analogous to that of Proposition 3.2.1, and we leave it to the reader.

Proposition 3.5.1. *Let F be a function defined on $\mathcal{BV}$, and select an $\eta \geq 0$. The extended real-valued functions $V_{*,\eta}F$ and $V_* F$ are Borel regular measures. Moreover, $V_{*,\eta}F = 0$ whenever $\eta \geq 1/(2m)$.*

Note the measures $V_{*,\eta}F$ and V_*F vanish on every thin set, however, Example 3.5.6 below shows they need not be absolutely continuous.

Proposition 3.5.2. *Let F be a function defined on $\mathcal{BV}$, let $E \subset \mathbb{R}^m$, and let $\eta \geq 0$. The following statements are true.*

- $(V_{*,\eta}F) \mathsf{L}\, E = (V_{ess,\eta}F) \mathsf{L}\, E$ *whenever the measure $(V_{*,\eta}F) \mathsf{L}\, E$ is absolutely continuous.*
- $(V_*F) \mathsf{L}\, E = (V_{ess}F) \mathsf{L}\, E$ *whenever the measure $(V_*F) \mathsf{L}\, E$ is absolutely continuous.*

PROOF. Suppose the measure $(V_{*,\eta}F) \mathsf{L}\, E$ is absolutely continuous. In view of inequality (3.5.1), the first claim will be established by showing that $(V_{*,\eta}F) \mathsf{L}\, E \leq (V_{ess,\eta}F) \mathsf{L}\, E$. Seeking a contradiction, assume

$$V_{ess,\eta}F(B) < V_{*,\eta}F(B)$$

for a set B contained in E, and find an essential gage δ on B so that $F(\delta,\eta) < V_{*,\eta}F(B)$. Since $N := \{\delta = 0\}$ is a negligible subset of B, our assumption implies $V_{*,\eta}F(N) = 0$. Thus given $\varepsilon > 0$, there is a gage σ on N such that $F(\sigma,\eta) < \varepsilon$. The formula

$$\rho(x) := \begin{cases} \delta(x) & \text{if } x \in B - N, \\ \sigma(x) & \text{if } x \in N, \end{cases}$$

defines a gage ρ on B. If P is a ρ-fine partition, then

$$\{(A,x) \in P : x \in B - N\} \quad \text{and} \quad \{(A,x) \in P : x \in N\}$$

are δ-fine and σ-fine partitions, respectively. Therefore

$$V_{*,\eta}F(B) \leq F(\rho,\eta) \leq F(\delta,\eta) + F(\sigma,\eta) < F(\delta,\eta) + \varepsilon,$$

and the arbitrariness of ε yields a contradiction.

If the measure $(V_*F) \mathsf{L}\, E$ is absolutely continuous, then so is the measure $(V_{*,\eta}F) \mathsf{L}\, E$ for each $\eta > 0$, and the proposition follows from the first part of the proof. $\square$

Proposition 3.5.3. *Let F be a charge, and let A be a bounded BV set. If $0 < \eta < 1/(2m)$, then*

$$VF(A) \leq V_{*,\eta}F(cl_*A) \leq V_*F(cl_*A)$$

and the equalities occur whenever A is essentially open.

PROOF. Assuming $V_{*,\eta}F(\mathrm{cl}_*A) < VF(A)$, find a gage δ on cl_*A and a disjoint family $\{A_1, \dots, A_p\} \subset \mathcal{BV}(A)$ so that

$$F(\delta, \eta) < \sum_{k=1}^{p} |F(A_k)|.$$

Choose an $\varepsilon > 0$, and let $\delta_k := \delta \upharpoonright \mathrm{cl}_*A_k$. According to Theorem 2.6.7, there are η-regular δ_k-fine partitions $P_k := \{(A_1^k, x_1^k), \dots, (A_{p_k}^k, x_{p_k}^k)\}$ such that $[P_k] \subset A_k$ and

$$\sum_{i=1}^{p_k} |F(A_i^k)| \geq \left|F([P_k])\right| > |F(A_k)| - \frac{\varepsilon}{p}.$$

Since $P := \bigcup_{k=1}^{p} P_k$ is an η-regular δ-fine partition, we obtain

$$F(\delta, \eta) \geq \sum_{k=1}^{p} \sum_{i=1}^{p_k} |F(A_i^k)| > \sum_{k=1}^{p} |F(A_k)| - \varepsilon,$$

and a contradiction follows from the arbitrariness of ε. Thus

$$V(A) \leq V_{*,\eta}F(\mathrm{cl}_*A) \leq V_*F(\mathrm{cl}_*A).$$

Now assume int_*A is open. Select a gage δ on cl_*A so that $\delta(x) = 0$ when $x \in \partial_*A$, and $B[x, \delta(x)] \subset \mathrm{int}_*A$ when $x \in \mathrm{int}_*A$. For each δ-fine partition $\{(A_1, x_1), \dots, (A_p, x_p)\}$, we obtain

$$\sum_{i=1}^{p} |F(A_i)| = \sum_{i=1}^{p} |F(A \cap A_i)| \leq VF(A),$$

since every A_i is equivalent to $A \cap A_i$. Therefore

$$V_{*,\eta}F(\mathrm{cl}_*A) \leq F(\delta, \eta) \leq VF(A)$$

and as η is arbitrary, the proposition follows. $\square$

Remark 3.5.4. In Proposition 3.5.3 it may not be possible to replace $V_*F(\mathrm{cl}_*A)$ by $V_*F(A)$. However, if F is a charge, then Corollary 1.5.3 implies

$$VF(A) = VF(\mathrm{int}_*A) = VF(\mathrm{cl}_*A)$$

for each bounded BV set A.

Corollary 3.5.5. *Let F be a charge, and let E be a locally BV set. If $V_{*,\eta}F(\mathrm{cl}_*E) = 0$ for a positive $\eta < 1/(2m)$, then $F \llcorner E = 0$.*

PROOF. Since (3.1.1) yields

$$|F \llcorner E| \leq V(F \llcorner E) = (VF) \llcorner E,$$

the corollary follows immediately from Proposition 3.5.3. $\square$

Example 3.5.6. Let S be the charge associated with the devil's staircase s_B in a cell B (Example 3.2.3). According to Proposition 3.5.3

$$V_* S(B) = V S(B) = 1$$

and a direct calculation shows $V_* S(B - C_B) = 0$. Thus $V_* S(C_B) = 1$, which means $V_* S$ is not absolutely continuous.

Theorem 3.5.7. *Let F be a charge, and let E be a measurable set such that the measure $(V_* F) \, \mathsf{L} \, E$ is absolutely continuous. Then $(V_* F) \, \mathsf{L} \, E$ is σ-finite, F is derivable at almost all $x \in E$, and*

$$V_* F(B) = \int_B |DF(x)| \, dx$$

for each measurable set $B \subset E$.

PROOF. Choose a positive $\eta < 1/(2m)$. In view of Proposition 2.3.5,

$$N_\eta := \left\{ x \in E : \overline{D}_\eta |F|(x) = \infty \right\}$$

is a measurable set. Seeking a contradiction, suppose $|N_\eta| > 0$. For a number $s := (2m - 1)/2$, let $C \subset N_\eta$ and $D \subset C$ be, respectively, a compact essential set and a G_δ set satisfying conditions (i) and (ii) of Theorem 3.4.1. Since D is a negligible subset of E, our assumption yields $V_* F(D) = 0$. Choose a gage δ on D, and observe $\mathcal{H}^s(\{\delta = 0\}) = 0$. By condition (ii) of Theorem 3.4.1, there is a compact essential set $K \subset C - \{\delta = 0\}$ such that $D \cap K$ is a dense subset of K. Applying Baire's category theorem to the G_δ set $D \cap K$, find a $t > 0$ and an open set U so that $D \cap K \cap U \neq \emptyset$ and the set

$$D_t := \left\{ x \in D \cap K \cap U : \delta(x) > t \right\}$$

is dense in $D \cap K \cap U$, and hence in $K \cap U$. According to inequality (2.3.1) and Lemma 2.3.4, the family of all essentially clopen BV sets B with

$$B = \mathrm{cl}_* B, \quad d(B) < t, \quad r(B) > \eta, \quad \text{and} \quad |F(B)| > \frac{|B|}{|K \cap U|}$$

is a Vitali cover of $K \cap U$. By Vitali's theorem, this family has a countable disjoint subfamily $\mathcal{B}$ such that almost all of $K \cap U$ is contained in $\bigcup \mathcal{B}$. Since Observation 1.5.8 implies ∂B is a negligible set (in fact, a thin set, but we shall not need this) for each $B \in \mathcal{B}$, almost all of $K \cap U$ is contained in $\bigcup_{B \in \mathcal{B}} \mathrm{int} \, B$. Making $\mathcal{B}$ smaller, we may assume

$(K \cap U) \cap \operatorname{int} B \neq \emptyset$ whenever $B \in \mathcal{B}$. Consequently, each $B \in \mathcal{B}$ contains a point $x_B \in D_t$. Now

$$\sum_{B \in \mathcal{B}} |F(B)| > \frac{1}{|K \cap U|} \sum_{B \in \mathcal{B}} |B| \geq 1,$$

and there is a finite family $\mathcal{A} \subset \mathcal{B}$ such that $\sum_{A \in \mathcal{A}} |F(A)| > 1$. As the collection $\{(A, x_A) : A \in \mathcal{A}\}$ is an η-regular δ-fine partition, we have $F(\delta, \eta) > 1$, and the arbitrariness of δ implies a contradiction:

$$V_* F(D) \geq V_{*,\eta} F(D) \geq 1.$$

It follows $N := \bigcup_{k=2m+1}^{\infty} N_{1/k}$ is a negligible set. As F is almost derivable at each $x \in E - N$, it is derivable at almost all $x \in E$ by Theorem 2.4.3. Given a measurable set $B \subset E$, the equality

$$V_* F(B) = \int_B |DF(x)| \, dx$$

follows from Proposition 3.5.2 and Theorem 3.2.4. Up to a negligible set $Z \subset E$, the sets

$$E_n := \Big\{ x \in E \cap B(n) : |DF(x)| < n \Big\}, \quad n = 1, 2, \ldots,$$

cover E, and $V_* F(E_n) \leq n |B(n)| < \infty$. As $V_* F(Z) = 0$, the measure $(V_* F) \, \llcorner \, E$ is σ-finite. $\square$

An alternative but equally complicated proof of Theorem 3.5.7 can be found in [9, Theorem 2].

Proposition 3.5.8. *Let F be a charge, and let $E \subset \mathbb{R}^m$. If $V_* F(E) < \infty$, then $E \subset S \cup \bigcup_{n=1}^{\infty} K_n$ where S is a Borel set with $\mathcal{H}^{m-1}(S) = 0$, and each K_n is a compact set with $V_* F(K_n) < \infty$.*

PROOF. The main idea of the proof is similar to that we employed in proving the Borel regularity of $V_{\mathrm{ess}} F$ (Proposition 3.2.1). Choose a positive $\eta < 1/(2m)$, and find a gage δ on E so that $F(\delta, \eta) < V_* F(E) + 1$. For an integer $k \geq 1$, let

$$E_k := \big\{ x \in E : \delta(x) > 1/k \big\}.$$

Further let $\rho(x) := 1/k$ for each $x \in E_k$, and $\sigma(x) := 1/k$ for each $x \in \operatorname{cl} E_k$. Given an η-regular σ-fine partition $\{(B_1, y_1), \ldots, (B_p, y_p)\}$, there is an η-regular ρ-fine partition $\{(B_1, x_1), \ldots, (B_p, x_p)\}$ obtained by moving the point $y_i \in \operatorname{cl} E_k$ to a nearby point $x_i \in E_k$. This and (3.2.1) yield

$$V_{*,\eta} F(\operatorname{cl} E_k) \leq F(\sigma, \eta) = F(\rho, \eta) \leq F(\delta, \eta) < V_* F(E) + 1,$$

and hence $V_* F(\operatorname{cl} E_k) \leq V_* F(E) + 1 < \infty$ by the arbitrariness of η.

Now $\{\delta = 0\} \subset \bigcup_{n=1}^{\infty} T_n$ where $\mathcal{H}^{m-1}(T_n) < \infty$ for all n. As $\mathcal{H}^{m-1}$ is a Borel regular measure, we may assume each T_n is a Borel set. Fix an integer $n \geq 1$, and observe $\mathcal{H}^{m-1} \llcorner T_n$ is a Radon measure [22, Section 1.1, Theorem 3]. Thus

$T_n = S_n \cup \bigcup_{i=1}^{\infty} C_{n,i}$ where S_n is a Borel set with $\mathcal{H}^{m-1}(S_n) = 0$ and each $C_{n,i}$ is a compact set [22, Section 1.1, Theorem 4]. Clearly $V_* F(C_{n,i}) = 0$ for $i = 1, 2, \ldots$, and $S := \bigcup_{n=1}^{\infty} S_n$ is Borel set with $\mathcal{H}^{m-1}(S) = 0$. Since

$$E \subset \{\delta = 0\} \cup \left(\bigcup_{k=1}^{\infty} \operatorname{cl} E_k \right) \subset S \cup \left(\bigcup_{n,i=1}^{\infty} C_{n,i} \right) \cup \left(\bigcup_{k=1}^{\infty} \operatorname{cl} E_k \right),$$

and since each $\operatorname{cl} E_k$ is a σ-compact set, the proposition is proved. $\square$

An immediate consequence of Theorem 3.5.7 and Proposition 3.5.8 is the next corollary.

Corollary 3.5.9. *Let F be a charge such that $V_* F$ is absolutely continuous. Then $\mathbb{R}^m = S \cup \bigcup_{n=1}^{\infty} K_n$ where S is a Borel set with $\mathcal{H}^{m-1}(S) = 0$, and each K_n is a compact set with $V_* F(K_n) < \infty$.*

Remark 3.5.10. While it is not easy to prove that $V_* F$ is σ-finite whenever F is a charge and $V_* F$ is absolutely continuous, the result is not unexpected. Indeed, if the measure $V_* F$ were not σ-finite, there would be a Borel set B such that $V_* F(B) = \infty$ and $V_* F(C) = 0$ or $V_* F(C) = \infty$ for each $C \subset B$ — an unlikely behavior for the measure so intimately connected with a charge. To substantiate our claim, assume μ is an absolutely continuous Borel regular measure such that each Borel set B with $\mu(B) = \infty$ contains a subset C with $0 < \mu(C) < \infty$; as μ is Borel regular, we may assume C is Borel. If $\mu(\mathbb{R}^m) = \infty$, use Zorn's lemma to find a maximal disjoint family $\mathcal{C}$ of Borel sets such that $0 < \mu(C) < \infty$ for each $C \in \mathcal{C}$. Since μ is absolutely continuous and $\mathcal{L}^m$ is σ-finite, it is easy to see the family $\mathcal{C}$ is countable. The maximality of $\mathcal{C}$ implies $\mu(\mathbb{R}^m - \bigcup \mathcal{C}) = 0$, which means μ is σ-finite.

Example 3.5.11. We show that not every σ-finite absolutely continuous Borel regular measure is the critical variation of a charge. Let $\{t_1, t_2, \ldots\}$ be a countable dense subset of $\mathbb{R}$, and let $\{r_i\}$ be a sequence of positive numbers with $\sum_{i=1}^{\infty} r_i < \infty$. For $i = 1, 2, \ldots$, let $B_i := \{t \in \mathbb{R} : 0 < |t - t_i| < r_i\}$ and

$$f_i(t) := \begin{cases} |t - t_i|^{-1} & \text{if } t \in B_i, \\ 0 & \text{if } t \in \mathbb{R} - B_i. \end{cases}$$

The function $f := \sum_{i=1}^{\infty} f_i$ is nonnegative, measurable, and finite almost everywhere. Indeed, $f(t) = \infty$ if and only if t belongs to infinitely many sets B_i, or alternatively, to a negligible set $\bigcap_{k=1}^{\infty} \bigcup_{i=k}^{\infty} B_i$. For $E \subset \mathbb{R}$, let

$$\mu(E) := \inf_B \int_B f(t) \, dt$$

where $B \subset \mathbb{R}$ is a Borel set containing E. It is easy to verify that $\mu : E \mapsto \mu(E)$ is a σ-finite absolutely continuous Borel regular measure in $\mathbb{R}$, and that the product measure $\nu := \mu \times \mathcal{L}^{m-1}$ in $\mathbb{R}^m$ has the same properties.

Now assume there is a charge F with $V_* F = \nu$. In view of Corollary 3.5.9, we can find a Borel set S and a sequence $\{K_n\}$ of compact sets so that $\mathcal{H}^{m-1}(S) = 0$, $\nu(K_n) < \infty$ for $n = 1, 2, \ldots$, and $\mathbb{R}^m = S \cup \bigcup_{n=1}^{\infty} K_n$. Given a set $E \subset \mathbb{R}^m$ and a point $\xi \in \mathbb{R}^{m-1}$, let $E^\xi := \{t \in \mathbb{R} : (t, \xi) \in E\}$. It follows from [23, Theorem 5.8] that $S^\xi = \emptyset$ for all $\xi \in B$ where $B \subset \mathbb{R}^{m-1}$ differs from $\mathbb{R}^{m-1}$ by an $\mathcal{L}^{m-1}$-negligible set; in particular $\mathcal{L}^{m-1}(B) > 0$. According to Baire's category theorem, for each

$\xi \in B$, there is an integer $n_\xi \geq 1$ for which the set $(K_{n_\xi})^\xi$ has nonempty interior. Since

$$B = \bigcup_{n=1}^{\infty} \{\xi \in B : n_\xi = n\},$$

there is an integer $p \geq 1$ such that the set $A := \{\xi \in B : n_\xi = p\}$ has positive $\mathcal{L}^{m-1}$ measure. If $\xi \in A$, then K_p^ξ contains some B_i, and hence

$$\mu(K_p^\xi) = \int_{K_p^\xi} f(t)\, dt \geq \int_{B_i} f_i(t)\, dt = \infty.$$

On the other hand, Fubini's theorem implies $\mu(K_p^\xi) < \infty$ for $\mathcal{L}^{m-1}$-almost all ξ in $\mathbb{R}^{m-1}$, a contradiction.

Example 3.5.11 raises an interesting problem: characterize all σ-finite absolutely continuous Borel regular measures in $\mathbb{R}^m$ that are critical variations of charges. The work of Thomson [85] suggests the necessary condition of Corollary 3.5.9 may also be sufficient.

3.6. AC$_*$ charges

We begin by showing that AC charges can be characterized by means of critical variation.

Proposition 3.6.1. *Let F be a charge, and let E be a locally BV set. If the measure $(V_*F) \llcorner \mathrm{cl}_* E$ is absolutely continuous and locally finite, then F is AC in E.*

PROOF. Choose an $\varepsilon > 0$, and use Lemma 3.1.6 to find a $\delta > 0$ so that $V_*F(B) < \varepsilon$ for each set $B \subset B[1/\varepsilon] \cap \mathrm{cl}_* E$ with $|B| < \delta$. Let A be a BV subset of $B(1/\varepsilon) \cap E$ and $|A| < \delta$. Since $\mathrm{cl}_* A \subset B[1/\varepsilon] \cap \mathrm{cl}_* E$ and $|\mathrm{cl}_* A| < \delta$, Proposition 3.5.3 implies

$$\left| F(A) \right| \leq VF(A) \leq V_*F(\mathrm{cl}_* A) < \varepsilon.$$

Consequently F is AC in E.   □

Theorem 3.6.2. *A charge F in a locally BV set E is AC if and only if $(V_*F) \llcorner \mathrm{cl}_* E$ is absolutely continuous and locally finite; in which case*

$$V_*F(\mathbb{R}^m - \mathrm{cl}_* E) = 0, \quad V_*F = V_{\mathrm{ess}}F, \quad \text{and} \quad (V_*F) \upharpoonright \mathcal{BV} = VF.$$

PROOF. As the converse follows from Proposition 3.6.1, suppose F is AC. The measure V_*F is locally finite, since

$$V_*F(B[r]) = VF(B[r]) < \infty$$

for each $r > 0$ by Propositions 3.5.3 and 3.1.4.

Choose a bounded negligible set N and an $\varepsilon > 0$ so small that $N \subset B(1/\varepsilon)$. By Proposition 3.1.4, there is a $\delta > 0$ such that $VF(B) < \varepsilon$ for each BV set $B \subset B(1/\varepsilon)$ with $|B| < \delta$. Find an open set $U \subset B(1/\varepsilon)$ for which $N \subset U$ and $|U| < \delta$, and observe $\sigma : x \mapsto \mathrm{dist}\,(x, \partial U)$ is a positive gage on N. If $P := \{(A_1, x_1), \ldots, (A_p, x_p)\}$ is a σ-fine partition, then $[P]$ is a BV subset of U, and consequently

$$\sum_{i=1}^{p} |F(A_i)| \leq VF([P]) < \varepsilon.$$

As ε is arbitrary, $V_*F(N) = 0$ and we conclude the measure V_*F is absolutely continuous. Propositions 3.5.2 and 3.2.11 imply $V_*F = V_{\mathrm{ess}}F$ and $V_*F \restriction \mathcal{BV} = VF$. Moreover, since VF is a charge in E,

$$V_*F\left(\left[\mathbb{R}^m - \mathrm{cl}_*E\right] \cap B[n]\right) = V_{\mathrm{ess}}F\left(\left[\mathbb{R}^m - \mathrm{cl}_*E\right] \cap B[n]\right)$$
$$= V_{\mathrm{ess}}F\left(\left[\mathbb{R}^m - E\right] \cap B[n]\right)$$
$$= VF\left(\left[\mathbb{R}^m - E\right] \cap B[n]\right) = 0$$

for $n = 1, 2, \ldots$. We infer $V_*F(\mathbb{R}^m - \mathrm{cl}_*E) = 0$. $\qquad\square$

For charges that are not AC, the critical variation allows us to define a weaker concept of absolute continuity. Indeed, Proposition 3.6.1 suggests the following definition.

Definition 3.6.3. A charge F is called AC_* *in a locally BV set* E whenever the measure $(V_*F) \llcorner \mathrm{cl}_*E$ is absolutely continuous. The family of all charges in E that are AC_* in E is denoted by $CH_*(E)$.

In view of inequality (3.5.2), the family $CH_*(E)$ is a linear space. It follows from Theorem 3.6.2, the linear space $CH_{AC}(E)$ is a subspace of $CH_*(E)$. Example 3.3.3 shows the inclusion $CH_{AC}(E) \subset CH_*(E)$ is generally proper. Our goal is to demonstrate that the larger space $CH_*(E)$ retains many important properties of $CH_{AC}(E)$. However, the next proposition warns against drawing automatic analogies between the spaces $CH_{AC}(E)$ and $CH_*(E)$ — cf. Remark 6.1.2 below.

Proposition 3.6.4. *For locally BV sets* $A \subset E$ *the following claims are true.*

(i) *If* $F \in CH_{AC}(A)$, *then* $F \in CH_{AC}(E)$ *and* $DF(x) = 0$ *for almost all* $x \in E - A$.

(ii) *If* $F \in CH_*(A)$ *and* cl_*A *is a relatively closed subset of* cl_*E, *then* $F \in CH_*(E)$ *and* $DF(x) = 0$ *for almost all* $x \in E - A$.

PROOF. (i) If $F \in CH_{AC}(A)$, then $F \in CH_{AC}(E)$ according to Remark 3.1.3, (2). Since $F \llcorner (E - A) = 0$, the statement follows from equality (3.1.1), Proposition 3.2.2, and Corollary 3.2.5.

(ii) By our assumption, the function $\delta : x \mapsto \text{dist}(x, \text{cl}_* A)$ is a positive gage on $\text{cl}_* E - \text{cl}_* A$. If $\{(A_1, x_1), \ldots, (A_p, x_p)\}$ is a δ-fine partition, then each A_i meets A in a negligible set. Thus $F(A_i) = 0$ for $i = 1, \ldots, p$, which implies $V_* F(\text{cl}_* E - \text{cl}_* A) = 0$. Since the measure $(V_* F) \llcorner \text{cl}_* A$ is absolutely continuous, so is $V_* F \llcorner \text{cl}_* E$. By a similar argument, $DF(x) = 0$ for each x in $\text{cl}_* E - \text{cl}_* A$. □

Our concept of AC_* differs from that employed in [75, Chapter 7, Section 8]. Nonetheless, there is a close relationship between a variant of the critical variation and functions ACG_* defined ibid. For the details we refer to [67].

Proposition 3.6.5. *Let E be a locally BV set. If $F \in CH_*(E)$ and $VF < \infty$, then F is AC.*

PROOF. Propositions 3.5.2 and 3.2.2 imply

$$V_* F\big(B[n] \cap \text{cl}_* E\big) = V_{\text{ess}} F\big(B[n] \cap \text{cl}_* E\big) \leq VF\big(B[n] \cap \text{cl}_* E\big) < \infty$$

for $n = 1, 2, \ldots$. Thus $(V_* F) \llcorner E$ is locally finite, and F is AC by Theorem 3.6.2. □

Let F be a charge in a locally BV set E. If F is singular in E, then $V_{\text{ess}} F(E) = 0$ by Corollary 3.2.5. If F is also AC_* in E, then

$$V_* F(\text{cl}_* E) = V_{\text{ess}} F(\text{cl}_* E) = V_{\text{ess}} F(E) = 0$$

according to Proposition 3.5.2. Now Corollary 3.5.5 implies $F = 0$. Thus we have established the equality

$$CH_*(E) \cap CH_S(E) = \{0\}$$

for each locally BV set E. The next theorem improves on this result.

Theorem 3.6.6. *Let E be a locally BV set, and let $F \in CH_*(E)$. If $\underline{D}F(x) \geq 0$ for almost all $x \in E$, then $F \geq 0$.*

PROOF. Choose an $A \in \mathcal{BV}(E)$, an $\varepsilon > 0$, and a positive $\eta < 1/(2m)$. Since the measure $V_* F \llcorner \text{cl}_* E$ is absolutely continuous, and since

$$N := \big\{x \in \text{cl}_* A : \underline{D}F(x) < 0\big\}$$

is a negligible subset of $\mathrm{cl}_* E$, there is a gage σ on N with $F(\sigma, \eta) < \varepsilon$. For each $x \in \mathrm{cl}_* A - N$ find a $\rho_x > 0$ so that $F(C) > -\varepsilon|C|$ for each BV set C with $d(C \cup \{x\}) < \rho_x$ and $r(C \cup \{x\}) > \eta$. The formula

$$\delta(x) := \begin{cases} \sigma(x) & \text{if } x \in N, \\ \rho_x & \text{if } x \in \mathrm{cl}_* A - N, \end{cases}$$

defines a gage δ on $\mathrm{cl}_* A$. By Theorem 2.6.7, there is an η-regular δ-fine partition $P := \{(C_1, x_1), \dots, (C_p, x_p)\}$ such that $[P] \subset A$ and

$$\left| F(A - [P]) \right| < \varepsilon.$$

As $\{(C_i, x_i) : x_i \in N\}$ is an η-regular σ-fine partition, we obtain

$$F(A) = F(A - [P]) + \sum_{x_i \in N} F(C_i) + \sum_{x_i \notin N} F(C_i)$$

$$> -\varepsilon - F(\sigma, \eta) - \varepsilon \sum_{x_i \notin N} |C_i| > -\varepsilon(2 + |A|),$$

and $F(A) \geq 0$ by the arbitrariness of ε. $\square$

Theorem 3.6.7. *Let F be a charge, and let E be a locally BV set. If F is almost derivable at each $x \in \mathrm{cl}_* E - T$ where T is a thin set, then F is AC_* in E.*

PROOF. Choose an $\eta > 0$, and for each $x \in \mathrm{cl}_* E - T$ find positive numbers δ_x and c_x so that $|F(B)| < c_x|B|$ for each BV set B with

$$d(B \cup \{x\}) < \delta_x \quad \text{and} \quad r(B \cup \{x\}) > \eta.$$

Let $N \subset \mathrm{cl}_* E$ be a negligible set, and let $N_k := \{x \in N - T : c_x < k\}$ for $k = 1, 2, \dots$. Given $\varepsilon > 0$, find an open set U so that $N_k \subset U$ and $|U| < \varepsilon/k$. The function $\delta : x \mapsto \min\{\delta_x, \mathrm{dist}\,(x, \partial U)\}$ is a positive gage on N_k. If $P := \{(B_1, x_1), \dots, (B_p, x_p)\}$ is an η-regular δ-fine partition, then $[P] \subset U$, and hence

$$\sum_{i=1}^p |F(B_i)| < \sum_{i=1}^p c_{x_i}|B_i| < k|U| < \varepsilon.$$

It follows $V_{*,\eta} F(N_k) \leq F(\delta, \eta) \leq \varepsilon$. As ε is arbitrary, $V_{*,\eta} F(N_k) = 0$. Since $N - T = \bigcup_{k=1}^\infty N_k$, and since the measure $V_{*,\eta} F$ vanishes on thin sets, $V_{*,\eta} F(N) = 0$. The arbitrariness of η yields $V_* F(N) = 0$. $\square$

Theorem 3.6.8. *Let E be a locally BV set. If $F \in CH_*(E)$, then F is derivable almost everywhere in E, and $F = \int DF \, d\mathcal{L}^m$ whenever $DF \in L^1_\ell(E)$.*

PROOF. If $F \in CH_*(E)$, then F is derivable almost everywhere in E according to Theorem 3.5.7. If $DF \in L^1_\ell(E)$, then the indefinite integral $G := \int DF \, d\mathcal{L}^m$ of DF is an AC charge by Proposition 3.1.7. Thus $F-G$ is an AC_* charge in E, which is singular in E in view of Theorem 3.1.9. Now it follows from Theorem 3.6.6 that $F = G$. $\qquad\square$

The next theorem summarizes important properties of AC_* functions. It follows directly from Theorems 3.6.2, 3.6.6, 3.6.7, and 3.6.8.

Theorem 3.6.9. *Let E be a locally BV set. Then*

$$CH_{AC}(E) + CH_{AD}(E) \subset CH_*(E) \subset CH_D(E)$$

and $CH_*(E) \cap CH_S(E) = \{0\}$.

Let F be a function defined on $\mathcal{BV}$, and let E be a locally BV set. For $\eta \geq 0$ and a nonnegative function δ defined on a set $B \subset \mathbb{R}^m$, let

$$F(\delta, \eta; E) := \sup_P \sum_{i=1}^{p} |F(A_i)|$$

where $P := \{(A_1, x_1), \ldots, (A_p, x_p)\}$ is an η-regular δ-fine partition such that $[P] \subset E$.

Proposition 3.6.10. *Let E be a locally BV set, let $B \subset \mathrm{cl}_* E$, and let F be a function defined on $\mathcal{BV}$. If $\theta > \eta > 0$, then*

$$V_{*,\theta}(F \llcorner E)(B) \leq \inf_\delta F(\delta, \eta; E) \leq V_{*,\eta}(F \llcorner E)(B)$$

where δ is a gage on B. Moreover,

$$V_{*,\theta}(F \llcorner E)(B) \leq V_{*,\eta}F(B) \quad \text{and} \quad V_*(F \llcorner E)(B) \leq V_*F(B).$$

PROOF. Let $c := \inf_\delta F(\delta, \eta; E)$ where δ is a gage on B. Since

$$c \leq F(\delta, \eta; E) = (F \llcorner E)(\delta, \eta; E) \leq (F \llcorner E)(\delta, \eta)$$

for each gage δ on B, we have $c \leq V_{*,\eta}(F \llcorner E)(B)$. Seeking a contradiction, assume $c < V_{*,\theta}(F \llcorner E)(B)$, and find a gage δ on B so that $F(\delta, \eta; E) < V_{*,\theta}(F \llcorner E)(B)$. Let Δ be a gage on $\mathrm{cl}_* E$ associated with the pair (η, θ) according to Lemma 2.6.6. Define a gage on B by letting $\sigma(x) := \min\{\delta(x), \Delta(x)\}$ for each $x \in B$. Since

$$F(\delta, \eta; E) < V_{*,\theta}(F \llcorner E)(B) \leq (F \llcorner E)(\sigma, \theta),$$

there is an θ-regular σ-fine partition $\{(B_1, x_1), \ldots, (B_p, x_p)\}$ such that

$$F(\delta, \eta; E) < \sum_{i=1}^{p} |(F \llcorner E)(B_i)| = \sum_{i=1}^{p} |F(E \cap B_i)|.$$

Clearly the partition $P := \{(E \cap B_1, x_1), \ldots, (E \cap B_p, x_p)\}$ is δ-fine, and by Lemma 2.6.6, it is also η-regular. Since $[P] \subset E$, we have a contradiction. Next observe that for each gage δ on B,

$$(F \, L \, E)(\delta, \eta; E) = F(\delta, \eta; E) \leq F(\delta, \eta).$$

Thus by the first part of the proof, $V_{*,\theta}(F \, L \, E)(B) \leq V_{*,\eta} F(B)$ and the lemma follows from the arbitrariness of η and θ. $\qquad \Box$

Corollary 3.6.11. *Let E be a locally BV set, and let $F \in CH_*(E)$. Then $F \, L \, A$ belongs to $CH_*(A)$ for each locally BV set $A \subset E$.*

Remark 3.6.12. Let $A \subset E$ be locally BV sets, and let $F \in CH_*(E)$. To prove that $F \, L \, A$ belongs to $CH_*(E)$ (Theorem 4.5.7 below) is appreciably harder than proving $F \, L \, A$ belongs to $CH_*(A)$. The claim $F \, L \, A$ belongs to $CH_*(E)$ cannot be deduced from Corollary 3.6.11, because $CH_*(A)$ is generally not contained in $CH_*(E)$ — cf. Proposition 6.1.1 and Remark 6.1.2, (1) below. However, Proposition 3.6.4, (ii) shows that $CH_*(A) \subset CH_*(E)$ whenever $\mathrm{cl}_* A$ is relatively closed in $\mathrm{cl}_* E$.

The following result is another version of the Gauss-Green theorem.

Theorem 3.6.13. *Let E be a locally BV set, and let $v : \mathrm{cl}_* E \to \mathbb{R}^m$ be a charging vector field. Suppose the flux of v is almost derivable at each $x \in \mathrm{cl}_* E - T$ where T is a thin set. Then $\mathfrak{div}\, v : x \mapsto \mathfrak{div}\, v(x)$ is defined almost everywhere in E, and*

$$\int_A \mathfrak{div}\, v \, d\mathcal{L}^m = \int_{\partial_* A} v \cdot \nu_A \, d\mathcal{H}^{m-1}$$

for each bounded BV set $A \subset E$ for which $\mathfrak{div}\, v \in L^1(A)$.

PROOF. The function $\mathfrak{div}\, v$ is defined almost everywhere in E by Theorem 2.4.3. According to Theorem 3.6.7, the flux of v belongs to $CH_*(E)$. An application of Corollary 3.6.11 and Theorem 3.6.8 completes the proof. $\qquad \Box$

We close this section by presenting yet another variational measure, which is useful for arbitrary additive functions on $\mathcal{BV}$. We shall apply it in Section 5.7 below.

Given a function F defined on $\mathcal{BV}$ and a set $E \subset \mathbb{R}^m$, we let

$$V_{\#} F(E) := \inf_\delta \sup_P \sum_{i=1}^p |F(A_i)|$$

where δ is a *positive* function on E and $P = \{(A_1, x_1), \ldots, (A_p, x_p)\}$ is an *arbitrary* δ-fine partition. Following the proof of Proposition 3.2.1, it is easy to show that the obviously defined extended real-valued function $V_{\#}F$ is a Borel regular measure, called the *sharp variation* of F. We have

$$V_*F \le V_{\#}F \quad \text{and} \quad VF(A) \le V_{\#}F(\text{cl }A) \tag{3.6.1}$$

for each $A \in \mathcal{BV}$. The first inequality is obvious, and using Proposition 2.6.3, the proof of the second one is similar to that of Proposition 3.5.3.

Proposition 3.6.14. *Let F be an additive function on $\mathcal{BV}$ such that $V_{\#}F$ is absolutely continuous. Then $V_{\#}F$ is locally finite and $VF < \infty$.*

PROOF. Choose an $x \in \mathbb{R}^m$. Since $V_{\#}F(\{x\}) = 0$, there is a $\delta_x > 0$ such that $\sum_{i=1}^{p}|F(A_i)| < 1$ for each disjoint collection $\{A_1, \dots, A_p\}$ of BV subsets of $B(x, \delta_x)$; indeed, $\{(A_1, x), \dots, (A_p, x)\}$ is a δ_x-fine partition. Thus each $x \in \mathbb{R}^m$ has a neighborhood B_x with $VF(B_x) \le 1$. Given $r > 0$, the compact ball $B[r+1]$ is covered by a finite collection $\{B_{x_1}, \dots, B_{x_k}\}$ of these neighborhoods, and an easy calculation reveals

$$V_{\#}F\big(B[r]\big) \le VF\big(B[r+1]\big) \le \sum_{i=1}^{k} VF(B_{x_i}) \le k \, .$$

The proposition follows from the arbitrariness of r. □

We mention without proof that the sharp variation $V_{\#}F$ of a nonnegative additive function F on $\mathcal{BV}$ is the usual *Lebesgue-Stieltjes measure* measure F^* defined in [75, Chapter 3, Section 5]. For details we refer to [65, Chapters 3 and 8], and in a more general setting to [68].

A final comment is in order. For a charge F, among the variational measures

$$V_{\text{ess}}F \le V_*F \le V_{\#}F$$

the critical variation V_*F is the most *informative*, since it can be independently absolutely continuous and locally finite. In contrast, the essential variation $V_{\text{ess}}F$ is always absolutely continuous (Proposition 3.2.1), and the sharp variation $V_{\#}F$ is locally finite whenever it is absolutely continuous (Proposition 3.6.14).

3.7. Essentially clopen sets

Lemma 2.3.4 shows that the derivates of a charge can be defined by means of closed essentially clopen BV sets, and this fact was used to demonstrate the measurability of derivates. We prove that variations of charges can also be defined by means of closed essentially clopen BV sets.

Proposition 3.7.1. *Let F be a charge, and let A be a figure. Then*

$$VF(A) = \sup_{\mathcal{A}} \sum_{i=1}^{p}|F(A_i)|$$

where $\mathcal{A} := \{A_1, \dots, A_p\}$ is a disjoint subfamily of $\mathcal{F}(A)$.

PROOF. Denoting by a the right side of the desired equality, we have $a \le VF(A)$. Proceeding toward a contradiction, suppose $a < VF(A)$ and find a disjoint collection $\{B_1, \ldots, B_q\} \subset \mathcal{BV}(A)$ so that $a < \sum_{j=1}^{q} |F(B_j)|$. It follows from Proposition 1.10.3 there is a figure $A_1 \subset A$ such that

$$a < |F(A_1)| + \sum_{j=2}^{q} |F(B_j - A_1)|.$$

Applying Proposition 1.10.3 again, find a figure $A_2 \subset \operatorname{cl}(A - A_1)$ so that

$$a < |F(A_1)| + |F(A_2)| + \sum_{j=3}^{q} \left| F\big(B_j - [A_1 \cup A_2]\big) \right|.$$

Inductively, construct nonoverlapping figures $A_1, \ldots, A_q$ so that $a < \sum_{i=1}^{p} |F(A_i)|$. Making each A_i slightly smaller, we obtain a contradiction. □

Proving a result similar to Proposition 3.7.1 for the variational measures $V_{\mathrm{ess}} F$ and $V_* F$ requires some preliminary work.

Lemma 3.7.2. *Let μ be a measure, and let $E \subset \mathbb{R}^m$ be a measurable set. Suppose $\{f_i\}$ is a sequence in $L^1(E, \mu)$ such that $\lim \int_E f_i \, d\mu = 0$ and $\liminf f_i(x) \ge 0$ for μ-almost all $x \in E$. If $\inf f_i \ge g$ for a $g \in L^1(E, \mu)$, then $\{f_i\}$ has a subsequence that converges to zero μ-almost everywhere in E.*

PROOF. Our assumptions imply that $\lim f_i^-(x) = 0$ for μ-almost all $x \in E$. Since $0 \le f_i^- \le g^-$ for $i = 1, 2, \ldots$, we have $\lim \int_E f_i^- \, d\mu = 0$. Thus

$$\lim \int_E |f_i| \, d\mu = \lim \int_E f_i \, d\mu + 2 \int_E f_i^- \, d\mu = 0 \,,$$

and the lemma follows from [73, Theorem 3.12]. □

Proposition 3.7.3. *Let A be a BV set, and let $\{A_i\}$ be a sequence of BV sets such that $\lim |A_i \triangle A| = 0$ and $\lim \|A_i\| = \|A\|$. If B is a BV set and $|A \cap B| = 0$, then*

$$\lim \|B - A_i\| = \|B\| \,.$$

PROOF. It suffices to show that there is a subsequence $\{A_{i_k}\}$ of $\{A_i\}$ with

$$\lim \|B - A_{i_k}\| = \|B\| \,.$$

Indeed, in this case each subsequence of $\{\|B - A_i\|\}$ has a subsequence converging to $\|B\|$, and the proposition follows. We shall use the notation introduced in Section 1.9. A result completely analogous to Theorem 1.7.1, (1) yields $\|A\|_{\mathbf{e}} \le \liminf \|A_i\|_{\mathbf{e}}$ for each $\mathbf{e} \in \partial B(1)$. We apply Lemma 3.7.2 to the sequence $\{\|A_i\|_{\mathbf{e}} - \|A\|_{\mathbf{e}}\}$ and the function $g : \mathbf{e} \mapsto -\|A\|_{\mathbf{e}}$. As $\lim \|A_i\| = \|A\|$, Theorem 1.9.3 shows there is a subsequence of $\{A_i\}$, still denoted by $\{A_i\}$, such that

$$\lim \|A_i\|_{\mathbf{e}} = \|A\|_{\mathbf{e}} < \infty$$

for $\mathcal{H}^{m-1}$-almost all $\mathbf{e} \in \partial B(1)$. Fix such an $\mathbf{e} \in \partial B(1)$. Since $\lim |A_i \triangle A| = 0$, Fubini's theorem shows there is a subsequence of $\{A_i\}$, still denoted by $\{A_i\}$, with $\lim \mathcal{H}^1 \big[l_x \cap (A_i \triangle A) \big] = 0$ for $\mathcal{H}^{m-1}$-almost all $x \in \Pi_{\mathbf{e}}$. By Theorem 1.7.1, (1),

$$\|l_x \cap A\| \le \liminf \|l_x \cap A_i\|$$

for $\mathcal{H}^{m-1}$-almost all $x \in \Pi_e$. Next we apply Lemma 3.7.2 to the sequence

$$\big\{ \|l_x \cap A_i\| - \|l_x \cap A\| \big\}$$

and the function $g : x \mapsto -\|l_x \cap A\|$. Using equality (1.9.3), we obtain a subsequence of $\{A_i\}$, still denoted by $\{A_i\}$, such that

$$\lim \|l_x \cap A_i\| = \|l_x \cap A\| < \infty$$

for $\mathcal{H}^{m-1}$-almost all $x \in \Pi_e$. Since both $\|l_x \cap A_i\|$ and $\|l_x \cap A\|$ are integers for $\mathcal{H}^{m-1}$-almost all $x \in \Pi_e$, we have $\|l_x \cap A_i\| = \|l_x \cap A\|$ whenever i is sufficiently large. As $|l_x \cap A \cap B| = 0$ for $\mathcal{H}^{m-1}$-almost all $x \in \Pi_e$, it is easy to see that

$$\lim \big\| l_x \cap (B - A_i) \big\| = \|l_x \cap B\|$$

for $\mathcal{H}^{m-1}$-almost all $x \in \Pi_e$.

Select an $e \in \partial B(1)$ so that $\lim \|A_i\|_e = \|A\|_e < +\infty$ and $\|B\|_e < +\infty$; note these conditions are satisfied by $\mathcal{H}^{m-1}$-almost all $e \in \partial B(1)$. The functions

$$\varphi_i : x \mapsto \|l_x \cap B\| + \|l_x \cap A_i\| \quad \text{and} \quad \varphi : x \mapsto \|l_x \cap B\| + \|l_x \cap A\|$$

belong to $L^1(\Pi_e, \mathcal{H}^{m-1})$ by Theorem 1.9.1, and $\big\| l_x \cap (B - A_i) \big\| \leq \varphi_i(x)$. In view of equality (1.9.3),

$$\lim \int_{\Pi_e} \varphi_i \, d\mathcal{H}^{m-1} = \|B\|_e + \lim \|A_i\|_e = \|B\|_e + \|A\|_e = \int_{\Pi_e} \varphi \, d\mathcal{H}^{m-1}.$$

The generalized dominated convergence theorem [22, Section 1.3, Theorem 4] yields

$$\lim \|B - A_i\|_e = \lim \int_{\Pi_e} \big\| l_x \cap (B - A_i) \big\| \, d\mathcal{H}^{m-1}(x)$$

$$= \int_{\Pi_e} \|l_x \cap B\| \, d\mathcal{H}^{m-1}(x) = \|B\|_e.$$

Now $\|B - A_i\|_e \leq \|B\|_e + \|A_i\|_e$, and by Theorem 1.9.3

$$\lim \int_{\partial B(1)} \big(\|B\|_e + \|A_i\|_e \big) \, d\mathcal{H}^{m-1}(e) = 2\alpha(m-1) \lim \big(\|B\| + \|A_i\| \big)$$

$$= 2\alpha(m-1) \big(\|B\| + \|A\| \big)$$

$$= \int_{\partial B(1)} \big(\|B\|_e + \|A\|_e \big) \, d\mathcal{H}^{m-1}(e).$$

Another application of Theorem 1.9.3, and the generalized dominated convergence theorem completes the proof:

$$\lim \|B - A_i\| = \lim \frac{1}{2\alpha(m-1)} \int_{\partial B(1)} \|B - A_i\|_e \, d\mathcal{H}^{m-1}(e)$$

$$= \frac{1}{2\alpha(m-1)} \int_{\partial B(1)} \|B\|_e \, d\mathcal{H}^{m-1}(e) = \|B\|. \qquad \square$$

A partition $\big\{ (A_1, x_1), \ldots, (A_p, x_p) \big\}$ is called *special*, if for $i = 1, \ldots, p$, the set A_i is essentially clopen, $A_i = \mathrm{cl}_* A_i$, and $x_i \in \mathrm{int}\, A_i$. Let F be a function defined on $\mathcal{BV}$, and let $\eta \geq 0$. For a nonnegative function δ defined on a set $E \subset \mathbb{R}^m$, let

$$SF(\delta, \eta) := \sup_P \sum_{i=1}^p |F(A_i)|$$

where $P := \big\{ (A_1, x_1), \ldots, (A_p, x_p) \big\}$ is an η-regular δ-fine special partition.

Theorem 3.7.4. *Let F be a charge, and let $\eta \geq 0$. For each $E \subset \mathbb{R}^m$,*

$$V_{\text{ess},\eta}F(E) = \inf_{\delta} SF(\delta,\eta) \quad \text{and} \quad V_{*,\eta}F(E) = \inf_{\sigma} SF(\sigma,\eta)$$

where δ is an essential gage on E and σ is a gage on E.

PROOF. As $SF(\delta,\eta) \leq F(\delta,\eta)$, we have

$$\inf_{\delta} SF(\delta,\eta) \leq V_{\text{ess},\eta}F(E)$$

where δ is an essential gage on E. Proceeding toward a contradiction, suppose the previous inequality is strict, and find an essential gage δ on E so that

$$SF(\delta,\eta) < V_{\text{ess},\eta}F(E) \leq F(\delta,\eta).$$

In view of Lemma 2.6.10, we may assume δ has no strict local maximum. There is an η-regular δ-fine partition $P = \{(A_1, x_1), \ldots, (A_p, x_p)\}$ such that

$$SF(\delta,\eta) < \sum_{i=1}^{p} |F(A_i)|.$$

By our choice of δ, the partition P remains δ-fine when the points x_i are moved slightly. Thus with no loss of generality, we suppose that $x_1, \ldots, x_p$ are distinct points. Choose an $\varepsilon > 0$ and for $i = 1, \ldots, p$, let $D_i = A_i - \bigcup_{j=1}^{p} B(x_j, \varepsilon)$. Proposition 3.7.3 shows $\{(D_1, x_1), \ldots, (D_p, x_p)\}$ is an η-regular δ-fine partition with

$$SF(\delta,\eta) < \sum_{i=1}^{p} |F(D_i)|$$

whenever ε is sufficiently small. So we assume from the onset that $\{x_1, \ldots, x_p\}$ and $\bigcup_{i=1}^{p} \operatorname{cl} A_i$ are disjoint sets. By Theorem 1.10.1 of Chapter 1 and Proposition 3.7.3, there is a compact BV set B_1' such that the following conditions are satisfied:

- $\partial B_1'$ is a C^∞ submanifold of $\mathbb{R}^m$ without boundary;
- B_1' contains no x_i;
- $\{(B_1', x_1), (A_2 - B_1', x_2) \ldots, (A_p - B_1', x_p)\}$ is an η-regular δ-fine partition;
- $SF(\delta,\eta) < |F(B_1')| + \sum_{i=2}^{p} |F(A_i - B_1')|.$

Since $\partial B_1'$ is a C^∞ submanifold of $\mathbb{R}^m$ without boundary, using the standard techniques of Riemannian geometry, it is not difficult to show that there are compact BV sets $S_k \subset \operatorname{int} B_1'$ such that ∂S_k is a C^∞ submanifold of $\mathbb{R}^m$ without boundary, $\lim |B_1' - S_k| = 0$, and $\lim \|S_k\| = \|B_1'\|$. Thus letting $B_1 = S_k$ for a sufficiently large k, we obtain

- $\{(B_1, x_1), (A_2 - B_1', x_2) \ldots, (A_p - B_1', x_p)\}$ is an η-regular δ-fine partition,
- $SF(\delta,\eta) < |F(B_1)| + \sum_{i=2}^{p} |F(A_i - B_1')|.$

Similarly, there are compact BV sets B_2 and B_2' such that

- ∂B_2 and $\partial B_2'$ are C^∞ submanifold of $\mathbb{R}^m$ without boundary,
- $B_2 \subset \operatorname{int} B_2'$ and $B_1 \cap B_2' = \emptyset$,
- B_2' contains no x_i,
- $\{(B_1, x_1), (B_2, x_2), (A_3 - [B_1' \cup B_2'], x_3), \ldots, (A_p - [B_1' \cup B_2'], x_p)\}$ is an η-regular δ-fine partition,
- $SF(\delta,\eta) < |F(B_1)| + |F(B_2)| + \sum_{i=3}^{p} |F(A_i - [B_1' \cup B_2'])|.$

Inductively, we construct an η-regular δ-fine partition $\{(B_1, x_1), \ldots, (B_p, x_p)\}$ such that each B_i is a compact BV set whose boundary ∂B_i is a C^∞ submanifold of $\mathbb{R}^m$ without boundary, no x_j is contained in any B_i, and the inequality

$$SF(\delta, \eta) < \sum_{i=1}^{p} |F(B_i)|$$

holds. Let $r > 0$ be such that no $B[x_j, r]$ meets any B_i, and let $C_i = B_i \cup B[x_i, r]$. If r is sufficiently small, then $\{(C_1, x_1), \ldots, (C_p, x_p)\}$ is an η-regular δ-fine special partition with

$$SF(\delta, \eta) < \sum_{i=1}^{p} |F(C_i)|.$$

This contradiction establishes the desired equality for $V_{\text{ess},\eta} F(E)$. The proof of the equality for $V_{*,\eta} F(E)$ is identical. $\qquad\square$

Observation 3.7.5. *Let F be a charge, let δ be a nonnegative function defined on a set $E \subset \mathbb{R}^m$, and let $\eta \geq 0$. There is a $\beta = \beta(m) > 1$ such that*

$$SF(\delta, \beta\eta) \leq \sup_P \sum_{i=1}^{p} |F(A_i)| \leq SF(\delta, \eta)$$

where $P = \{(A_1, x_1), \ldots, (A_p, x_p)\}$ is an η-regular δ-fine dyadic partition.

Remark 3.7.6. Observation 3.7.5 follows easily from Lemma 1.10.2. In conjunction with Theorem 3.7.4, it shows that the essential and critical variations of charges can be defined by means of dyadic figures. In view of Observation 2.3.7, the same is true for the derivates of charges. From this and Proposition 2.6.11, it is easy to infer that a large portion of the material discussed in this book can be developed by means of figures (dyadic, if one cares) without any references to BV sets.

While using figures instead of BV sets may have a pedagogical merit, this approach obliterates some important details, such as Proposition 3.6.4, or Proposition 5.1.8 below. Since it is not possible to avoid BV sets in the long run (cf. the next chapter), it appears that working with them from the onset is a better strategy.

4

Charges and BV functions

We show that with each charge F in $\mathbb{R}^m$ we can, in a natural way, associate a charge $F \times \mathcal{L}^1$ in $\mathbb{R}^{m+1}$, which corresponds to our intuitive concept of a product charge. Indeed, if F is the restriction of a measure μ in $\mathbb{R}^m$, then the product $F \times \mathcal{L}^1$ is the restriction of the product measure $\mu \times \mathcal{L}^1$ in $\mathbb{R}^{m+1}$. Among other applications, the charges $F \times \mathcal{L}^1$ are used to establish a duality between the spaces $CH(\mathbb{R}^m)$ and $BV_{\text{loc}}^{\infty}(\mathbb{R}^m)$.

4.1. The charge $F \times \mathcal{L}^1$

In this section, we shall consider BV sets not only in $\mathbb{R}^m$ but also in $\mathbb{R}^{m+1}$. To avoid confusion, we employ the notation

$$\mathcal{BV}(\mathbb{R}^m), \ \mathcal{BV}_n(\mathbb{R}^m), \ \mathcal{F}(\mathbb{R}^m), \ CH(\mathbb{R}^m),$$

$$\mathcal{BV}(\mathbb{R}^{m+1}), \ \mathcal{BV}_n(\mathbb{R}^{m+1}), \ \mathcal{F}(\mathbb{R}^{m+1}), \ CH(\mathbb{R}^{m+1})$$

whose meaning is obvious. The elements of $\mathcal{F}(\mathbb{R}^m)$ and $\mathcal{F}(\mathbb{R}^{m+1})$ are referred to as m-dimensional and $(m+1)$-dimensional figures, respectively. When the context allows no ambiguity, we still write $|E|$, $\|E\|$, $CH(E)$, etc., irrespective whether E is a subset of $\mathbb{R}^m$ or of $\mathbb{R}^{m+1}$.

If $E \in \mathcal{BV}(\mathbb{R}^{m+1})$, then it follows from Theorem 1.9.1 the set

$$E^t := \left\{ x \in \mathbb{R}^m : (x, t) \in E \right\}$$

belongs to $\mathcal{BV}(\mathbb{R}^m)$ for $\mathcal{L}^1$-almost all $t \in \mathbb{R}$. In view of Fubini's theorem, given a Borel measure μ in $\mathbb{R}^m$, we can define the product measure $\mu \times \mathcal{L}^1$ by the formula

$$(\mu \times \mathcal{L}^1)(E) := \int_{\mathbb{R}} \mu(E^t) \, dt$$

for each Borel set $E \subset \mathbb{R}^{m+1}$. The following theorem shows that for a set $E \in \mathcal{BV}(\mathbb{R}^{m+1})$, the same formula can be used to define the product charge $F \times \mathcal{L}^1$ in $\mathbb{R}^{m+1}$.

Theorem 4.1.1. *Let F be a charge in $\mathbb{R}^m$. If $E \in \mathcal{BV}(\mathbb{R}^{m+1})$, then the function $t \mapsto F(E^t)$, defined for $\mathcal{L}^1$-almost all $t \in \mathbb{R}$, belongs to $L^1(\mathbb{R})$. Letting*

$$(F \times \mathcal{L}^1)(E) := \int_{\mathbb{R}} F(E^t)\, dt$$

for each $E \in \mathcal{BV}(\mathbb{R}^{m+1})$ defines a charge $F \times \mathcal{L}^1$ in $\mathbb{R}^{m+1}$. The map

$$F \mapsto F \times \mathcal{L}^1 : \big(CH(\mathbb{R}^m), \mathcal{S}\big) \to \big(CH(\mathbb{R}^{m+1}), \mathcal{S}\big)$$

is linear and continuous.

PROOF. If $A \in \mathcal{F}(\mathbb{R}^{m+1})$, then $t \mapsto F(A^t)$ is an $\mathcal{L}^1$-measurable function with compact support that has only finitely many values. Clearly,

$$H : A \mapsto \int_{\mathbb{R}} F(A^t)\, dt : \mathcal{F}(\mathbb{R}^{m+1}) \to \mathbb{R}$$

is an additive function on $\mathcal{F}(\mathbb{R}^{m+1})$. Choose an $\varepsilon > 0$ and real numbers $a < b$. According to Proposition 2.2.6, there is a $\theta > 0$ such that

$$\big|F(B)\big| < \theta|B| + \varepsilon\big(\|B\| + 1\big)$$

for each BV set $B \subset [a,b]^m$. If $A \subset [a,b]^{m+1}$ is an $(m+1)$-dimensional figure, then

$$|H(A)| \leq \int_a^b \big|F(A^t)\big|\, dt \leq \theta \int_a^b |A^t|\, dt + \varepsilon \int_a^b \big(\|A^t\| + 1\big)\, dt \tag{$*$}$$
$$\leq \theta|A| + \varepsilon\big(\|A\| + b - a\big)$$

by Fubini's theorem, Theorem 1.9.1 and inequality (1.9.1). We infer H has a unique extension to a charge in $\mathbb{R}^{m+1}$, denoted by $F \times \mathcal{L}^1$.

Select a BV set $E \subset \mathbb{R}^{m+1}$ with $\operatorname{cl} E \subset (a,b)^{m+1}$. By Proposition 1.10.3, there is a sequence $\{A_i\}$ of $(m+1)$-dimensional figures contained in $[a,b]^{m+1}$ such that $\{A_i\} \to E$. For $i = 1, 2, \ldots$ and $t \in \mathbb{R}$, let $f_i(t) := F(A_i^t)$ and observe

$$\big|f_i(t) - f_j(t)\big| \leq \big|F(A_i^t - A_j^t)\big| + \big|F(A_j^t - A_i^t)\big|.$$

Integrating and applying inequality $(*)$, we obtain

$$\int_{\mathbb{R}} \big|f_i(t) - f_j(t)\big|\, dt = \int_a^b \big|f_i(t) - f_j(t)\big|\, dt \leq$$
$$\theta|A_i \triangle A_j| + 2\varepsilon\big[\|A_i\| + \|A_j\| + (b-a)\big].$$

As ε is arbitrary and $\{A_i\} \to E$, the previous inequality shows $\{f_i\}$ is a Cauchy sequence in $L^1(\mathbb{R})$, and hence it converges in $L^1(\mathbb{R})$ to an $f \in L^1(\mathbb{R})$. It follows $\{f_i\}$ has a subsequence, still denoted by $\{f_i\}$, such that $\lim f_i(t) = f(t)$ for all $t \in \mathbb{R} - N$ where $N \subset \mathbb{R}$ is an $\mathcal{L}^1$-negligible set. In view of Fubini's theorem, we may assume $\lim |A_i^t \bigtriangleup E^t| = 0$ for each $t \in \mathbb{R} - N$. As Fatou's lemma yields

$$\int_{\mathbb{R}} \liminf \|A_i^t\| \, dt \le \liminf \int_{\mathbb{R}} \|A_i^t\| \, dt \le \limsup \|A_i\| < \infty \,,$$

we may further assume $\liminf \|A_i^t\| < \infty$ for every $t \in \mathbb{R} - N$. Select a point $t \in \mathbb{R} - N$, and find a subsequence $\{A_{i_k}\}$ of $\{A_i\}$ such that $\limsup \|A_{i_k}^t\| < \infty$. Then $\{A_{i_k}^t\} \to E^t$ and we have

$$f(t) = \lim f_{i_k}(t) = \lim F(A_{i_k}^t) = F(E^t) \,,$$

since F is a charge in $\mathbb{R}^m$. As $\{f_i\}$ converges in $L^1(\mathbb{R})$ to f,

$$(F \times \mathcal{L}^1)(E) = \lim H(A_i) = \lim \int_{\mathbb{R}} F(A_i^t) \, dt$$

$$= \lim \int_{\mathbb{R}} f_i(t) \, dt = \int_{\mathbb{R}} f(t) \, dt = \int_{\mathbb{R}} F(E^t) \, dt \,.$$

Now choose an integer $n \ge 1$ and a set $E \in \mathcal{BV}_n(\mathbb{R}^{m+1})$. Observe $E^t \subset B[n]$ for each $t \in \mathbb{R}$, and let

$$k := \max\{n, 1 + 4(2n)^{m-1}\} \,,$$

$$J_- := \{t \in [-n, n] : \|E^t\| \le k\}, \quad J_+ := \{t \in [-n, n] : \|E^t\| > k\} \,.$$

The first part of the proof and Corollary 2.2.3 imply

$$|(F \times \mathcal{L}^1)(E)| \le \int_{J_-} |F(E^t)| \, dt + \int_{J_+} |F(E^t)| \, dt$$

$$\le \mathcal{L}^1(J_-) \|F\|_k + 2\|F\|_k \int_{J_+} \|E^t\| \, dt$$

$$\le \Big(\mathcal{L}^1(J_-) + 2\|E\|\Big) \|F\|_k \le 4n\|F\|_k \,,$$

and hence $\|F \times \mathcal{L}^1\|_n \le 4n\|F\|_k$. As the map $F \mapsto F \times \mathcal{L}^1$ is obviously linear, it is also continuous. $\square$

Remark 4.1.2. If E is a locally BV set in $\mathbb{R}^m$, then $E \times \mathbb{R}$ is a locally BV set in $\mathbb{R}^{m+1}$, and a direct calculation reveals

$$(F \, \mathsf{L} \, E) \times \mathcal{L}^1 = (F \times \mathcal{L}^1) \, \mathsf{L} \, (E \times \mathbb{R})$$

for each charge F in $\mathbb{R}^m$. In particular, if F is a charge in E, then $F \times \mathcal{L}^1$ is a charge in $E \times \mathbb{R}$.

We use the product charge $F \times \mathcal{L}^1$ to establish a duality between the linear spaces $CH(\mathbb{R}^m)$ and $BV_c^\infty(\mathbb{R}^m)$. If g is a nonnegative function defined on a set E, we call the set

$$\Sigma_g := \big\{(x,t) \in E \times \mathbb{R} : 0 < t < g(x)\big\}$$

the *subgraph* of g. Observe

$$\Sigma_g^t = \begin{cases} \{g > t\} & \text{if } t > 0, \\ \emptyset & \text{if } t \le 0. \end{cases}$$

If $g \in BV_c^\infty(\mathbb{R}^m)$ and $g \ge 0$, then Fubini's theorem, the coarea theorem, Theorem 1.9.1, and inequality (1.9.1) imply

$$|\Sigma_g| = |g|_1 \quad \text{and} \quad \|\Sigma_g\| \le \|g\| + 2\big|\{g \ne 0\}\big| ; \qquad (4.1.1)$$

in particular Σ_g belongs to $\mathcal{BV}(\mathbb{R}^{m+1})$. In view of Theorem 4.1.1, the real number

$$\langle F, g \rangle : = (F \times \mathcal{L}^1)(\Sigma_{g^+}) - (F \times \mathcal{L}^1)(\Sigma_{g^-})$$

$$= \int_0^\infty F(\{g > t\})\, dt - \int_0^\infty F(\{-g > t\})\, dt . \qquad (4.1.2)$$

is defined for each pair (F, g) in $CH(\mathbb{R}^m) \times BV_c^\infty(\mathbb{R}^m)$.

A *simple BV function* in $\mathbb{R}^m$ is a linear combination

$$g := \sum_{i=1}^p c_i \chi_{E_i}$$

where $c_1, \ldots, c_p$ are real numbers and $E_1, \ldots, E_p$ belong to $\mathcal{BV}(\mathbb{R}^m)$. The collection of all simple BV functions in $\mathbb{R}^m$ is a linear subspace of $BV_c^\infty(\mathbb{R}^m)$, denoted by $BVS(\mathbb{R}^m)$. The role of the space $BVS(\mathbb{R}^m)$ is purely auxiliary: it will help us to establish the linearity of the functional

$$g \mapsto \langle F, g \rangle : BV_c^\infty(\mathbb{R}^m) \to \mathbb{R} ,$$

and prove other results below.

The space $BVS(\mathbb{R}^m)$ of simple BV functions should not be confused with the space $SBV(\mathbb{R}^m)$ of *special BV functions* discussed in [2].

Lemma 4.1.3. *If F is a charge in $\mathbb{R}^m$ and $g = \sum_{i=1}^p c_i \chi_{E_i}$ where $c_1, \ldots, c_p$ are real numbers and $E_1, \ldots, E_p$ belong to $\mathcal{BV}(\mathbb{R}^m)$, then*

$$\langle F, g \rangle = \sum_{i=1}^p c_i F(E_i) .$$

In particular, the map $g \mapsto \langle F, g \rangle$ is a linear functional on $BVS(\mathbb{R}^m)$.

PROOF. If the sets $E_1, \ldots, E_p$ are disjoint, then

$$\int_0^\infty F(\{g > t\}) \, dt = \int_0^\infty F\left(\bigcup \{E_i : c_i > t\}\right) dt$$
$$= \int_0^\infty \left[\sum_{c_i > t} F(E_i)\right] dt = \sum_{c_i > 0} c_i F(E_i) \, ;$$

since for each $t \geq 0$, we have

$$\sum_{c_i > t} F(E_i) = \sum_{c_i > 0} \chi_{[0, c_i)}(t) F(E_i) \, .$$

Similarly $\int_0^\infty F(\{-g > t\}) \, dt = -\sum_{c_i < 0} c_i F(E_i)$, and hence

$$\langle F, g \rangle = \sum_{i=1}^p c_i F(E_i) \, .$$

In the general case, proceeding by induction on p, find a disjoint sub-collection $\mathcal{B} := \{B_1, \ldots, B_q\}$ of $\mathcal{BV}(\mathbb{R}^m)$ so that $\bigcup_{i=1}^p E_i = \bigcup \mathcal{B}$ and each $E_i = \bigcup \mathcal{B}_i$ for some $\mathcal{B}_i \subset \mathcal{B}$. Since

$$\chi_{E_i} = \sum_{j=1}^q c_{i,j} \chi_{B_j}$$

where each $c_{i,j}$ equals zero or one,

$$g = \sum_{i=1}^p c_i \left(\sum_{j=1}^q c_{i,j} \chi_{B_j}\right) = \sum_{j=1}^q \left(\sum_{i=1}^p c_i c_{i,j}\right) \chi_{B_j} \, .$$

By the first part of the proof,

$$\langle F, g \rangle = \sum_{j=1}^q \left(\sum_{i=1}^p c_i c_{i,j}\right) F(B_j) = \sum_{i=1}^p c_i \left[\sum_{j=1}^q c_{i,j} F(B_j)\right]$$
$$= \sum_{i=1}^p c_i \langle F, \chi_{E_i} \rangle = \sum_{i=1}^p c_i F(E_i) \, . \qquad \square$$

Lemma 4.1.4. *For each* $g \in BV_c^\infty(\mathbb{R}^m)$, *there is a sequence* $\{g_k\}$ *in* $BVS(\mathbb{R}^m)$ *having the following properties:*

(i) $\{g_k \neq 0\} \subset \{g \neq 0\}$ *for* $k = 1, 2, \ldots$;

(ii) $\{g_k\}$ *converges uniformly to* g;

(iii) $\{g_k\}$ $\mathcal{T}$-*converges to* g.

In particular, $BVS(\mathbb{R}^m)$ *is dense in* $\left(BV_c^\infty(\mathbb{R}^m), \mathcal{T}\right)$.

PROOF. In view of Theorem 1.8.12, it suffices to consider a function $g \in BV_c^\infty(\mathbb{R}^m)$ with $0 \le g \le 1$. Select an integer $k \ge 1$, and using the coarea theorem, find numbers $0 = t_0 < \cdots < t_{k+1} = 1$ so that $(i-1)/k < t_i < i/k$, the sets $E_i := \{g > t_i\}$ are in $\mathcal{BV}$, and

$$\frac{1}{k}\|E_i\| \le \int_{(i-1)/k}^{i/k} \|\{g > t\}\| \, dt$$

for $i = 1, \ldots, k$. Letting $E_0 := \mathbb{R}^m$ and $E_{k+1} := \emptyset$, the function

$$g_k := \sum_{i=1}^{k+1} t_{i-1}\chi_{(E_{i-1}-E_i)} = \sum_{i=1}^{k} (t_i - t_{i-1})\chi_{E_i}$$

belongs to $BVS(\mathbb{R}^m)$. Moreover, $0 \le g_k \le g$ and

$$g(x) - g_k(x) \le t_i - t_{i-1} \le 2/k$$

for all $x \in E_{i-1} - E_i$ and $i = 1, \ldots, k+1$. The first inequality yields condition (i). As $\mathbb{R}^m = \bigcup_{i=1}^{k+1}(E_{i-1} - E_i)$, the second inequality implies condition (ii), and therefore $\lim |g - g_k|_1 = 0$. Now condition (iii) follows from our choice of the numbers t_i and the coarea theorem:

$$\|g_k\| \le \sum_{i=1}^{k}(t_i - t_{i-1})\|E_i\| \le \sum_{i=1}^{k} \frac{2}{k} \cdot k \int_{(i-1)/k}^{i/k} \|\{g > t\}\| \, dt$$

$$= 2\int_0^1 \|\{g > t\}\| \, dt = 2\|g\|. \qquad \square$$

For a locally BV set $E \subset \mathbb{R}^m$, let

$$BV_b^\infty(E) := \{g \in BV_c^\infty(\mathbb{R}^m) : \{g \ne 0\} \subset E\} \qquad (4.1.3)$$

and observe that $BV_b^\infty(E)$ as a closed linear subspace of the locally convex space $(BV_c^\infty(\mathbb{R}^m), \mathcal{T})$.

Theorem 4.1.5. *Let $E \subset \mathbb{R}^m$ be a locally BV set. The pairing*

$$(F, g) \mapsto \langle F, g \rangle$$

is a nondegenerate bilinear functional on $CH(E) \times BV_b^\infty(E)$, and

$$\lim \langle F_i, g_i \rangle = \langle F, g \rangle$$

for each sequence $\{(F_i, g_i)\}$ in $CH(E) \times BV_b^\infty(E)$ that $(\mathcal{S} \times \mathcal{T})$-converges to (F, g) in $CH(E) \times BV_b^\infty(E)$.

PROOF. Choose a charge F in E. Since $\langle F, \chi_B \rangle = F(B)$ for each bounded BV set $B \subset E$, we see that $F = 0$ whenever $\langle F, g \rangle = 0$ for all $g \in BV_b^\infty(E)$. On the other hand, if g is a nonzero element of $BV_b^\infty(E)$, we may assume by symmetry that so is g^+. Observe that $G := \mathcal{L}^m \llcorner \{g > 0\}$ is a charge in E and

$$\langle G, g \rangle = \int_0^\infty \left| \{g > t\} \right| dt = \int_E g^+(x)\, dx > 0$$

by Fubini's theorem. It follows $(F, g) \mapsto \langle F, g \rangle$ is a nondegenerate pairing on $CH(E) \times BV_b^\infty(E)$.

Given $g \in BV_c^\infty(\mathbb{R}^m)$, the linearity and continuity of the map

$$F \mapsto \langle F, g \rangle : \big(CH(\mathbb{R}^m), \mathcal{S}\big) \to \mathbb{R}$$

is a direct consequence of Theorem 4.1.1.

Let F be any charge. Select a sequence $\{g_i\}$ in $BV_c^\infty(\mathbb{R}^m)$ that $\mathcal{T}$-converges to a function $g \in BV_c^\infty(\mathbb{R}^m)$, and note

$$\lim \left| \Sigma_{g_i^\pm} \triangle \Sigma_{g^\pm} \right| \leq \lim |g_i - g|_1 = 0.$$

Moreover, there is an integer $k \geq 1$ such that

$$\|g_i\| + \|g_i\|_\infty < k \quad \text{and} \quad \operatorname{supp} g_i \subset B(k)$$

for $i = 1, 2, \ldots$. Thus $\Sigma_{g_i^\pm} \subset B(k) \times (-k, k)$, and (4.1.1) implies

$$\left\| \Sigma_{g_i^\pm} \right\| \leq 2|B(k)| + \|g_i\| \leq 2|B(k)| + k.$$

It follows the sequences $\{\Sigma_{g_i^\pm}\}$ $\mathcal{T}$-converge to $\Sigma_{g^\pm}$. Therefore

$$\lim \langle F, g_i \rangle = \lim (F \times \mathcal{L}^1)(\Sigma_{g_i^+}) - \lim (F \times \mathcal{L}^1)(\Sigma_{g_i^-})$$
$$= (F \times \mathcal{L}^1)(\Sigma_{g^+}) - (F \times \mathcal{L}^1)(\Sigma_{g^-}) = \langle F, g \rangle$$

by Theorem 4.1.1. As $\mathcal{T}$ is a sequential topology, the map

$$g \mapsto \langle F, g \rangle : \big(BV_c^\infty(\mathbb{R}^m), \mathcal{T}\big) \to \mathbb{R}$$

is continuous, and hence linear according to Lemmas 4.1.3 and 4.1.4.

We have established that the pairing

$$(F, g) \mapsto \langle F, g \rangle : \big(CH(\mathbb{R}^m), \mathcal{S}\big) \times \big(BV_c^\infty(\mathbb{R}^m), \mathcal{T}\big) \to \mathbb{R}$$

is a bilinear separately continuous functional. As $\big(CH(\mathbb{R}^m), \mathcal{S}\big)$ is a Fréchet space, the proposition follows from the Banach-Steinhaus theorem [74, Theorem 2.17]. $\qquad \Box$

If E is a locally BV set, $F \in CH(E)$, and $g \in BV_b^\infty(E)$, then it follows from Theorem 4.1.5 that the maps

$$\Theta_E(F) : g \mapsto \langle F, g \rangle : \left(BV_b^\infty(E), \mathcal{T}\right) \to \mathbb{R},$$

$$\Phi_E(g) : F \mapsto \langle F, g \rangle : \left(CH(E), \mathcal{S}\right) \to \mathbb{R} \tag{4.1.4}$$

are continuous linear functionals. If $E = \mathbb{R}^m$, we write $\Theta(F)$ and $\Phi(g)$ instead of $\Theta_{\mathbb{R}^m}(F)$ and $\Phi_{\mathbb{R}^m}(g)$, respectively.

4.2. The space $\left(CH_*(E), \mathcal{S}\right)$

While the space $\left(CH_*(E), \mathcal{S}\right)$ is typically not complete, we show that it is barrelled. Our proof relies heavily on ideas of Thomson [83].

Lemma 4.2.1. *If B be a BV set and $y \in \mathbb{R}^m$, then*

$$\left|(B + y) \bigtriangleup B\right| \leq |y| \cdot \|B\|.$$

PROOF. If $g \in BV(\mathbb{R}^m) \cap C^1(\mathbb{R}^m)$, then

$$g(x - y) - g(x) = \int_0^1 \frac{d}{dt} g(x - ty)\, dt = - \int_0^1 y \cdot \nabla g(x - ty)\, dt$$

for all $x \in \mathbb{R}^m$, and Schwartz's inequality yields

$$\int_{\mathbb{R}^m} \left|g(x - y) - g(x)\right| dx \leq \int_{\mathbb{R}^m} \left(\int_0^1 |y| \cdot |\nabla g(x - ty)|\, dt \right) dx$$

$$= |y| \int_0^1 \left(\int_{\mathbb{R}^m} |\nabla g(x - ty)|\, dx \right) dt$$

$$= |y| \cdot |g|.$$

By Theorem 1.7.1, (2), there is a sequence $\{g_i\}$ in $BV(\mathbb{R}^m) \cap C^1(\mathbb{R}^m)$ with $\lim \|g_i\| = \|B\|$ and $\lim |g_i - \chi_B|_1 = 0$. Since

$$\left|(B + y) \bigtriangleup B\right| = \int_{\mathbb{R}^m} \left|\chi_B(x - y) - \chi_B(x)\right| dx$$

$$\leq \int_{\mathbb{R}^m} \left|\chi_B(x - y) - g_i(x - y)\right| dx$$

$$+ \int_{\mathbb{R}^m} \left|g_i(x - y) - g_i(x)\right| dx$$

$$+ \int_{\mathbb{R}^m} \left|g_i(x) - \chi_B(x)\right| dx$$

$$\leq 2|g_i - \chi_B|_1 + |y| \cdot |g_i|$$

for $i = 1, 2, \ldots$, the lemma follows. $\qquad \square$

Lemma 4.2.1 has an interesting geometric consequence. If B is a bounded BV set and $|y| > d(B)$, then the sets B and $B + y$ are disjoint. Thus

$$2|B| = \big|(B + y) \,\triangle\, B\big| \leq |y| \cdot \|B\|,$$

which implies $|B| \leq \frac{1}{2} d(B) \|B\|$. A stronger result follows from (2.3.2), however, a complete proof of inequality (2.3.2) is harder than that of Lemma 4.2.1.

The *convolution* of functions $f, g \in L_c^\infty(\mathbb{R}^m)$ is a function $f * g$ defined on $\mathbb{R}^m$ by the formula

$$(f * g)(x) := \int_{\mathbb{R}^m} f(x - y) g(y) \, dy.$$

Note $f * g$ is a measurable function [73, Theorem 8.14], and $f * g = g * f$. For $\varphi \in C_c^1(\mathbb{R}^m)$ and $g \in BV_c^\infty$, the following inequalities are obtained by routine calculations:

$$
\begin{aligned}
|\varphi * g|_\infty &\leq |\varphi|_1 \cdot |g|_\infty, \\
|\varphi * g|_1 &\leq |\varphi|_1 \cdot |g|_1, \\
\|\varphi * g\| &\leq \min\big\{ |\varphi|_1 \cdot \|g\|, \|\varphi\| \cdot |g|_1 \big\}.
\end{aligned}
\tag{4.2.1}
$$

Define a function $\eta \in C_c^\infty(\mathbb{R}^m)$ by letting

$$
\eta(x) := \begin{cases} \gamma \exp\left(\frac{1}{|x|^2 - 1} \right) & \text{if } |x| < 1, \\ 0 & \text{if } |x| \geq 1, \end{cases}
$$

where the constant γ is chosen so that $\int_{\mathbb{R}^m} \eta(x) \, dx = 1$. Given $\varepsilon > 0$, the function

$$\eta_\varepsilon : x \mapsto \frac{1}{\varepsilon^m} \eta\left(\frac{x}{\varepsilon} \right) \tag{4.2.2}$$

is called the *standard mollifier* [22, Section 4.2].

Convolutions and mollifiers are useful in many areas of analysis. We shall apply them in proving the next proposition.

Proposition 4.2.2. *The space $CH_{AC}(\mathbb{R}^m)$ is dense in $\big(CH(\mathbb{R}^m), \mathcal{S}\big)$.*

PROOF. Let F be a charge, and let $\varphi \in C_c^1(\mathbb{R}^m)$ be a nonnegative function. For a bounded BV set B, let

$$F_\varphi(B) := \Theta(F)(\varphi * \chi_B) = \langle F, \varphi * \chi_B \rangle$$

where $\Theta(F)$ is the map defined in (4.1.4). We show that F_φ is an AC charge which is $\mathcal{S}$-close to F for a suitable choice of φ.

Given $\varepsilon > 0$, use Proposition 2.2.6 to find a $\theta > 0$ so that

$$\big|F(B)\big| < \theta|B| + \varepsilon\big(\|B\| + 1\big)$$

for each BV set $B \subset B(1/\varepsilon)$. Select such a set B and let $b = |\varphi * \chi_B|_\infty$. Fubini's theorem, the coarea theorem, and inequalities (4.2.1) yield

$$\left| F_\varphi(B) \right| = \left| (F \times \mathcal{L}^1)(\Sigma_{\varphi*\chi_B}) \right| \leq \int_0^b \left| F(\Sigma_{\varphi*\chi_B}^t) \right| dt$$

$$\leq \theta \int_0^b \left| \Sigma_{\varphi*\chi_B}^t \right| dt + \varepsilon \int_0^b \left(\left\| \Sigma_{\varphi*\chi_B}^t \right\| + 1 \right) dt$$

$$= \theta |\varphi * \chi_B|_1 + \varepsilon \left(\|\varphi * \chi_B\| + b \right)$$

$$\leq \theta |B| \cdot |\varphi|_1 + \varepsilon \left(\|\varphi\| \cdot |B| + |\varphi|_1 \right),$$

and we conclude F_φ is an AC charge.

For $i = 1, 2, \ldots$, let $\varphi_i := \eta_{1/i}$ be the standard mollifier, and write F_i instead of F_{φ_i}. It suffices to show that the sequence $\{F_i\}$ of AC charges converges to F uniformly on each $\mathcal{BV}_n$. To this purpose, select an integer $n \geq 1$ and a set $B \in \mathcal{BV}_n$. Clearly $\operatorname{supp}(\varphi_i * \chi_B) \subset B[n+1]$, and since $|\varphi_i|_1 = 1$, the inequalities (4.2.1) imply

$$|\varphi_i * \chi_B|_\infty \leq 1 \quad \text{and} \quad \|\varphi_i * \chi_B\| \leq \|B\| \leq n.$$

According to Fubini's theorem and Lemma 4.2.1,

$$|\chi_B - \varphi_i * \chi_B|_1 =$$

$$\int_{\mathbb{R}^m} \left| \chi_B(x) \int_{\mathbb{R}^m} \varphi_i(y)\, dy - \int_{\mathbb{R}^m} \chi_B(x-y)\varphi_i(y)\, dy \right| dx \leq$$

$$\int_{\mathbb{R}^m} \varphi_i(y) \left(\int_{\mathbb{R}^m} |\chi_B(x) - \chi_B(x-y)|\, dx \right) dy =$$

$$\int_{B[1/i]} \varphi_i(y)\, |(B+y) \bigtriangleup B|\, dy \leq$$

$$\int_{B[1/i]} \varphi_i(y)\, |y| \cdot \|B\|\, dy \leq \frac{1}{i} \|B\| \leq \frac{n}{i}.$$

Thus $\{\varphi_i * \chi_B\}$ is a sequence in $BV_c^\infty(\mathbb{R}^m)$ that $\mathcal{T}$-converges to χ_B uniformly with respect to $B \in \mathcal{BV}_n$. It follows from Theorem 4.1.5 that the linear functional $\Theta(F)$ is $\mathcal{T}$-continuous on $BV_c^\infty(\mathbb{R}^m)$. Consequently

$$F(B) = \Theta(F)(\chi_B) = \lim \Theta(F)(\varphi_i * \chi_B) = \lim F_i(B)$$

where the convergence is uniform with respect to $B \in \mathcal{BV}_n$.   □

Corollary 4.2.3. *For a locally BV set $E \subset \mathbb{R}^m$, the following claims are true.*

   (i) *The space $CH_{AC}(E)$ is dense in $\big(CH(E), \mathbf{S}\big)$.*

   (ii) *The space $\big(CH_*(E), \mathbf{S}\big)$ is not complete whenever $\operatorname{int} E \neq \emptyset$.*

PROOF. (i) Let F be a charge in E. By Proposition 4.2.2, there is a sequence $\{F_i\}$ in $CH_{AC}(\mathbb{R}^m)$ that $\mathbf{S}$-converges to F. Clearly, $\{F_i \mathsf{L} E\}$ is a sequence in $CH_{AC}(E)$ that also $\mathbf{S}$-converges to F.

(ii) Suppose int $E \neq \emptyset$. Assertion (i) together with Observation 3.3.1 and Theorem 3.6.9 imply $CH_*(E)$ is a proper dense subset of the complete space $(CH(E), \mathbf{S})$. □

In the rest of this section we show that for each locally BV set E, the space $CH_*(E)$ is barrelled. The proof, which is rather technical, begins with a few lemmas.

Lemma 4.2.4. *Let E be a locally BV set, and let C be the union of non-overlapping essentially closed locally BV sets $C_1, \ldots, C_p$ whose perimeters are finite. Assume $V \subset CH_*(E \cap C)$ is a convex set such that for $i = 1, \ldots, p$, the intersection $V \cap CH_*(E \cap C_i)$ is a neighborhood of 0 in $CH_*(E \cap C_i)$. Then V is a neighborhood of 0 in $CH_*(E \cap C)$.*

PROOF. There are $\varepsilon_i > 0$ and integers $n_i \geq 1$ such that

$$\{G \in CH_*(E \cap C_i) : \|G\|_{n_i} < \varepsilon_i\} \subset V$$

for $i = 1, \ldots, p$. Find integers $k_i > \|C_i\|$, and let

$$\varepsilon := \min\{\varepsilon_1, \ldots, \varepsilon_p\} \quad \text{and} \quad n := \max\{n_1 + k_1, \ldots, n_p + k_p\}.$$

Choose an $F \in CH_*(E \cap C)$ with $\|F\|_n < \varepsilon/p$, and observe

$$F = \sum_{i=1}^p (F \mathsf{L} C_i).$$

According to Remark 3.6.12, each $F \mathsf{L} C_i$ belongs to $CH_*(E \cap C)$, and Proposition 1.8.4 yields

$$\|F \mathsf{L} C_i\|_{n_i} = \sup_{B \in \mathcal{BV}_{n_i}} |F(B \cap C_i)| \leq \|F\|_{n_i + k_i} \leq \|F\|_n < \frac{\varepsilon_i}{p}.$$

Thus each $p(F \mathsf{L} C_i)$ belongs to V, and since V is convex, it contains the charge $F = (1/p)\sum_{i=1}^p p(F \mathsf{L} C_i)$. □

Recall that Fréchet spaces, barrels, and barrelled spaces are defined in Section 1.2.

Lemma 4.2.5. *Let C be a convex, symmetric, bounded, and complete subset of a metrizable topological vector space X. If V is a barrel in X, then V absorbs C.*

PROOF. Since $C = \bigcup_{n=1}^{\infty} \left[C \cap (nV) \right]$, the Baire category theorem implies there is an integer $p \geq 1$ such that some $x \in C \cap (pV)$ is an interior point of $C \cap (pV)$ relative to C. As $0 = (x - x)/2$ is also an interior point of $C \cap (pV)$ relative to C, we can find a neighborhood U of 0 with $C \cap U = C \cap (pV)$. Now C is bounded, and so $C \subset rU$ for an $r > 0$. Select an $x \in C$ and observe x/r belongs to $C \cap U \subset pV$. We conclude $C \subset (rp)V$, and the lemma is proved.   □

Lemma 4.2.6. *Let X be a Fréchet space whose topology is induced by an increasing sequence $\{\|\cdot\|_n\}$ of seminorms, and let $\{x_i\}$ be a sequence in X such that $\sum_{i=1}^{\infty} \|x_i\|_i < \infty$. Then we can define a set*

$$X\big[\{x_i\}\big] := \left\{ \sum_{i=1}^{\infty} t_i x_i : |t_i| \leq 1 \text{ for } i = 1, 2, \dots \right\},$$

which is convex, symmetric, and compact.

PROOF. By Tychonoff's theorem [21, Theorem 3.2.4], the set I of all sequences $\{t_i\}$ in $[-1, 1]$ equipped with the product topology is a compact space. If $\{t_i\}$ is a sequence in $[-1, 1]$ and $q \geq p \geq n \geq 1$ are integers, then

$$\left\| \sum_{i=p}^{q} t_i x_i \right\|_n \leq \sum_{i=p}^{q} |t_i| \cdot \|x_i\|_n \leq \sum_{i=p}^{q} \|x_i\|_i.$$

Our condition on $\{x_i\}$ implies the sequence $\left\{ \sum_{i=1}^{k} t_i x_i \right\}$ is Cauchy, and hence

$$\sum_{i=1}^{\infty} t_i x_i = \lim_{k \to \infty} \sum_{i=1}^{k} t_i x_i$$

exists by the completeness of X. It follows the set $X\big[\{x_i\}\big]$ is defined, and it is clearly convex and symmetric. A standard argument shows the surjective map

$$\{t_i\} \mapsto \sum_{i=1}^{\infty} t_i x_i : I \to X\big[\{x_i\}\big]$$

is continuous, and hence $X\big[\{x_i\}\big]$ is compact.   □

Theorem 4.2.7. *If E is a locally BV set, then the space $\big(CH_*(E), S\big)$ is barrelled.*

PROOF. Select a barrel V in $\big(CH_*(E), S\big)$, and proceeding toward a contradiction, suppose V is not a neighborhood of 0 in $CH_*(E)$.

Claim 1. There is a decreasing sequence of locally BV sets $\{C_i\}$ such that for $i = 1, 2, \dots$, the intersection $V \cap CH_*(E \cap C_i)$ is not a neighborhood of 0 in $CH_*(E \cap C_i)$ and $\partial_* C_i = \partial C_i$. Moreover, the set $\bigcap_{i=1}^{\infty} C_i$ contains at most one point.

Proof. Assume there is an integer $n \geq 1$ such that

$$V \cap CH_*\big(E \cap K[n]\big)$$

is not a neighborhood of 0 in $CH_*\big(E \cap K[n]\big)$. Let $C := K[n]$, and use Lemma 4.2.4 to find a dyadic cube $C_1 \subset C$ of diameter $\sqrt{m}/2$ so that $V \cap CH_*(E \cap C_1)$ is not a neighborhood of 0 in $CH_*(E \cap C_1)$. Proceeding inductively, construct a sequence $\{C_i\}$ of dyadic cubes such that $d(C_i) = \sqrt{m}/2^i$ and $V \cap CH_*(E \cap C_i)$ is not a neighborhood of 0 in $CH_*(E \cap C_i)$. Clearly $\partial_* C_i = \partial C_i$ and $\bigcap_{i=1}^{\infty} C_i$ is a singleton.

Conversely, assume that for $i = 1, 2, \ldots$,

$$V \cap CH_*\big(E \cap K[i]\big)$$

is a neighborhood of 0 in $CH_*\big(E \cap K[i]\big)$. If $C_i := \mathbb{R}^m - K[i]$, then $\partial_* C_i = \partial C_i$, $\bigcap_{i=1}^{\infty} C_i = \emptyset$, and it follows from Lemma 4.2.4 that no $V \cap CH_*(E \cap C_i)$ is a neighborhood of 0 in $CH_*(E \cap C_i)$.

By Claim 1, there are $F_i \in CH_*(E \cap C_i)$ such that $\|F_i\|_i \leq 2^{-i}$ and $F_i \notin iV$ for $i = 1, 2, \ldots$. According to Remark 3.6.12, each charge F_i belongs to $CH_*(E)$. Using Lemma 4.2.6, define a compact convex symmetric set $Z := CH(E)\big[\{F_i\}\big]$.

Claim 2. $Z \subset CH_*(E)$.

Proof. If $U_0 := \mathbb{R}^m - \operatorname{cl} C_1$ and $U_k := \operatorname{int} C_k - \operatorname{cl} C_{k+1}$ for $k = 1, 2, \ldots$, then $S := \mathbb{R}^m - \bigcup_{k=0}^{\infty} U_k$ is a thin set. Select $F := \sum_{j=1}^{\infty} t_j F_j$ in Z and a negligible set $N \subset \operatorname{cl}_* E$. Observe $F_j \, \llcorner \, U_k = 0$ whenever $j > k$. Given $\varepsilon > 0$ and an integer $k \geq 1$, find a gage δ_k on $N \cap U_k$, so that

$$\sum_{i=1}^{p} |F_j(A_i)| < \frac{\varepsilon}{k 2^k}, \quad j = 1, \ldots, k,$$

for each ε-regular δ_k-fine partition $\{(A_1, x_1), \ldots, (A_p, x_p)\}$. Making δ_k smaller we can assume $B\big[x, \delta_k(x)\big] \subset U_k$. The formula

$$\delta(x) := \begin{cases} \operatorname{dist}(x, \operatorname{cl} C_1) & \text{if } x \in N \cap U_0, \\ \delta_k(x) & \text{if } x \in N \cap U_k, \ k = 1, 2, \ldots, \\ 0 & \text{if } x \in N \cap S, \end{cases}$$

defines a gage on N. If $\{(A_1, x_1), \ldots, (A_p, x_p)\}$ is an ε-regular δ-fine partition, then

$$\sum_{i=1}^{p} |F(A_i)| \leq \sum_{i=1}^{p} \sum_{j=1}^{\infty} |F_j(A_i)| = \sum_{k=0}^{\infty} \sum_{x_i \in U_k} \sum_{j=1}^{\infty} |F_j(A_i)|$$

$$= \sum_{k=1}^{\infty} \sum_{x_i \in U_k} \sum_{j=1}^{k} |F_j(A_i)| = \sum_{k=1}^{\infty} \sum_{j=1}^{k} \sum_{x_i \in U_k} |F_j(A_i)|$$

$$< \sum_{k=1}^{\infty} \sum_{j=1}^{k} \frac{\varepsilon}{k2^k} = \varepsilon \sum_{k=1}^{\infty} 2^{-k} = \varepsilon,$$

and consequently, $V_{*,\varepsilon} F(N) \leq F(\delta, \varepsilon) \leq \varepsilon$. Letting $\varepsilon \to 0$ yields $V_* F(N) = 0$, and we conclude $F \in CH_*(E)$.

Since Z is compact by Lemma 4.2.6, it is a complete and bounded subset of $CH_*(E)$. As Z is also convex and symmetric, Lemma 4.2.5 implies $Z \subset nV$ for an integer $n \geq 1$. A contradiction follows, since the charge F_n belongs to $Z - nV$. □

4.3. Duality

The *dual space* of a topological vector space X (abbreviated as the *dual* of X) is the linear space X^* of all continuous linear functionals $T : X \to \mathbb{R}$. The w^*-topology is the smallest topology $\mathcal{W}$ in X^* such that for each $x \in X$, the *evaluation map*

$$T \mapsto T(x) : (X^*, \mathcal{W}) \to \mathbb{R}$$

is continuous. The basic facts about dual spaces and w^*-topologies can be found in [74, Chapter 3].

The w^*-topology can be defined in the dual of any topological vector space X. However, for a particular X, it is often advantageous to consider stronger topologies in X^*. What follows is a case in point.

Let E be a locally BV set. It follows from Section 1.7 that

$$BV_n(E) := BV_n \cap BV_b^\infty(E), \quad n = 1, 2, \ldots,$$

are compact subsets of $\left(BV_b^\infty(E), \mathcal{T} \right)$. Thus letting

$$\|T\|_n := \sup\left\{ |T(g)| : g \in BV_n(E) \right\} \tag{4.3.1}$$

for each $T \in BV_b^\infty(E)^*$, we obtain a separating family of seminorms $\left\{ \|\cdot\|_n : n = 1, 2, \ldots \right\}$ which induces a Fréchet topology $\mathcal{N}$ in $BV_b^\infty(E)^*$.

Theorem 4.3.1. *For $T \in BV_b^\infty(E)^*$, the function*

$$\Lambda_E(T) : A \mapsto T(\chi_{E \cap A}) : \boldsymbol{\mathcal{BV}} \to \mathbb{R}$$

is a charge in E, and the map

$$\Lambda_E : T \mapsto \Lambda_E(T) : \big(BV_b^\infty(E)^*, \boldsymbol{\mathcal{N}}\big) \to \big(CH(E), \boldsymbol{\mathcal{S}}\big)$$

is a surjective linear homeomorphism with $\Lambda_E^{-1} : F \mapsto \Theta_E(F)$.

PROOF. Since $F = \Lambda_E(T)$ is a function on $\boldsymbol{\mathcal{BV}}$ such that $F = F \, \mathsf{L} \, E$, and since $F \upharpoonright \boldsymbol{\mathcal{BV}}(E)$ is the composition of T and the map

$$\chi : A \mapsto \chi_A : \boldsymbol{\mathcal{BV}}(E) \to BV_b^\infty(E) \,,$$

it is easy to see that $\Lambda_E(T)$ is a charge in E. In view of (2.2.1),

$$\|\Lambda_E(T)\|_n = \sup_{B \in \boldsymbol{\mathcal{BV}}_n} \big|T(\chi_{E \cap B})\big| \leq \|T\|_n$$

for $n = 1, 2, \ldots$. It follows the map Λ_E, which is clearly linear, is also continuous and injective. Recall the continuous linear functional

$$\Theta_E(F) : g \mapsto \langle F, g \rangle : \big(BV_b^\infty(E), \boldsymbol{\mathcal{J}}\big) \to \mathbb{R}$$

was defined in (4.1.4), and let

$$\Theta_E : F \mapsto \Theta_E(F) : CH(E) \to BV_b^\infty(E)^* \,.$$

According to Lemma 4.1.3, for each charge F in E and each bounded BV set $A \subset E$, we obtain

$$(\Lambda_E \circ \Theta_E)(F)(A) = \Theta_E(F)(\chi_A) = \langle F, \chi_A \rangle = F(A) \,.$$

Hence $\Lambda_E \circ \Theta_E$ is the identity map on $CH(E)$, which implies Λ_E is surjective. By the closed graph theorem [74, Theorem 2.15], the map Λ_E is a linear homeomorphism and $\Lambda_E^{-1} = \Theta_E$. $\qquad\square$

The following corollary is a direct consequence of Theorem 4.3.1 and Proposition 2.2.4.

Corollary 4.3.2. *If A is a bounded BV set, then the topology $\boldsymbol{\mathcal{N}}$ in $BV_b^\infty(A)^*$ is induced by a Banach norm.*

Remark 4.3.3. Let $\mathcal{D} := \mathcal{D}(\mathbb{R}^m)$ be the space of Schwartz's test functions equipped with its usual direct limit topology [74, Definition 6.3]. Observing the inclusion map $\mathcal{D} \hookrightarrow BV_c^\infty(\mathbb{R}^m)$ is continuous, we deduce that for each charge F, the linear functional $T := \Theta(F) \upharpoonright \mathcal{D}$ is a distribution. We shall elaborate on this fact in Section 5.6 below.

Following ideas of De Pauw [59], we show next that for a locally BV set E, the linear space $BV_b^\infty(E)$ is linearly isomorphic to the dual space of any linear subspaces of $(CH(E), \mathcal{S})$ which contains $CH_{AC}(E)$.

Lemma 4.3.4. *If $T \in (CH_{AC}(E), \mathcal{S})^*$, then there is a $g \in L_c^\infty(\mathbb{R}^m)$ such that $\{g \neq 0\} \subset E$ and for each $F \in CH_{AC}(E)$,*

$$g\,DF \in L^1(\mathbb{R}^m) \quad \text{and} \quad T(F) = \int_{\mathbb{R}^m} g(x)DF(x)\,dx.$$

PROOF. Suppose first that $E := \mathbb{R}^m$, and let $B_k := B[k]$ for each positive integer k. The bijection

$$\Upsilon_k : f \mapsto \int f(x)\,dx : L^1(B_k) \to (CH_{AC}(B_k), \mathcal{S})$$

is continuous with respect to the L^1-topology: if $f \in L^1(B_k)$, then

$$\left|\Upsilon_k(f)(B)\right| = \left|\int_{B \cap B_k} f(x)\,dx\right| \leq \int_{B_k} |f(x)|\,dx = |f|_1$$

for each bounded BV set B; so $\|\Upsilon_k(f)\|_n \leq |f|_1$ for $n = 1, 2, \ldots$. It follows $T \circ \Upsilon_k$ is a continuous linear functional on $L^1(B_k)$. According to standard results about the duality of L^p spaces [73, Theorem 6.16], there is a function $g_k \in L^\infty(B_k)$ such that

$$T[\Upsilon_k(f)] = \int_{B_k} g_k(x)f(x)\,dx$$

for each $f \in L^1(B_k)$. It follows that

$$T(F) = T[\Upsilon_k(DF)] = \int_{B_k} g_k(x)DF(x)\,dx$$

for every $F \in CH_{AC}(B_k)$. As

$$\int_B g_k(x)\,dx = \int_{B_k} g_k(x)D(\mathcal{L}^m \llcorner B)(x)\,dx = T(\mathcal{L}^m \llcorner B)$$
$$= \int_{B_{k+1}} g_{k+1}(x)D(\mathcal{L}^m \llcorner B)(x)\,dx = \int_B g_{k+1}(x)\,dx$$

for each measurable set $B \subset B_k$, we see that $g_{k+1}(x) = g_k(x)$ for almost all $x \in B_k$. Consequently, there is a function g defined on $\mathbb{R}^m$ such that $g(x) = g_k(x)$ for almost all $x \in B_k$ and $k = 1, 2, \ldots$; in particular g is measurable.

Select an $F \in CH_{AC}(\mathbb{R}^m)$. Since DF belongs to $L_{\text{loc}}^1(\mathbb{R}^m)$, so does the function $|DF|\,\text{sign}\,g$ whose indefinite integral G is an AC charge.

As the sequence $\{G \mathop{\llcorner} B_k\}$ $\mathcal{S}$-converges to G, we obtain

$$T(G) = \lim T(G \mathop{\llcorner} B_k) = \lim \int_{B_k} g_k(x) D(G \mathop{\llcorner} B_k)(x)\, dx$$

$$= \lim \int_{B_k} |g(x) DF(x)|\, dx = \int_{\mathbb{R}^m} |g(x) DF(x)|\, dx\,.$$

Hence gDF belongs to $L^1(\mathbb{R}^m)$ and

$$T(F) = \lim T(F \mathop{\llcorner} B_k) = \lim \int_{B_k} g(x) DF(x)\, dx$$

$$= \int_{\mathbb{R}^m} g(x) DF(x)\, dx\,.$$

Now Proposition 3.1.7 and Theorem 3.1.9 imply that $gf \in L^1(\mathbb{R}^m)$ for each $f \in L^1_{\mathrm{loc}}(\mathbb{R}^m)$; in particular, $g \in L^1(\mathbb{R}^m)$. We complete the argument by showing that the equivalence class of g contains a bounded function with compact support.

Assuming $\big|\{g \neq 0\} - B_n\big| > 0$ for $n = 1, 2, \ldots$, it is easy to construct a sequence $\{A_i\}$ of disjoint bounded measurable subsets of $\{g \neq 0\}$ so that $|A_i| > 0$ for $i = 1, 2, \ldots$, and that each B_n meets only finitely many of the sets A_i. Let $\beta_i := \int_{A_i} |g(x)|\, dx$ and

$$f(x) := \begin{cases} \frac{1}{\beta_i} \operatorname{sign} g(x) & \text{if } x \in A_i,\ n = 1, 2, \ldots, \\ 0 & \text{if } x \in \mathbb{R}^m - \bigcup_{i=1}^{\infty} A_i. \end{cases}$$

By our choice of the sets A_i, the function f belongs to $L^1_{\mathrm{loc}}(\mathbb{R}^m)$ and

$$\int_{\mathbb{R}^m} g(x) f(x)\, dx = \sum_{i=1}^{\infty} \frac{1}{\beta_i} \int_{A_i} |g(x)|\, dx = \infty\,.$$

This contradiction shows that $\{g \neq 0\}$ is equivalent to a bounded set.

Assuming $|g|_\infty = \infty$, let $S_i := \big\{x \in \mathbb{R}^m : |g(x)| > 2^i\big\}$ and observe $0 < |S_i| < \infty$ for $i = 1, 2, \ldots$. As the function

$$f := \sum_{i=1}^{\infty} \frac{1}{2^i |S_i|} \chi_{S_i} \operatorname{sign} g$$

belongs to $L^1(\mathbb{R}^m)$, so does gf. A contradiction follows: $\lim |S_i| = 0$, while

$$\int_{S_i} f(x) g(x)\, dx \geq \frac{1}{2^i |S_i|} \int_{S_i} |g(x)|\, dx \geq 1$$

for $i = 1, 2, \ldots$. We conclude $|g|_\infty < \infty$, and the first part of the proof is completed.

Next suppose that E is an arbitrary locally BV set, and select integers $k_n > \|E \cap B_n\|$. In view of Proposition 1.8.4,

$$\|F \, \llcorner \, E\|_n = \sup_{B \in \mathcal{BV}_n} |F(B \cap B_n \cap E)| \leq \|F\|_{n+k_n}$$

for each charge F and $n = 1, 2, \ldots$. It follows the projection

$$\Pi : F \mapsto F \, \llcorner \, E : \big(CH_{AC}(\mathbb{R}^m), \mathcal{S}\big) \to \big(CH_{AC}(E), \mathcal{S}\big)$$

is continuous. By the first part of the proof, there is a $g \in L_c^\infty(\mathbb{R}^m)$ such that

$$gDF \in L^1(\mathbb{R}^m) \quad \text{and} \quad T \circ \Pi(F) = \int_{\mathbb{R}^m} g(x)DF(x)\,dx$$

for each $F \in CH_{AC}(\mathbb{R}^m)$. In particular, for each $F \in CH_{AC}(E)$,

$$gDF \in L^1(\mathbb{R}^m) \quad \text{and} \quad T(F) = \int_{\mathbb{R}^m} g(x)DF(x)$$

and this remains true when g is replaced by $g\chi_{\mathbb{R}^m - E}$. $\qquad\square$

Theorem 4.3.5. *Let E be a locally BV set, and let X be a linear subspace of $CH(E)$. Denote by X^* the dual space of $(X, \mathcal{S})$ equipped with the w^*-topology. If $CH_{AC}(E) \subset X$, then the map*

$$\Phi_E : g \mapsto \Phi_E(g) : \big(BV_b^\infty(E), \mathcal{T}\big) \to X^*$$

is a continuous linear bijection.

PROOF. For each $g \in BV_b^\infty(E)$, the continuous linear functional

$$\Phi_E(g) : F \mapsto \langle F, g \rangle : (CH(E), \mathcal{S}) \to \mathbb{R}$$

was defined in (4.1.4). In view of Theorem 4.1.5 and the definition of the w^*-topology in X^*, we only need to show that Φ_E is a surjective map. To this end, select a $T \in X^*$, and use Lemma 4.3.4 to find a function $g \in L_c^\infty(\mathbb{R}^m)$ such that $\{g \neq 0\} \subset E$ and

$$T(F) = \int_{\mathbb{R}^m} g(x)DF(x)\,dx$$

for each $F \in CH_{AC}(E)$. Now it suffices to prove that g is a BV function with $\Phi_E(g) = T$.

Employing the $\mathcal{S}$-continuity of T, find positive integers n and k so that $|T(F)| \leq k\|F\|_n$ for all $F \in X$. Let $c := n + \|B[n] \cap E\|$, and choose a vector field $v \in C_c^1(\mathbb{R}^m; \mathbb{R}^m)$ with $|v|_\infty \leq 1$. The charge

$$G := \int \chi_E(x)\mathrm{div}\,v(x)\,dx$$

belongs to $CH_{AC}(E)$ by Proposition 3.1.7. Given a $B \in \mathcal{BV}_n$, the Gauss-Green formula (1.8.1) and Proposition 1.8.4 imply

$$|G(B)| = \left| \int_{\partial_*(E \cap B)} v \cdot \nu_{E \cap B} \, d\mathcal{H}^{m-1} \right| \leq \|E \cap B\| \cdot |v|_\infty$$

$$\leq \|B[n] \cap E \cap B\| \leq \|B[n] \cap E\| + \|B\| \leq c.$$

Thus $\|G\|_n \leq c$, and Theorem 3.1.9 yields

$$\int_{\mathbb{R}^m} g(x)\mathrm{div}\, v(x)\, dx = \int_{\mathbb{R}^m} g(x)\chi_E(x)\mathrm{div}\, v(x)\, dx$$

$$= \int_{\mathbb{R}^m} g(x)DG(x)\, dx$$

$$= T(G) \leq k\|G\|_n \leq kc.$$

We infer $\|g\| < \infty$, which means g is a BV function. If $F \in CH_{AC}(E)$, then Fubini's theorem, Proposition 3.1.7, and Theorem 3.1.9 imply

$$T(F) = \int_{\mathbb{R}^m} g^+(x)DF(x)\, dx - \int_{\mathbb{R}^m} g^-(x)DF(x)\, dx$$

$$= \int_{\mathbb{R}^m} \left(\int_0^{g^+(x)} dt \right) DF(x)\, dx - \int_{\mathbb{R}^m} \left(\int_0^{g^-(x)} dt \right) DF(x)\, dx$$

$$= \int_0^\infty \left[\int_{\{g>t\}} DF(x)\, dx \right] dt - \int_0^\infty \left[\int_{\{-g>t\}} DF(x)\, dx \right] dt$$

$$= \int_0^\infty F(\{g>t\})\, dt - \int_0^\infty F(\{-g>t\})\, dt$$

$$= \langle F, g \rangle = \Phi_E(g)(F).$$

In view of Corollary 4.2.3, (i), and Theorem 4.1.5, the equality

$$T(F) = \Phi_E(g)(F)$$

extends to all $F \in X$, and the surjectivity of Φ_E is established. $\square$

Question 4.3.6. Is there a useful description of the topology $\mathcal{S}^*$ in X^* for which the map

$$\Phi_E : \left(BV_b^\infty(E), \mathcal{T} \right) \to (X^*, \mathcal{S}^*)$$

is a homeomorphism? Does such a description exist for any particular X, such as $CH_{AC}(E)$, $CH_*(E)$, or $CH(E)$?

4.4. More on BV functions

In this section, a few results are stated without proofs. References to standard texts are supplied instead.

Observation 4.4.1. *Let $g \in BV_{\text{loc}}(\mathbb{R}^m)$. There is a negligible set N and a positive function δ defined on $\mathbb{R}^m - N$ such that*

$$\int_{B(x,r)} |g(y) - g(x)| \, dy \leq r|B(x,r)|\left(1 + |\nabla g(x)|\right),$$

for each $x \in \mathbb{R}^m - N$ and $0 < r < \delta(x)$.

PROOF. According to Theorem 1.7.3, there is a negligible set N and a positive function δ defined on $\mathbb{R}^m - N$ such that

$$\int_{B(x,r)} |g(y) - g(x) - \nabla g(x) \cdot (y - x)| \, dy < r|B(x,r)|$$

for each $x \in \mathbb{R}^m - N$ and $0 < r < \delta(x)$. For such an x and r,

$$\int_{B(x,r)} |g(y) - g(x)| \, dy \leq r|B(x,r)| + \int_{B(x,r)} |\nabla g(x)| \cdot |y - x| \, dy$$

$$\leq r|B(x,r)|\left(1 + |\nabla g(x)|\right). \qquad \square$$

Notation 4.4.2. Given $g \in L^1_{\text{loc}}(\mathbb{R}^m)$, $x \in \mathbb{R}^m$, and $r > 0$, we let

$$(g)_{x,r} := \frac{1}{|B(x,r)|} \int_{B(x,r)} g(y) \, dy,$$

and if a finite limit

$$g^*(x) := \lim_{r \to 0+} (g)_{x,r}$$

exists, we call $g^*(x)$ the *precise value* of g at x.

Note that the precise values of a function $g \in L^1_{\text{loc}}(\mathbb{R}^m)$ depend only on the equivalence class of g. If G is the indefinite integral of g, then

$$g^*(x) = DG(x) = g(x)$$

for almost all $x \in \mathbb{R}^m$ according to Theorem 3.1.9.

Theorem 4.4.3. *Let $g \in BV_{\text{loc}}(\mathbb{R}^m)$. The precise value $g^*(x)$ of g is defined for $\mathcal{H}^{m-1}$-almost all $x \in \mathbb{R}^m$, and there is a thin set T such that*

$$\lim_{r \to 0+} \frac{1}{|B(x,r)|} \int_{B(x,r)} |g(y) - g^*(x)| \, dy = 0$$

for each $x \in \mathbb{R}^m - T$.

For $m \geq 2$, Theorem 4.4.3 follows from [22, Section 5.9, Theorem 3] by a standard application of Hölder's inequality (cf. Theorem 1.7.3); if $m = 1$, then Remark 1.7.2 facilitates a direct verification.

Corollary 4.4.4. *Let G be the indefinite integral of a $g \in BV_{\mathrm{loc}}(\mathbb{R}^m)$. The function g^* is $\mathcal{H}^{m-1}$-measurable, and there is a thin set T such that $DG(x) = g^*(x)$ for each $x \in \mathbb{R}^m - T$. If Ω is a Lipschitz domain, then*

$$\mathrm{Tr}\,(g \upharpoonright \Omega)(x) = g^*(x)$$

for $\mathcal{H}^{m-1}$-almost all $x \in \partial\Omega - T$.

PROOF. Enumerate all positive rationals as $r_1, r_2, \ldots$, and observe that g^* equals $\mathcal{H}^{m-1}$-almost everywhere to a Baire function

$$x \mapsto \lim_{n \to \infty} \sup_{r_i < 1/n} (g)_{x, r_i}\,.$$

Let T be a thin set associated with g according to Theorem 4.4.3. Select an $x \in \mathbb{R}^m - T$ and a sequence $\{C_i\}$ of measurable sets such that

$$\lim d(C_i \cup \{x\}) = 0 \quad \text{and} \quad s = \inf s(C_i \cup \{x\}) > 0\,.$$

Letting $r_i := d(C_i \cup \{x\})$ and $B_i := B[x, r_i]$, we obtain $C_i \subset B_i$ and $|B_i|/|C_i| \leq \alpha(m)/s$ for $i = 1, 2, \ldots$. Theorem 4.4.3 yields

$$\lim \left| \frac{G(C_i)}{|C_i|} - g^*(x) \right| = \lim \left| \frac{1}{|C_i|} \int_{C_i} [g(y) - g^*(x)]\,dy \right|$$

$$\leq \lim \left[\frac{|B_i|}{|C_i|} \cdot \frac{1}{|B_i|} \int_{B_i} |g(y) - g^*(x)|\,dy \right]$$

$$\leq \frac{\alpha(m)}{s} \lim \frac{1}{|B_i|} \int_{B_i} |g(y) - g^*(x)|\,dy = 0\,.$$

Thus $DG(x) = g^*(x)$ by (2.3.1) and Observation 2.3.6.

Finally, let Ω be a Lipschitz domain. By Theorem 1.7.4, the trace

$$\mathrm{Tr}\,(g \upharpoonright \Omega)(x) := \lim_{r \to 0+} \frac{1}{|\Omega \cap B(x, r)|} \int_{\Omega \cap B(x, r)} g(y)\,dy$$

is defined at $\mathcal{H}^{m-1}$-almost all $x \in \partial\Omega$. Select an $x \in \Omega - T$ so that $\mathrm{Tr}\,(g \upharpoonright \Omega)(x)$ is defined. Since $\partial\Omega = \partial_*\Omega$, there is a sequence $\{r_i\}$ of positive numbers such that $\lim r_i = 0$ and $\inf s[\Omega \cap B(x, r_i)] > 0$. Letting $C_i := \Omega \cap B(x, r_i)$, the first part of the proof shows

$$g^*(x) = \lim \frac{G(C_i)}{|C_i|} = \mathrm{Tr}\,(g \upharpoonright \Omega)(x)\,. \qquad \square$$

Lemma 4.4.5. *Let g be a locally BV function with $0 \leq g \leq 1$, and let A be a bounded BV set. If $0 \leq c \leq 1$, then each of the sets*

$$\Sigma_{g\chi_A} - \left(A \times [0,c]\right), \quad \left(A \times [0,c]\right) - \Sigma_{g\chi_A}, \quad \Sigma_{g\chi_A} \bigtriangleup \left(A \times [0,c]\right)$$

has perimeter smaller than or equal to

$$2|A| + \|Dg\|(\mathrm{cl}_* A) + \int_{\partial_* A} \left|g^*(x) - c\right| d\mathcal{H}^{m-1}(x).$$

PROOF. Observe

$$A_+ := \left\{(x,t) \in A \times (c,1] : t < g(x)\right\} = \Sigma_{g\chi_A} - \left(A \times [0,c]\right),$$
$$A_- := \left\{(x,t) \in A \times [0,c] : g(x) \leq t\right\} = \left(A \times [0,c]\right) - \Sigma_{g\chi_A}.$$

According to Theorem 1.9.1,

$$\max\left\{\|A_+\|_{m+1}, \|A_-\|_{m+1}, \|A_+ \cup A_-\|_{m+1}\right\} \leq 2|A|. \qquad (*)$$

For $t \in \mathbb{R}$ and a set $B \subset \mathbb{R}^{m+1}$, let $B^t := \left\{x \in \mathbb{R}^m : (x,t) \in B\right\}$. If $c < t \leq 1$ then $A_+^t = A \cap \{g > t\}$, and Proposition 1.8.5 implies

$$\|A_+^t\| \leq \|\mathcal{D}A\|\left(\mathrm{int}_*\{g > t\}\right) + \|\mathcal{D}\{g > t\}\|(\mathrm{cl}_* A)$$
$$\leq \|\mathcal{D}A\|\left(\mathrm{int}_*\{g \geq t\}\right) + \|\mathcal{D}\{g > t\}\|(\mathrm{cl}_* A).$$

By Theorem 4.4.3, there is an $\mathcal{H}^{m-1}$-negligible set N such that $g^*(x)$ is defined for each x in $\mathbb{R}^m - N$. Select an x in $\mathrm{int}_*\{g \geq t\} - N$ and let $0 < \beta < 1$. If $B_\varepsilon := B[x,\varepsilon]$, then

$$\frac{\left|\{g \geq t\} \cap B_\varepsilon\right|}{|B_\varepsilon|} > \beta$$

for all sufficiently small $\varepsilon > 0$. Consequently

$$g^*(x) = \lim_{\varepsilon \to 0+} \frac{1}{|B_\varepsilon|} \int_{B_\varepsilon} g(y)\, dy \geq \lim_{\varepsilon \to 0+} \frac{1}{|B_\varepsilon|} \int_{\{g \geq t\} \cap B_\varepsilon} g(y)\, dy$$
$$\geq t \lim_{\varepsilon \to 0+} \frac{\left|\{g \geq t\} \cap B_\varepsilon\right|}{|B_\varepsilon|} > t\beta,$$

and $g^*(x) \geq t$ by the arbitrariness of β. We conclude

$$\partial_* A \cap \mathrm{int}_*\{g \geq t\} \subset \left\{x \in \partial_* A : g^*(x) \geq t\right\} \cup N.$$

An application of Fubini's theorem yields

$$
I_+ := \int_c^1 \|\mathcal{D}A\|(\mathrm{int}_*\{g > t\})\, dt
$$

$$
\leq \int_c^1 \mathcal{H}^{m-1}\Big(\{x \in \partial_* A : g^*(x) \geq t\}\Big) dt
$$

$$
= \int_{\{g^*>c\}\cap\partial_* A} \left(\int_c^{g^*(x)} dt\right) d\mathcal{H}^{m-1}(x)
$$

$$
= \int_{\{g^*>c\}\cap\partial_* A} [g^*(x) - c]\, d\mathcal{H}^{m-1}(x).
$$

Now we apply symmetry. The function $h := 1 - g$ satisfies the same conditions as g, and the equality

$$
A_-^t = A \cap \{g \leq t\} = A \cap \{h \geq 1 - t\}
$$

holds whenever $0 \leq t \leq c$. Moreover, $h^*(x) = 1 - g^*(x)$ for each x in $\mathbb{R}^m - N$. Proposition 1.8.5 implies

$$
\|A_-^t\| \leq \|\mathcal{D}A\|(\mathrm{int}_*\{h \geq 1 - t\}) + \|\mathcal{D}\{g \leq t\}\|(\mathrm{cl}_* A),
$$

and by the first part of the proof,

$$
I_- := \int_0^c \|\mathcal{D}A\|(\mathrm{int}_*\{h \geq 1 - t\})\, dt
$$

$$
\leq \int_0^c \mathcal{H}^{m-1}\Big(\{x \in \partial_* A : h^*(x) \geq 1 - t\}\Big) dt
$$

$$
= \int_0^c \mathcal{H}^{m-1}\Big(\{x \in \partial_* A : g^*(x) \leq t\}\Big) dt
$$

$$
= \int_{\{g^*<c\}\cap\partial_* A} [c - g^*(x)]\, d\mathcal{H}^{m-1}(x).
$$

As $\partial_*\{g \leq t\} = \partial_*\{g > t\}$, Theorem 1.8.2, (3) yields

$$
\|\mathcal{D}\{g \leq t\}\| = \|\mathcal{D}\{g > t\}\|
$$

for $\mathcal{L}^1$-almost all $t \in \mathbb{R}$. In view of Theorem 1.9.1,

$$
\|A_+\|_{\mathbb{R}^m} = \int_c^1 \|A_+^t\|\, dt \leq I_+ + \int_c^1 \|\mathcal{D}\{g > t\}\|(\mathrm{cl}_* A)\, dt,
$$

$$
\|A_-\|_{\mathbb{R}^m} = \int_0^c \|A_-^t\|\, dt \leq I_- + \int_0^c \|\mathcal{D}\{g > t\}\|(\mathrm{cl}_* A)\, dt.
$$

Using Propositions 1.5.1 and 1.8.10, we obtain

$$\|A_+ \cup A_-\|_{\mathbb{R}^m} \leq \|A_+\|_{\mathbb{R}^m} + \|A_-\|_{\mathbb{R}^m}$$

$$\leq I_+ + I_- + \int_0^1 \|\mathcal{D}\{g > t\}\|(\mathrm{cl}_* A)\, dt$$

$$\leq \int_{\partial_* A} |g^*(x) - c|\, d\mathcal{H}^{m-1}(x) + \|Dg\|(\mathrm{cl}_* A),$$

and the lemma follows from inequalities (1.9.1) and ($*$).   □

Remark 4.4.6. For applications of Lemma 4.4.5, it is necessary to find an effective estimate of the integral

$$\int_{\partial_* A} |g^*(y) - g^*(x)|\, d\mathcal{H}^{m-1}(y).$$

This is easy if g is a Lipschitz function and the regularity of the set $A \cup \{x\}$ is bounded away from zero (cf. the proof of Theorem 5.6.1 below). For a general locally BV function g, the estimate may depend on the local geometry of A, not controlled by the regularity of $A \cup \{x\}$. We obtain a useful estimate only when $A := B(x, r)$ — a special case sufficient for our purposes.

The first step toward estimating the integral of Remark 4.4.6 is a theorem that follows from the *Poincaré inequality* [22, Section 5.6.1, Theorem 1, (ii)] by a routine application of Hölder's inequality (cf. Theorems 1.7.3 and 4.4.3).

Theorem 4.4.7. *Let $m \geq 2$. There is a constant $\gamma = \gamma(m)$ such that*

$$\int_B |g(y) - (g)_{x,r}|\, dy \leq \gamma\, r \|Dg\|(B)$$

for each ball $B := B(x, r)$ and each $g \in BV(B)$.

The following result is proved in [88, Theorem 5.10.7] for a general class of BV sets containing all Lipschitz domains.

Theorem 4.4.8. *For $r > 0$, there is a constant $\kappa_r = \kappa_r(m)$ such that*

$$\int_{\partial B} |\mathrm{Tr}\, h(y)|\, d\mathcal{H}^{m-1}(y) \leq \kappa_r \left(\|Dh\|(B) + \int_B |h(y)|\, dy \right)$$

for each ball $B := B(x, r)$ and each $h \in BV(B)$.

Note that letting $h := \chi_B$ in Theorem 4.4.8, we obtain $\|B\| \leq \kappa_r |B|$. Consequently $\kappa_r \geq m/r$ is *large* for small $r > 0$.

Corollary 4.4.9. *Let $m \geq 2$. There is a constant $\beta = \beta(m)$ having the following property: given a function $g \in BV_{\mathrm{loc}}(\mathbb{R}^m)$, we can find a countable set C so that*

$$\int_{\partial B} \left| g^*(y) - (g)_{x,r} \right| d\mathcal{H}^{m-1}(y) \leq \beta \|Dg\|(B)$$

for each ball $B := B(x,r)$ with $x \in \mathbb{R}^m - C$.

PROOF. Let $\beta := \kappa_1(1+\gamma)$ where κ_1 and γ are the constants from Theorems 4.4.8 and 4.4.7, respectively. Choose a function $g \in BV_{\mathrm{loc}}(\mathbb{R}^m)$, and denote by T a thin set associated with g according to Corollary 4.4.4. Let C be the set of all $x \in \mathbb{R}^m$ such that $\mathcal{H}^{m-1}\left(T \cap \partial B[x, r_x]\right) > 0$ for some $r_x > 0$.

Claim. The set C is countable.

Proof. Seeking a contradiction, assume C is uncountable and let $S_x := \partial B[x, r_x]$ for each $x \in C$. By definition, $T = \bigcup_{n=1}^{\infty} T_n$ where $\mathcal{H}^{m-1}(T_n) < \infty$ for $n = 1, 2, \ldots$. Since

$$C = \bigcup_{n,p=1}^{\infty} \left\{ x \in C : \mathcal{H}^{m-1}(T_n \cap S_x) > \frac{1}{p} \right\},$$

there are positive integers n and p such that $\mathcal{H}^{m-1}(T_n \cap S_x) > 1/p$ for infinitely many $x \in C$. If x and y are distinct points of C, then S_x and S_y are distinct spheres in $\mathbb{R}^m$ with $m \geq 2$. Hence $\mathcal{H}^{m-1}(S_x \cap S_y) = 0$, and a contradiction follows.

Select an $x \in \mathbb{R}^m - C$ and for $r > 0$, let $B_r := B(x, r)$. By the choice of C and Corollary 4.4.4,

$$g^*(y) = \mathrm{Tr}\,(g \restriction B_r)(y)$$

for $\mathcal{H}^{m-1}$-almost all $y \in \partial B_r$. In particular, $g^* \in L^1(\partial B_r, \mathcal{H}^{m-1})$ by Theorem 1.7.4. Applying Theorem 4.4.8 to the function $h := g - (g)_{x,1}$ and using Theorem 4.4.7, we obtain

$$\int_{\partial B_1} \left| g^*(y) - (g)_{x,1} \right| d\mathcal{H}^{m-1}(y) \leq$$

$$\kappa_1 \left(\|Dg\|(B_1) + \int_{B_1} \left| g(y) - (g)_{x,1} \right| dy \right) \leq$$

$$\kappa_1 \|Dg\|(B_1) + \kappa_1 \gamma \|Dg\|(B_1) = \beta \|Dg\|(B_1).$$

For a fixed $r > 0$, define an affine map

$$\Phi : y \mapsto x + r(y - x) : \mathbb{R}^m \to \mathbb{R}^m$$

and note the function $f := g \circ \Phi$ belongs to $BV_{\text{loc}}(\mathbb{R}^m)$. As $\Phi(B_1) = B_r$, it is easy to verify that $(f)_{x,1} = (g)_{x,r}$, and

$$f^*(y) = g^*\big[\Phi(y)\big] = \text{Tr}\,(f \upharpoonright B_1)(y)$$

for $\mathcal{H}^{m-1}$-almost all $y \in \partial B_1$. Thus

$$\int_{\partial B_1} \big|f^*(y) - (f)_{x,1}\big|\, d\mathcal{H}^{m-1}(y) \le \beta \|Df\|(B_1)$$

by applying the above inequality to the function f. Observe

$$\int_{\partial B_r} \big|g^*(y) - (g)_{x,r}\big|\, d\mathcal{H}^{m-1}(y) =$$

$$r^{m-1} \int_{\partial B_1} \big|f^*(y) - (f)_{x,1}\big|\, d\mathcal{H}^{m-1}(y) \le \qquad (*)$$

$$r^{m-1} \beta \|Df\|(B_1)\,.$$

If $g \upharpoonright B_r$ belongs to $C^1(B_r)$, then $f \upharpoonright B_1$ belongs to $C^1(B_1)$. The chain rule and area theorem imply

$$\|Df\|(B_1) = \int_{B_1} \nabla f(y)\, dy = r \int_{B_1} \nabla g\big[\Phi(y)\big]\, dy$$

$$= \frac{1}{r^{m-1}} \int_{B_r} \nabla g(y)\, dy = \frac{1}{r^{m-1}} \|Dg\|(B_r)\,,$$

and it follows from Theorem 1.7.1, (2) that the equality

$$\|Dg\|(B_r) = r^{m-1} \|Df\|(B_1)$$

remains true for an arbitrary $g \in BV_{\text{loc}}(\mathbb{R}^m)$. Combining the last equality with inequality $(*)$ completes the proof. $\square$

Recall from Remark 2.1.5 that in dimension one, the flux of any function $f : \mathbb{R} \to \mathbb{R}$ is defined. A proof of the following theorem can be found in [24, Theorem 4.5.9, (23)].

Theorem 4.4.10. *Let $m = 1$ and let $g \in BV_{\text{loc}}(\mathbb{R})$. The value $g^*(x)$ is defined for all $x \in \mathbb{R}$, and if G is the flux of $g^* : x \mapsto g^*(x)$, then*

$$VG\big([a,b]\big) \le \|Dg\|\big([a,b]\big)$$

for each cell $[a,b] \subset \mathbb{R}$.

Now we are ready to estimate the integral from Remark 4.4.6. The estimate will be applied in the proof of Proposition 4.5.6 below.

Proposition 4.4.11. *There is a constant $\kappa = \kappa(m)$ having the following property: given a function $g \in BV_{\mathrm{loc}}(\mathbb{R}^m)$, we can find a negligible set N and a positive function δ defined on $\mathbb{R}^m - N$ so that*

$$\int_{\partial B} \left| g^*(y) - g(x) \right| d\mathcal{H}^{m-1}(y) \leq \kappa \|Dg\|(B) + m|B|\left(1 + \left|\nabla g(x)\right|\right)$$

for each closed ball $B := B[x,r]$ with $x \in \mathbb{R}^m - N$ and $0 < r < \delta(x)$.

PROOF. Choose a negligible set $N \subset \mathbb{R}^m$ so that $g^*(x)$ is defined and equals $g(x)$ for each $x \in \mathbb{R}^m - N$.

Suppose $m = 1$, and denote by G the flux of g^*. If $B := B[x,r]$ where $x \in \mathbb{R} - N$ and $r > 0$, then Theorem 4.4.10 implies

$$\int_{\partial B} \left| g^*(y) - g(x) \right| d\mathcal{H}^0(y) = \left| g^*(x-r) - g^*(x) \right| + \left| g^*(x+r) - g^*(x) \right|$$

$$= \left| G([x-r,x]) \right| + \left| G([x,x+r]) \right|$$

$$\leq VG(B) \leq \|Dg\|(B),$$

and the lemma holds. We mention that this is the only place where a *closed* ball $B := B[x,r]$ is needed.

Suppose $m \geq 2$. Enlarging N, we may assume that it contains a countable set C associated with g according to Corollary 4.4.9. Select an $x \in \mathbb{R}^m - N$ and for $r > 0$, let $B_r := B(x,r)$. By Corollary 4.4.9,

$$\int_{\partial B_r} \left| g^*(y) - (g)_{x,r} \right| d\mathcal{H}^{m-1}(y) \leq \beta \|Dg\|(B_r)$$

where $\beta = \beta(m)$ is a constant. Enlarging N further, it follows from Observation 4.4.1 there is a positive function δ defined on $\mathbb{R}^m - N$ such that the inequality

$$\int_{\partial B_r} \left| (g)_{x,r} - g(x) \right| d\mathcal{H}^{m-1}(y) = \|B_r\| \cdot \left| (g)_{x,r} - g(x) \right|$$

$$= \frac{m}{r} \left| \int_{B_r} [g(y) - g(x)] \, dy \right|$$

$$\leq m|B_r|\left(1 + \left|\nabla g(x)\right|\right)$$

holds whenever $0 < r < \delta(x)$. Combining the last two inequalities completes the argument. □

We note that in Proposition 4.4.11 the negligible set N depends on a particular *function* $g \in BV_{\mathrm{loc}}(\mathbb{R}^m)$, and may change when g is replaced by a function h equal to g almost everywhere.

4.5. The charge $F \mathbin{\llcorner} g$

Let F be a charge, and let $g \in BV_{\mathrm{loc}}^{\infty}(\mathbb{R}^m)$. As $g\chi_B \in BV_c^{\infty}(\mathbb{R}^m)$ for each bounded BV set B, there is a function

$$F \mathbin{\llcorner} g : B \mapsto \langle F, g\chi_B \rangle : \mathcal{BV}(\mathbb{R}^m) \to \mathbb{R}.$$

It follows from Theorem 4.1.5 that $F \mathbin{\llcorner} g$ is a charge, called the *charge induced by F via g*.

Remark 4.5.1. For a charge F and $g \in BV_{\mathrm{loc}}^{\infty}(\mathbb{R}^m)$, the notation $F \mathbin{\llcorner} g$ is consistent with our previous usage of the symbol $\mathbin{\llcorner}$.

(i) If E is a locally BV set, then

$$(F \mathbin{\llcorner} \chi_E)(A) = \langle F, \chi_{E \cap A} \rangle = F(E \cap A) = (F \mathbin{\llcorner} E)(A)$$

for each bounded BV set A.

(ii) If F is the restriction of a Radon measure μ and A is a bounded BV set, then Fubini's theorem yields

$$(F \mathbin{\llcorner} g)(A) = \langle F, g\chi_A \rangle$$
$$= \int_0^{\infty} \mu\big(\{g^+\chi_A > t\}\big)\, dt - \int_0^{\infty} \mu\big(\{g^-\chi_A > t\}\big)\, dt$$
$$= \int_A g^+\, d\mu - \int_A g^-\, d\mu = \int_A g\, d\mu = (\mu \mathbin{\llcorner} g)(A).$$

Thus $F \mathbin{\llcorner} g$ is the restriction of the signed measure $\mu \mathbin{\llcorner} g$ defined in Remark 1.3.1.

It is clear the pairing

$$(F, g) \mapsto F \mathbin{\llcorner} g : CH(\mathbb{R}^m) \times BV_{\mathrm{loc}}^{\infty}(\mathbb{R}^m) \to CH(\mathbb{R}^m)$$

is a nondegenerate bilinear map, and $F \mathbin{\llcorner} g$ is a charge in a locally BV set E whenever either F is a charge in E, or $\{g \ne 0\} \subset E$.

Proposition 4.5.2. *If $\big\{(F_i, g_i)\big\}$ is a sequence in $CH(\mathbb{R}^m) \times BV_c^{\infty}(\mathbb{R}^m)$ that $(\mathcal{S} \times \mathcal{T})$-converges to (F, g) in $CH(\mathbb{R}^m) \times BV_c^{\infty}(\mathbb{R}^m)$, then the sequence $\{F_i \mathbin{\llcorner} g_i\}$ $\mathcal{S}$-converges to $F \mathbin{\llcorner} g$.*

PROOF. Let $g \in BV_c^{\infty}(\mathbb{R}^m)$, and for a positive integer n, select an $A \in \mathcal{BV}_n(\mathbb{R}^m)$. By (1.7.5), (4.1.1) and Theorem 1.8.12,

$$\big\|\Sigma_{g^{\pm}\chi_A}\big\| \le |g|_{\infty}\|A\| + \|g\| + 2\big|\{g\chi_A \ne 0\}\big| \le c\beta_n + \|g\| \qquad (*)$$

where $c := \max\big\{1, |g|_{\infty}\big\}$ and $\beta_n := n + 2|B(n)|$.

Select a sequence $\{g_i\}$ in $BV_c^\infty(\mathbb{R}^m)$ that $\mathcal{T}$-converging to zero, and note that $\lim |g_i|_1 = 0$,

$$\operatorname{supp} g_i \subset [-k,k]^m \quad \text{and} \quad \|g_i\| + |g_i|_\infty \le k$$

for a fixed integer $k \ge 1$ and $i = 1, 2, \dots$. If $A \in \mathcal{BV}_n(\mathbb{R}^m)$, then

$$\Sigma_{g_i^\pm \chi_A} \subset [-k,k]^{m+1}, \quad |\Sigma_{g_i^\pm \chi_A}| \le |g_i|_1, \quad \|\Sigma_{g_i^\pm \chi_A}\| \le k(\beta_n + 1),$$

where the last inequality follows from $(*)$. Let F be a charge, and choose an $\varepsilon > 0$. By Proposition 2.2.6, there is a $\theta > 0$ with

$$\big|(F \times \mathcal{L}^1)(B)\big| < \theta|B| + \varepsilon\big(\|B\| + 1\big)$$

for each $B \in \mathcal{BV}(\mathbb{R}^{m+1})$ contained in $[-k,k]^{m+1}$. Now

$$\begin{aligned}
\big|(F \mathrel{\llcorner} g_i)(A)\big| &= \big|\langle F, g_i\chi_A\rangle\big| \\
&\le \big|(F \times \mathcal{L}^1)(\Sigma_{g_i^+ \chi_A})\big| + \big|(F \times \mathcal{L}^1)(\Sigma_{g_i^- \chi_A})\big| \\
&< 2\theta|g_i|_1 + 2\varepsilon\big[k(\beta_n + 1) + 1\big],
\end{aligned}$$

and as $A \in \mathcal{BV}_n(\mathbb{R}^m)$ is arbitrary,

$$\|F \mathrel{\llcorner} g_i\|_n \le 2\theta|g_i|_1 + 2\varepsilon\big[k(\beta_n + 1) + 1\big].$$

From this and the arbitrariness of ε, we infer $\lim \|F \mathrel{\llcorner} g_i\|_n = 0$.

By $(*)$, there is a positive integer $p := p(g,n)$ such that $\Sigma_{g^\pm \chi_A}$ belongs to $\mathcal{BV}_p(\mathbb{R}^{m+1})$. Therefore

$$\begin{aligned}
\big|(F \mathrel{\llcorner} g)(A)\big| &= \big|\langle F, g\chi_A\rangle\big| \\
&\le \big|(F \times \mathcal{L}^1)(\Sigma_{g^+\chi_A})\big| + \big|(F \times \mathcal{L}^1)(\Sigma_{g^-\chi_A})\big| \\
&\le 2\|F \times \mathcal{L}^1\|_p.
\end{aligned}$$

Hence $\|F \mathrel{\llcorner} g\|_n \le 2\|F \times \mathcal{L}^1\|_p$, which implies $F \times \mathcal{L}^1 \mapsto F \mathrel{\llcorner} g$ is a continuous map. Since the map $F \mapsto F \times \mathcal{L}^1$ is continuous by Theorem 4.1.1, we have shown that the pairing

$$(F,g) \mapsto F \mathrel{\llcorner} g : \big(CH(\mathbb{R}^m),\mathcal{S}\big) \times \big(BV_c^\infty(\mathbb{R}^m),\mathcal{T}\big) \to \big(CH(\mathbb{R}^m),\mathcal{S}\big)$$

is a separately continuous bilinear map. As $\big(CH(\mathbb{R}^m),\mathcal{S}\big)$ is a Fréchet space, the proposition follows from the Banach-Steinhaus theorem [74, Theorem 2.17]. $\square$

Proposition 4.5.3. *If F is a charge and $g, h \in BV_{\mathrm{loc}}^\infty(\mathbb{R}^m)$, then*

$$(F \mathrel{\llcorner} g) \mathrel{\llcorner} h = F \mathrel{\llcorner} (gh) = (F \mathrel{\llcorner} h) \mathrel{\llcorner} g.$$

PROOF. By Remark 4.5.1, the proposition holds when g and h are the indicators of bounded BV sets. Linearity implies the proposition still holds when g and h are simple BV functions.

Let $g, h \in BV_c^\infty(\mathbb{R}^m)$. Use Lemma 4.1.4 to find sequences $\{g_i\}$ and $\{h_i\}$ of simple BV functions that $\mathfrak{T}$-converge to g and h, respectively. By Proposition 4.5.2, the sequence $\{F \mathsf{L} g_i\}$ $\mathcal{S}$-converges to $F \mathsf{L} g$, and so by the same proposition,

$$\{(F \mathsf{L} g_i) \mathsf{L} h_j\} = \{F \mathsf{L} (g_i h_j)\}$$

$\mathcal{S}$-converges to both $(F \mathsf{L} g) \mathsf{L} h_j$ and $F \mathsf{L} (gh_j)$. Therefore $(F \mathsf{L} g) \mathsf{L} h_j = F \mathsf{L} (gh_j)$ for $j = 1, 2, \ldots$, and we infer $(F \mathsf{L} g) \mathsf{L} h = F \mathsf{L} (gh)$.

Finally, let $g, h \in BV_{\mathrm{loc}}^\infty(\mathbb{R}^m)$, and choose a bounded BV set A. Since $g\chi_A$ and $h\chi_A$ belong to $BV_c^\infty(\mathbb{R}^m)$, we can apply the result of the previous paragraph. By linearity, we may assume that g and h are nonnegative functions and calculate

$$[(F \mathsf{L} g) \mathsf{L} h](A) = \langle F \mathsf{L} g, h\chi_A \rangle = \int_0^\infty (F \mathsf{L} g)(\{h\chi_A > t\}) \, dt$$

$$= \int_0^\infty \left\langle F, g\chi_{\{h\chi_A > t\}} \right\rangle dt = \int_0^\infty \left\langle F, g\chi_A \chi_{\{h\chi_A > t\}} \right\rangle dt$$

$$= \int_0^\infty [F \mathsf{L} (g\chi_A)] (\{h\chi_A > t\}) \, dt = \langle F \mathsf{L} (g\chi_A), (h\chi_A)\chi_A \rangle$$

$$= \left([F \mathsf{L} (g\chi_A)] \mathsf{L} [h\chi_A] \right)(A) = F \mathsf{L} (gh\chi_A)(A)$$

$$= \langle F, gh\chi_A \rangle = [F \mathsf{L} (gh)](A) . \qquad \square$$

If g and h are functions defined on the sets A and B, respectively, we let

$$(g \otimes h)(x, y) := g(x)h(y)$$

for all $(x, y) \in A \times B$. We denote by 1_E the function defined on a set E such that $1_E(x) = 1$ for every $x \in E$. With this notation,

$$(g \otimes 1_E)(x, y) = g(x) \quad \text{and} \quad (1_E \otimes h)(y, z) = h(z)$$

for each $(x, y) \in A \times E$ and each $(y, z) \in E \times B$.

Proposition 4.5.4. *If F is a charge and $g \in BV_{\mathrm{loc}}^\infty(\mathbb{R}^m)$, then the function $g \otimes 1_\mathbb{R}$ belongs to $BV_{\mathrm{loc}}^\infty(\mathbb{R}^{m+1})$ and*

$$(F \mathsf{L} g) \times \mathcal{L}^1 = (F \times \mathcal{L}^1) \mathsf{L} (g \otimes 1_\mathbb{R}) .$$

PROOF. The first claim follows from the coarea theorem, Theorem 1.9.1, and inequality (1.9.1). Let χ_C be the indicator in $\mathbb{R}$ of a cell $C \subset \mathbb{R}$. By Remark 4.1.2,

$$[(F \mathsf{L} g) \times \mathcal{L}^1] \mathsf{L} (1_{\mathbb{R}^m} \otimes \chi_C) = (F \times \mathcal{L}^1) \mathsf{L} (g \otimes \chi_C)$$

whenever g is the indicator of locally BV sets $B \subset \mathbb{R}^m$. In view of linearity, the previous equality still holds when g is a simple BV function.

Let $g \in BV_{\mathrm{loc}}^\infty(\mathbb{R}^m)$ and $A \in \mathcal{BV}(\mathbb{R}^{m+1})$. Choose an integer $k \geq 1$ so that $A \subset [-k, k]^{m+1}$, and let $B := [-k, k]^m$ and $C := [-k, k]$. Proceeding as in the proof of Proposition 4.5.3, it is easy to show

$$[(F \, \llcorner \, g) \times \mathcal{L}^1](A) = \Big([(F \, \llcorner \, g\chi_B) \times \mathcal{L}^1] \, \llcorner \, (1_{\mathbb{R}^m} \times \chi_C) \Big)(A),$$
$$[(F \times \mathcal{L}^1) \, \llcorner \, (g \otimes 1_{\mathbb{R}})](A) = [(F \times \mathcal{L}^1) \, \llcorner \, (g\chi_B \otimes \chi_C)](A). \tag{*}$$

Select a sequence $\{g_i\}$ in $BVS(\mathbb{R}^m)$ that $\mathcal{T}$-converges to $g\chi_B$, and observe the sequence $\{g_i \otimes \chi_C\}$ $\mathcal{T}$-converges to $g\chi_B \otimes \chi_C$. According to Theorem 4.1.1 and Proposition 4.5.2, the sequence

$$\Big\{ [(F \, \llcorner \, g_i) \times \mathcal{L}^1] \, \llcorner \, (1_{\mathbb{R}^m} \otimes \chi_C) \Big\} = \big\{ (F \times \mathcal{L}^1) \, \llcorner \, (g_i \otimes \chi_C) \big\}$$

converges to both $[(F \, \llcorner \, g\chi_B) \times \mathcal{L}^1] \, \llcorner \, (1_{\mathbb{R}^m} \otimes \chi_C)$ and $(F \times \mathcal{L}^1) \, \llcorner \, (g\chi_B \otimes \chi_C)$. This and equalities ($*$) complete the proof. $\square$

Proposition 4.5.5. *Let $g \in BV_{\text{loc}}^\infty(\mathbb{R}^m)$, and let F be a charge in a locally BV set E. If F belongs to $CH_*(E)$, then so does $F \, \llcorner \, g$.*

PROOF. In view of Theorem 1.8.12, we may assume with no loss of generality that $0 \leq g \leq 1$. Denote by G the charge $F \, \llcorner \, g$, and by I the interval $[0, 1]$. For $t \in I$ and any set $C \subset \mathbb{R}^m$, let

$$C^t := \{g > t\} \cap C,$$

and observe $G(B) = \int_I F(B^t) \, dt$ whenever B is a bounded BV set. It suffices to show $V_* G(N) = 0$ for each bounded negligible set $N \subset \text{cl}_* E$. Choose such a set N, and find an $\varepsilon > 0$ with $N \subset B(1/\varepsilon)$. Let

$$\eta := \frac{\varepsilon}{6 + 2\|Dg\|\big(B(1/\varepsilon)\big)},$$

and applying Proposition 2.2.6, find a $\theta > 0$ so that

$$|F(B)| < \theta|B| + \eta\big(\|B\| + 1\big) \tag{*}$$

for every BV set $B \subset B(1/\eta)$. As F is AC_* in $\text{cl}_* E$, there is a gage δ on N such that

$$\sum_{i=1}^{p} |F(A_i)| < \eta$$

for each $(\eta/6)$-regular δ-fine partition $\big\{(A_1, x_1), \ldots, (A_p, x_p)\big\}$. Choose an open set $U \subset B(1/\varepsilon)$ with $N \subset U$ and $|U| < \eta/\theta$. Making δ smaller, we may assume $B[x, \delta(x)]$ is contained in U for every $x \in N$.

Select an η-regular δ-fine partition $\big\{(B_1, x_1), \ldots, (B_q, x_q)\big\}$, and note that U contains the closure of each B_i by the choice of δ; in particular $B_i^t = U^t \cap B_i$. By the coarea theorem and Proposition 1.8.5,

$$\|\mathcal{D}U^t\|(\text{int}_* B_i) \leq \|B_i^t\| \leq \|\mathcal{D}U^t\|(\text{int}_* B_i) + \|B_i\| \tag{**}$$

for $\mathcal{L}^1$-almost all $t \in I$ and $i = 1, \ldots, q$. Letting

$$S_i := \left\{ t \in I : \|B_i^t\| \le 2\|B_i\| \right\}$$

and $T_i := I - S_i$, we distinguish two cases.

Case 1. Let $S_i' := \left\{ t \in S_i : |B_i| \le 2|B_i^t| \right\}$, $S_i'' := S_i - S_i'$, and

$$A_{i,t} := \begin{cases} B_i^t & \text{if } t \in S_i', \\ B_i - B_i^t & \text{if } t \in S_i''. \end{cases}$$

Since $\|A_{i,t}\| \le \|B_i\| + \|B_i^t\| \le 3\|B_i\|$ and $|B_i| \le 2|A_{i,t}|$, we have

$$r\bigl(A_{i,t} \cup \{x_i\}\bigr) \ge \frac{1}{2} \cdot \frac{|B_i|}{d\bigl(B_i \cup \{x_i\}\bigr)\|B_i\|} \cdot \frac{\|B_i\|}{\|A_{i,t}\|} > \frac{\eta}{6}$$

for each $t \in S_i$.

Case 2. By inequality $(**)$, for almost all $t \in T_i$,

$$\|B_i^t\| \le \|\mathcal{D}U^t\|(\text{int}_* B_i) + \|B_i\| < \|\mathcal{D}U^t\|(\text{int}_* B_i) + \frac{1}{2}\|B_i^t\|,$$

and hence $\|B_i^t\| < 2\|\mathcal{D}U^t\|(\text{int}_* B_i)$.

Using Cases 1 and 2, we estimate separately each of the sums

$$\alpha := \sum_{i=1}^q \int_{S_i} \bigl|F(B_i^t)\bigr|\, dt \quad \text{and} \quad \beta := \sum_{i=1}^q \left| \int_{T_i} F(B_i^t)\, dt \right|.$$

According to Case 1, the collection $P_t := \left\{ (A_{i,t}, x_i) : t \in S_i \right\}$ is an $(\eta/6)$-regular δ-fine partition for each $t \in I$. Thus

$$\alpha = \sum_{i=1}^q \left(\int_{S_i'} \bigl|F(A_{i,t})\bigr|\, dt + \int_{S_i''} \bigl|F(B_i - A_{i,t})\bigr|\, dt \right)$$

$$\le \sum_{i=1}^q \left(\int_{S_i} \bigl|F(A_{i,t})\bigr|\, dt + \int_{S_i''} \bigl|F(B_i)\bigr|\, dt \right)$$

$$\le \int_I \left(\sum_{i=1}^q \chi_{S_i}(t)\bigl|F(A_{i,t})\bigr| \right) dt + \int_I \left(\sum_{i=1}^q \bigl|F(B_i)\bigr| \right) dt$$

$$< \int_I \left(\sum_{t \in S_i} \bigl|F(A_{i,t})\bigr| \right) dt + \eta < 2\eta.$$

Given a set $J \subset \{1, \ldots, q\}$ and a $t \in I$, let $J_t := \{i \in J : t \in T_i\}$ and $B_t := \bigcup_{i \in J_t} B_i^t$. Using $(*)$ and Case 2, observe

$$F(B_t) \leq \theta|B_t| + \eta\big(\|B_t\| + 1\big) \leq \theta|U| + \eta \sum_{i \in J_t} \|B_i^t\| + \eta$$

$$< 2\eta + \eta \sum_{i \in J_t} \|\mathcal{D}U^t\|(\mathrm{int}_* B_i) \leq 2\eta + \eta\|\mathcal{D}U^t\|(U)$$

for $\mathcal{L}^1$-almost all $t \in I$. By the coarea theorem

$$\sum_{i \in J} \int_{T_i} F(B_i^t)\, dt = \int_I \bigg(\sum_{i \in J} \chi_{T_i}(t) F(B_i^t)\bigg) dt$$

$$= \int_I \bigg(\sum_{i \in J_t} F(B_i^t)\bigg) dt = \int_I F(B_t)\, dt$$

$$\leq 2\eta + \eta \int_I \|\mathcal{D}U^t\|\, dt \leq 2\eta + \eta\|Dg\|(U).$$

As J is arbitrary, $\beta \leq 4\eta + 2\eta\|Dg\|(U)$. Adding the estimates yields

$$\sum_{i=1}^q |G(B_i)| = \sum_{i=1}^q \bigg|\int_I F(B_i^t)\, dt\bigg| < \alpha + \beta \leq \eta\Big[6 + 2\|Dg\|(U)\Big] < \varepsilon.$$

Since $\eta < \varepsilon$, we conclude

$$V_{*,\varepsilon} G(N) \leq G(\delta, \varepsilon) \leq G(\delta, \eta) \leq \varepsilon,$$

and the proposition follows by letting $\varepsilon \to 0$. $\square$

Proposition 4.5.6. *Let* $g \in BV_{\mathrm{loc}}^\infty(\mathbb{R}^m)$, *and let* F *be a charge. There is a negligible set* N, *depending only on* g, *such that the following condition is satisfied: given* $\varepsilon > 0$, *we can find a positive function* δ *defined on* $B(1/\varepsilon) - N$ *so that*

$$\sum_{i=1}^p \Big|g(x_i)F(A_i) - (F \mathbin{\llcorner} g)(A_i)\Big| < \varepsilon$$

whenever $A_i = B[x_i, r_i]$ *are disjoint balls with* $x_i \in B(1/\varepsilon) - N$ *and* $r_i < \delta(x_i)$ *for* $i = 1, \ldots, p$.

PROOF. In view of Theorem 1.8.12, no generality is lost by assuming $0 \leq g \leq 1$. Choose an $\varepsilon > 0$, and let $U := B(1/\varepsilon)$, $H := F \times \mathcal{L}^1$, and

$$\rho := 3 + (3 + m)|U| + (3 + \kappa)\|Dg\|(U)$$

where $\kappa := \kappa(m)$ is the constant from Proposition 4.4.11. According to Propositions 2.2.6, given $\eta := \varepsilon/\rho$, we can find a $\theta > 0$ so that

$$\big|H(C)\big| < \theta|C| + \eta\big(\|C\| + 1\big) \tag{$*$}$$

for every BV set $C \subset U \times [0,1]$. As $\nabla g \in L^1_{\text{loc}}(\mathbb{R}^m)$, Proposition 3.1.8 implies there is a positive function σ on U such that

$$\sum_{i=1}^{p} \left| |\nabla g(x_i)| \cdot |A_i| - \int_{A_i} |\nabla g(x)| \, dx \right| < 1$$

for each σ-fine partition $\{(A_1, x_1), \ldots, (A_p, x_p)\}$. Making σ smaller, we can assume $B[x, \sigma(x)] \subset U$ for each $x \in U$. Observe

$$\sum_{i=1}^{p} |\nabla g(x_i)| \cdot |A_i| < 1 + \sum_{i=1}^{p} \int_{A_i} |\nabla g(x)| \, dx \qquad (**)$$

$$\leq 1 + \sum_{i=1}^{p} \|Dg\|(A_i) \leq 1 + \|Dg\|(U)$$

for each σ-fine partition $\{(A_1, x_1), \ldots, (A_p, x_p)\}$. Let N and δ be, respectively, a negligible set and a positive function associated with g according to Proposition 4.4.11. Making δ smaller, we may assume

$$\delta(x) \leq \min\{\eta/\theta, \sigma(x)/2\}$$

for each $x \in U - N$. Let $A_i := B[x_i, r_i]$ be disjoint balls such that $x_i \in U - N$ and $r_i < \delta(x_i)$ for $i = 1, \ldots, p$. By the choice of δ, the sets

$$C_i^1 := A_i \times [0, g(x_i)] \quad \text{and} \quad C_i^2 := \Sigma_{g\chi_{A_i}}$$

are contained in $U \times [0,1]$. For $j, k = 1, 2$ and $I \subset \{1, \ldots, p\}$, Observation 4.4.1, $(**)$, Lemma 4.4.5, and Proposition 4.4.11 yield

$$\left| \bigcup_{i \in I} (C_i^j - C_i^k) \right| = \sum_{i \in I} |C_i^j - C_i^k| \leq \sum_{i=1}^{p} \int_{A_i} |g(y) - g(x_i)| \, dy \leq$$

$$\sum_{i=1}^{p} r_i \left(1 + |\nabla g(x_i)| \right) \cdot |A_i| < \frac{\eta}{\theta} \left[1 + |U| + \|Dg\|(U) \right],$$

$$\left\| \bigcup_{i \in I} (C_i^j - C_i^k) \right\| \leq \sum_{i \in I} \|C_i^j - C_i^k\| \leq$$

$$\sum_{i=1}^{p} \left[2|A_i| + \|Dg\|(A_i) + \int_{\partial A_i} |g^*(y) - g(x_i)| \, d\mathcal{H}^{m-1}(y) \right] <$$

$$2|U| + (1 + \kappa)\|Dg\|(U) + m \sum_{i=1}^{p} \left(1 + |\nabla g(x_i)| \right) \cdot |A_i| <$$

$$1 + (2 + m)|U| + (2 + \kappa)\|Dg\|(U) \right].$$

From these inequalities and $(*)$, we obtain

$$\sum_{i \in I} H(C_i^j - C_i^k) = H\left[\bigcup_{i \in I}(C_i^j - C_i^k)\right] < \eta\rho = \varepsilon .$$

The arbitrariness of I and the inequality

$$\left|H(C_i^1) - H(C_i^2)\right| \le \left|H(C_i^1 - C_i^2)\right| + \left|H(C_i^2 - C_i^1)\right|$$

imply

$$\sum_{i=1}^{p}\left|H(C_i^1) - H(C_i^2)\right| < 4\varepsilon .$$

Observing that $H(C_i^1) = g(x_i)F(A_i)$ and $H(C_i^2) = (F \mathbin{\text{L}} g)(A_i)$ completes the argument. □

Theorem 4.5.7. *Let $g \in BV_{\mathrm{loc}}^{\infty}(\mathbb{R}^m)$, and let F be a charge in a locally BV set E. If F belongs to $CH_*(E)$, then so does $F \mathbin{\text{L}} g$ and*

$$D(F \mathbin{\text{L}} g)(x) = DF(x)g(x)$$

for almost all $x \in E$.

PROOF. The charge $G := F \mathbin{\text{L}} g$ belongs to $CH_*(E)$ by Proposition 4.5.5. Let N be a negligible set that is associated with g according to Proposition 4.5.6. In view of Theorem 3.6.9, enlarging N, we may assume that the charges F and G are derivable at each $x \in E - N$.

Claim. Given $\varepsilon > 0$, there is a positive function δ defined on $B(1/\varepsilon) \cap (E - N)$ such that

$$\sum_{i=1}^{p}\left|DF(x_i)g(x_i)|A_i| - G(A_i)\right| < \varepsilon$$

whenever $A_i := B[x_i, r_i]$ are disjoint balls with $x_i \in B(1/\varepsilon) \cap (E - N)$, and $r_i \le \delta(x_i)$ for $i = 1, \dots, p$.

Proof. Given $\varepsilon > 0$, let $U := B(1/\varepsilon)$ and $c := 1 + |g\chi_U|_\infty$. By Proposition 4.5.6, there is a positive function δ defined on $U \cap (E - N)$ such that

$$\sum_{i=1}^{p}\left|g(x_i)F(A_i) - G(A_i)\right| < \frac{\varepsilon}{2}$$

whenever $A_i := B[x_i, r_i]$ are disjoint balls with $x_i \in U \cap (E - N)$ and $r_i \le \delta(x_i)$ for $i = 1, \dots, p$. Making δ smaller, we may assume that $B[x, \delta(x)] \subset U$ for each x in $U \cap (E - N)$ and that

$$\left|DF(x)|A| - F(A)\right| < \frac{\varepsilon}{2c|U|}|A|$$

for every ball $A := B[x, r]$ with $x \in U \cap (E - N)$ and $r \le \delta(x)$. Thus

$$\sum_{i=1}^{p} \Big| DF(x)g(x_i)|A_i| - G(A_i) \Big| \le$$

$$\sum_{i=1}^{p} |g(x_i)| \cdot \Big| DF(x_i)|A_i| - F(A_i) \Big| + \sum_{i=1}^{p} \Big| g(x_i)F(A_i) - G(A_i) \Big| <$$

$$\frac{\varepsilon}{2|U|} \sum_{i=1}^{p} |A_i| + \frac{\varepsilon}{2} \le \varepsilon ,$$

and the claim is proved.

Denote by S_k the set of all $x \in (E - N) \cap B(k)$ such that

$$\left| DF(x)g(x) - \lim_{r \to 0+} \frac{G(B[x, r])}{|B[x, r]|} \right| > \frac{1}{k} , \qquad (*)$$

$k = 1, 2, \ldots$. Fix an integer $k \ge 1$, and select a positive $\varepsilon < 1/k$. There is a positive function δ defined on $B(1/\varepsilon) \cap (E - N)$ such that the claim holds with ε replaced by ε/k. Let $\mathcal{B}$ be the collection of all balls $B[x, r]$ such that $x \in S_k$, $r < \delta(x)$, and

$$\Big| DF(x)g(x)|B[x, r]| - G(B[x, r]) \Big| > \frac{1}{k} |B[x, r]| .$$

In view of inequality $(*)$, the family $\mathcal{B}$ covers S_k. According to Theorem 1.4.1, there is a countable family of disjoint balls $B_i := B[x_i, r_i]$ in $\mathcal{B}$ for which $S_k \subset \bigcup_i B[x_i, 5r_i]$. Observe

$$\sum_i |B_i| < k \sum_i \Big| DF(x_i)g(x_i)|B_i| - G(B_i) \Big| \le \varepsilon$$

by the choice of δ, and hence $|S_k| < 5^m \varepsilon$. By the arbitrariness of ε, the set S_k is negligible, and so is $S := \bigcup_{k=1}^{\infty} S_k$. As $G'(x) = DF(x)g(x)$ for each $x \in E - (N \cup S)$, the theorem is proved. $\qquad \square$

For a locally BV set E, we define an algebra

$$BV_\ell^\infty(E) := \big\{ g \upharpoonright E : g \in BV_{\mathrm{loc}}^\infty(\mathbb{R}^m) \big\}, \qquad (4.5.1)$$

Since $\overline{f \upharpoonright E} = f\chi_E$ for each function f defined on $\mathbb{R}^m$ and each $E \subset \mathbb{R}^m$, a function g defined on a locally BV set E belongs to $BV_\ell^\infty(E)$ if and only if its zero extension $\bar{g}$ belongs to $BV_{\mathrm{loc}}^\infty(\mathbb{R}^m)$. If F is a charge and $g \in BV_\ell^\infty(E)$, we define a charge $F \,\mathsf{L}\, g$ in E by the formula

$$F \,\mathsf{L}\, g := F \,\mathsf{L}\, \bar{g} . \qquad (4.5.2)$$

Let E be a locally BV set. In view of Propositions 4.5.3 and 4.5.5,

$$\mathop{\mathsf{L}} : (F, g) \mapsto F \mathop{\llcorner} g : CH_*(E) \times BV_\ell^\infty(E) \to CH_*(E)$$

is a linear action of the algebra $BV_\ell^\infty(E)$ on the space $CH_*(E)$. By pointwise multiplication, the algebra $BV_\ell^\infty(E)$ acts, also linearly, on the space $M(E)$ of all measurable functions defined on E. According to Theorem 4.5.7 there is a commutative diagram

$$
\begin{array}{ccc}
CH_*(E) \times BV_\ell^\infty(E) & \xrightarrow{\ \ \mathsf{L}\ \ } & CH_*(E) \\
{\scriptstyle D \times id} \downarrow & & \downarrow {\scriptstyle D} \\
M(E) \times BV_\ell^\infty(E) & \xrightarrow{\ \ \cdot\ \ } & M(E)
\end{array}
$$

where D is the derivation, id is the identity map of $BV_\ell^\infty(E)$, and $\cdot$ denotes the pointwise multiplication.

Given a charge F, we define a Borel regular measure μ_F by a technique similar to the construction of the Hausdorff measures. For $E \subset \mathbb{R}^m$ and $\delta > 0$, let

$$\mu_F^\delta(E) := \inf_{\mathcal{B}} \sum_{B \in \mathcal{B}} |F(B)|$$

where $\mathcal{B}$ is a countable family of balls $B := B[x, r]$ with $r < \delta$ covering E. Letting

$$\mu_F(E) := \sup_{\delta > 0} \mu_F^\delta(E),$$

it is well known, and easy to prove, that the extended real-valued functions

$$\mu_F^\delta : E \mapsto \mu_F^\delta(E) \quad \text{and} \quad \mu_F : E \mapsto \mu_F(E)$$

defined for all $E \subset \mathbb{R}^m$ are measures [71, Chapter 1, Section 4, Theorem 15]. Moreover, μ_F is a Borel regular measure by [71, Chapter 1, Section 4, Theorem 20].

If F and G are charges, we say G is *derivable with respect to* F at $x \in \mathbb{R}^m$ if $F(B[x, r]) \neq 0$ for all sufficiently small $r > 0$, and there exists a finite limit

$$D^F G(x) := \lim_{r \to 0+} \frac{G(B[x, r])}{F(B[x, r])},$$

called the *derivate* of G at x *with respect to* F. If G is derivable at x, then it is derivable at x with respect to the charge $\mathcal{L}^m \mathop{\restriction} \mathcal{BV}$, but not vice versa.

Proposition 4.5.8. *Let* $g \in BV_{\text{loc}}^\infty(\mathbb{R}^m)$ *and let* F *be a charge. There is a negligible set* N, *depending only on* g, *such that for* μ_F-*almost all* $x \in \mathbb{R}^m - N$, *the charge* $F \mathop{\llcorner} g$ *is derivable at* x *with respect to* F *and*

$$D^F(F \mathop{\llcorner} g)(x) = g(x).$$

PROOF. Let $G := F \mathop{\llcorner} g$, and let N be a negligible set associated with g according to Proposition 4.5.6. Denote by S the set of all $x \in \mathbb{R}^m - N$ at which either $D^F G(x)$ does not exist or $D^F G(x) \neq g(x)$. For each $x \in S$ there is a $\beta(x) > 0$ such that given $\gamma > 0$, we can find a positive $r < \gamma$ with

$$\left| G(B[x, r]) - g(x) F(B[x, r]) \right| \geq \beta(x) \left| F(B[x, r]) \right|. \qquad (*)$$

Fix an integer $k \geq 1$, and let $S_k := \{x \in B(k) : \beta(x) \geq 1/k\}$. Choose a positive $\varepsilon < 1/k$ and $\delta > 0$. Using Proposition 4.5.6, find a positive function $\sigma \leq \delta$ defined on $B(1/\varepsilon) - N$ so that

$$\sum_{i=1}^{p} \left| g(x_i) F(A_i) - G(A_i) \right| < \frac{\varepsilon}{k} \qquad (**)$$

whenever $A_i := B[x_i, r_i]$ are disjoint balls with $x_i \in B(1/\varepsilon) - N$ and $r_i < \sigma(x_i)$ for $i = 1, \ldots, p$. By Theorem 1.4.5, there is a disjoint countable collection $B_i = B[y_i, s_i]$ with $y_i \in S_k$ and $s_i < \sigma(y_i)$ such that each $B[y_i, s_i]$ satisfies inequality $(*)$, and if $B := \bigcup_i B_i$ then $\mu_F(S_k - B) = 0$. In view of inequality $(**)$,

$$\sum_i \left| F(B_i) \right| \leq k \sum_i \left| g(y_i) F(B_i) - G(B_i) \right| \leq \varepsilon,$$

and consequently $\mu_F^\delta(S_k) \leq \mu_F^\delta(B) \leq \varepsilon$. From the arbitrariness of ε and δ we obtain $\mu_F(S_k) = 0$. Since $S = \bigcup_{k=1}^{\infty} S_k$, the proposition follows. $\square$

Fix an integer $n \geq 1$. Given $E \subset \mathbb{R}^{m+n}$ and $y \in \mathbb{R}^n$, let

$$E^y := \{x \in \mathbb{R}^m : (x, y) \in E\}.$$

If $h \in L_{\text{loc}}^\infty(\mathbb{R}^n)$ and $\mu := \mathcal{L}^n \llcorner h$ (see Remark 1.3.1), then an argument analogous to the proof of Theorem 4.1.1 shows that the function

$$F \times \mu : E \mapsto \int_{\mathbb{R}^n} F(E^y) \, d\mu(y) : \mathcal{BV}(\mathbb{R}^{m+n}) \to \mathbb{R}$$

is a charge in $\mathbb{R}^{m+n}$.

Conjecture 4.5.9. *Let $h \in L_{\text{loc}}^\infty(\mathbb{R}^n)$ and $\mu = \mathcal{L}^n \llcorner h$. If E is a locally BV set in $\mathbb{R}^m$ and $F \in CH_*(E)$, then the charge $F \times \mu$ belongs to $CH_*(E \times \mathbb{R}^n)$.*

While Conjecture 4.5.9 remains unproved at the time of this writing, I was informed by Buczolich that the affirmative answer is imminent. Thus it is worthwhile to mention a simple consequence.

Proposition 4.5.10. *Let E be a locally BV set in $\mathbb{R}^m$, let $F \in CH_*(E)$, and let $h \in L_{\text{loc}}^\infty(\mathbb{R}^n)$. If Conjecture 4.5.9 is correct, then*

$$D\big[F \times (\mathcal{L}^n \llcorner h)\big](x, y) = \big[(DF) \otimes h\big](x, y)$$

for $\mathcal{L}^{m+n}$-almost all points (x, y) in $E \times \mathbb{R}^n$.

PROOF. Let $H := \mathcal{L}^n \llcorner h$ and $G := F \times H$. By Theorems 3.1.9 and 3.6.9, there is an $\mathcal{L}^{m+n}$-negligible set $N \subset E \times \mathbb{R}^n$ such that $DF(x)$, $DG(x, y)$, and $DH(y) = h(y)$ exist for each (x, y) in $A := (E \times \mathbb{R}^n) - N$. Choose an $(x, y) \in A$ and for an $r > 0$, give $K[y, r]$ and $K\big[(x, y), r\big]$ the obvious meaning analogous to $K[x, r]$. Then

$$
\begin{aligned}
DG(x, y) &= \lim_{r \to 0+} \frac{G\Big(K\big[(x, y), r\big]\Big)}{\mathcal{L}^{m+n}\Big(K\big[(x, y), r\big]\Big)} \\[2mm]
&= \lim_{r \to 0+} \frac{F(K[x, r])}{|K[x, r]|} \cdot \lim_{r \to 0+} \frac{H(K[y, r])}{\mathcal{L}^n(K[y, r])} \\[2mm]
&= DF(x) DH(y) = DF(x) h(y).
\end{aligned}
$$
$\square$

4.6. Lipeomorphisms

If $\phi : E \to \mathbb{R}^m$ is a Lipschitz map of a set $E \subset \mathbb{R}^m$, we denote by $\bar{\phi}$ the unique extension of ϕ to $\mathrm{cl}\, E$ (cf. Theorem 1.6.5) and employ the following notation: for $x \in \mathrm{cl}\, E$ and $B \subset \mathrm{cl}\, E$, we let

$$x^* := \bar{\phi}(x) \quad \text{and} \quad B^* := \bar{\phi}(B).$$

There is no danger that this notation will be confused with our notation for dual spaces introduced in Section 4.3.

Suppose E is a locally BV set and $\phi : E \to \mathbb{R}^m$ is a lipeomorphism. According to Theorem 1.8.6, if $B \subset \mathrm{cl}\, E$ is a BV set or a locally BV set, then so is B^*, respectively. Given $x, y \in \mathrm{cl}\, E$ and a measurable set $B \subset \mathrm{cl}\, E$, inequality (1.6.1) and Theorem 1.6.5 yield

$$c^{-1}|x - y| \leq |x^* - y^*| \leq c|x - y|,$$
$$c^{-m}|B| \leq |B^*| \leq c^m|B|, \tag{4.6.1}$$
$$c^{1-m}\|B\| \leq \|B^*\| \leq c^{m-1}\|B\|,$$

for any positive $c \geq \max\{\mathrm{Lip}(\phi), \mathrm{Lip}(\phi^{-1})\}$; in particular

$$r(B^*) \geq c^{-2m} r(B) \tag{4.6.2}$$

whenever $B \in \mathcal{BV}(E)$. Given a function F defined on $\mathcal{BV}$, we let

$$\phi^* F(A) := F\big[(A \cap E)^*\big] \tag{4.6.3}$$

for each $A \in \mathcal{BV}$. Observe $\phi^* F = (\phi^* F) \, \mathsf{L}\, E = \phi^*(F \, \mathsf{L}\, E^*)$. If F is additive, then so is $\phi^* F$. Moreover, $\phi^* F$ is a charge, necessarily in E, whenever F is a charge (see Proposition 1.8.4).

Proposition 4.6.1. *Let E be a locally BV set, and let $\phi : E \to \mathbb{R}^m$ be a lipeomorphism. Suppose F is a function defined on $\mathcal{BV}$ that is derivable almost everywhere in $\phi(E)$. Then $\phi^* F$ is derivable almost everywhere in E and for almost all $x \in E$,*

$$D(\phi^* F)(x) = DF\big[\phi(x)\big] J_\phi(x)$$

PROOF. Define a charge H in E by letting $H(B) := \big|(E \cap B)^*\big|$ for each bounded BV set B. By the area theorem and Theorem 3.1.9,

$$DH(x) = J_\phi(x)$$

for almost all $x \in E$. Choose an $x \in \mathrm{int}_c E$ so that $DH(x) = J_\phi(x)$ and F is derivable at x^*. If $\{E_i\}$ is a sequence in $\mathcal{BV}(E)$ tending to x, then

$\{E_i^*\}$ is a sequence in $\mathcal{BV}$ tending to x^* according to (4.6.1) and (4.6.2). In view of Observation 2.3.6,

$$\lim \frac{\phi^* F(E_i)}{|E_i|} = \lim \left(\frac{F(E_i^*)}{|E_i^*|} \cdot \frac{H(E_i)}{|E_i|} \right) = DF(x^*) J_\phi(x).$$

Since $\phi^* F = (\phi^* F) \mathbin{\text{L}} E$, an application of Proposition 2.5.3 completes the demonstration.   □

Proposition 4.6.1 shows that the short exact sequence (3.3.1) is functorial on the category **Lip** of all locally BV sets and their lipeomorphisms. Indeed, given a locally BV set E and a lipeomorphism $\phi : E \to \mathbb{R}^m$, let

$$\phi^*(f)(x) := f\big[\phi(x)\big] J_\phi(x)$$

for each function f defined on $\phi(E)$, and observe that the following diagram of linear spaces and linear maps commutes:

$$
\begin{array}{ccccc}
CH_S\big[\phi(E)\big] & \xrightarrow{\ \subset\ } & CH_D\big[\phi(E)\big] & \xrightarrow{\ D\ } & PI\big[\phi(E)\big] \\
\phi^* \downarrow & & \phi^* \downarrow & & \phi^* \downarrow \\
CH_S(E) & \xrightarrow{\ \subset\ } & CH_D(E) & \xrightarrow{\ D\ } & PI(E)
\end{array}
$$

The next proposition shows that ϕ^* maps $CH_*\big[\phi(E)\big]$ into $CH_*(E)$.

Proposition 4.6.2. *Let $\phi : E \to \mathbb{R}^m$ be a lipeomorphism of a locally BV set E. Suppose F is a function defined on $\mathcal{BV}$, and let B be a subset of $\mathrm{cl}_* E$. If $F = F \mathbin{\text{L}} \phi(E)$, then*

$$V_*(\phi^* F)(B) = V_* F\big[\bar{\phi}(B)\big].$$

PROOF. Seeking a contradiction, suppose $V_* F(B^*) < V_*(\phi^* F)(B)$ and select a positive $c \geq \max\{\mathrm{Lip}(\phi), \mathrm{Lip}(\phi^{-1})\}$. By Proposition 3.6.10, there are an $\eta > 0$ and a gage δ^* on B^* such that

$$F(\delta^*, \eta^*; E^*) < (\phi^* F)(\delta, \eta; E)$$

where $\eta^* := \eta/c^{2m}$ and $\delta : x \mapsto \delta^*(x^*)/c$ is a gage on B; indeed, the set $\{\delta = 0\} = \phi^{-1}\big(\{\delta^* = 0\}\big)$ is thin by (1.6.1). Find an η-regular δ-fine partition $P := \big\{(A_1, x_1), \ldots, (A_p, x_p)\big\}$ with $[P] \subset E$ and

$$F(\delta^*, \eta^*; E^*) < \sum_{i=1}^p \big|\phi^* F(A_i)\big| = \sum_{i=1}^p \big|F(A_i^*)\big|;$$

since $F = F \mathbin{\text{L}} E^*$ and $(A_i \cap E)^* = A_i^* \cap E^*$. A contradiction follows: (4.6.1) and (4.6.2) imply that $P^* := \big\{(A_1^*, x_1^*), \ldots, (A_p^*, x_p^*)\big\}$ is an

η^*-regular δ^*-fine partition with $[P^*] \subset E^*$. Therefore

$$V_*(\phi^*F)(B) \leq V_*F(B^*),$$

and the reverse inequality is obtained by symmetry. $\qquad\square$

Using symmetry and combining Propositions 4.6.1 and 4.6.2 with Theorem 3.6.9, we obtain the following theorem.

Theorem 4.6.3. *Let $\phi : E \to \mathbb{R}^m$ be a lipeomorphism of a locally BV set E. A charge F belongs to $CH_*[\phi(E)]$ if and only if the charge ϕ^*F belongs to $CH_*(E)$, in which case*

$$D(\phi^*F)(x) = DF[\phi(x)]J_\phi(x)$$

for almost all $x \in E$.

Theorem 4.6.3 is equivalent to the commutativity of the diagram

$$\begin{array}{ccc}
CH_*[\phi(E)] & \xrightarrow{\;D\;} & M[\phi(E)] \\
{\scriptstyle\phi^*}\big\downarrow & & \big\downarrow{\scriptstyle\phi^*} \\
CH_*(E) & \xrightarrow{\;D\;} & M(E)
\end{array} \qquad (4.6.4)$$

where the derivation D is an injective map.

Definition 4.6.4. A Lipschitz map $\phi : E \to \mathbb{R}^m$ of a set $E \subset \mathbb{R}^m$ is called a *local lipeomorphism* if it satisfies the following conditions:

(i) each $x \in \mathrm{cl}\, E$ has a neighborhood U such that $\phi : E \cap U \to \mathbb{R}^m$ is a lipeomorphism;

(ii) ϕ is a *proper map*, i.e., $\bar{\phi}^{-1}(K)$ is compact whenever $K \subset \mathbb{R}^m$ is compact.

Note that the definition of a local lipeomorphism ϕ, which is convenient for our purposes, is not entirely local. Indeed, it combines a local condition (i) with two global conditions: ϕ is Lipschitz and proper. Clearly, if ϕ is a local lipeomorphism, then so is $\bar{\phi}$.

It is easy to see that **Lip** is a subcategory of the category **Lip**$_{\mathrm{loc}}$ of all locally BV sets and their local lipeomorphisms. Our immediate goal is to extend Theorem 3.7.4 from **Lip** to **Lip**$_{\mathrm{loc}}$, and give it a clear geometric interpretation.

Observation 4.6.5. *Let $\phi : E \to \mathbb{R}^m$ be a local lipeomorphism of a locally BV set E. For each compact set $K \subset \mathrm{cl}\, E$ there is an integer $n \geq 1$ such that $\bar{\phi} \upharpoonright K$ is at most an n-to-one map. In particular, $\bar{\phi}$ is a finite-to-one map.*

PROOF. By condition (i) of Definition 4.6.4, a compact set $K \subset \mathrm{cl}\, E$ is the union of sets $A_1, \ldots, A_n$ such that each restriction $\bar{\phi} \restriction A_i$ is a one-to-one map. As ϕ is proper, the set $\bar{\phi}^{-1}(y) \subset \mathrm{cl}\, E$ is compact for every $y \in \mathbb{R}^m$. □

Proposition 4.6.6. *Let $\phi : E \to \mathbb{R}^m$ be a local lipeomorphism of a set $E \subset \mathbb{R}^m$. If $B \subset E$ is a bounded BV set or a locally BV set, then so is $\phi(B)$, respectively.*

PROOF. Let $B \subset E$ be a bounded BV set. By condition (i) of Definition 4.6.4, there are BV sets $B_1, \ldots, B_p$ such that $B = \bigcup_{i=1}^p B_i$ and $\phi : B_i \to \mathbb{R}^m$ is a lipeomorphism. Thus $\phi(B) = \bigcup_{i=1}^p \phi(B_i)$ is a BV set according to Theorem 1.8.6 and Proposition 1.8.4. As ϕ is Lipschitz, the set $\phi(B)$ is bounded.

Let $B \subset E$ be a locally BV set. Observe that $\psi := \phi \restriction B$ is again a local lipeomorphism, and choose an $r > 0$. Since ψ is proper, the set $\psi^{-1}[\psi(B) \cap B(r)]$ is contained in a bounded BV set $B \cap B(s)$ for an $s > 0$. According to Proposition 1.8.5 and the previous paragraph, $A := \psi[B \cap B(s)]$ is a BV subset of $\psi(B)$ containing $\psi(B) \cap B(r)$. Therefore

$$\psi(B) \cap B(r) = A \cap \psi(B) \cap B(r) = A \cap B(r)$$

is a BV set, and the proposition follows from the arbitrariness of r. □

If $\phi : E \to \mathbb{R}^m$ is a local lipeomorphism of a locally BV set E and F is a charge, then Proposition 4.6.6 allows us to define $\phi^* F$ by formula (4.6.3). Unfortunately, the function $\phi^* F$ need not be additive, since ϕ is generally not injective. To obtain the correct transformation formula for charges (a *pullback*), we must interpret bounded BV sets as *currents* — a very special case of the general theory developed by Federer and Fleming [25]. In what follows we **make no attempt to define currents**. We merely present some calculations which should enhance the intuition of readers unfamiliar with this area of geometric measure theory. A comprehensive discussion of currents can be found in [24, Chapter 4]; for a less complete but probably more accessible treatment we refer to [56, Chapter 4] and [77, Chapter 6].

As usual, we denote by $dx_1, \ldots, dx_m$ the dual base of the standard base $\mathbf{e}_1, \ldots, \mathbf{e}_m$ in $\mathbb{R}^m$. For an integer k with $1 \leq k \leq m$, the linear space generated by all elements $dx_{i_1} \wedge \cdots \wedge dx_{i_k}$ where $i_j = 1, \ldots, m$ and $j = 1, \ldots, k$ is denoted by $\wedge^k \mathbb{R}^m$. Let $\wedge^0 \mathbb{R}^m := \mathbb{R}$, and denote by

dx the generator $dx_1 \wedge \cdots \wedge dx_m$ of $\wedge^m \mathbb{R}^m$. The graded algebra

$$\wedge^* \mathbb{R}^m := \bigoplus_{k=0}^{m} \wedge^k \mathbb{R}^m$$

is discussed briefly in [79, Chapter 4]; more details can be found in [24, Chapter 1]. As the space $\mathbb{R}^m$ remains fixed throughout the exposition, we abbreviate $\wedge^k \mathbb{R}^m$ and $\wedge^* \mathbb{R}^m$ as $\wedge^k$ and $\wedge^*$, respectively.

Let E be a measurable set. For $k = 0, \ldots, m$, the linear spaces

$$\wedge_M^k(E) := M(E; \wedge^k) \quad \text{and} \quad \wedge_\infty^k(E) := L^\infty(E; \wedge^k)$$

are defined according to Convention 1.1.1, since $\wedge^k$ is a finite-dimensional vector space. Equipped with the discrete topology, $\wedge_\infty^k(E)$ is a locally convex space, whose dual is denoted by $\wedge_\infty^k(E)^*$ — cf. Section 4.3. Observe that elements of $\wedge_M^m(E)$ are forms

$$\omega := a\, dx = a\, dx_1 \wedge \cdots \wedge dx_m$$

where a is a measurable function defined on E.

Clearly, the map $a \mapsto a\, dx$ is an isomorphism from $M(\mathbb{R}^m)$ onto $\wedge_M^m(\mathbb{R}^m)$. Notwithstanding, there is a conceptual difference between a and $a\, dx$ demonstrated, e.g., by the transformation formula (4.6.5) below.

Let $\phi : E \to \mathbb{R}^m$ be a Lipschitz map of a measurable set E. Motivated by differentiable forms [79, Theorem 4-9], we introduce a linear map $\phi^\# : \wedge_M^m(\mathbb{R}^m) \to \wedge_M^m(E)$ defined by the formula

$$\phi^\#(a\, dx) := (a \circ \phi) \det D\phi\, dx \qquad (4.6.5)$$

for every $a\, dx \in \wedge_M^m(\mathbb{R}^m)$. Note $\phi^\#$ maps $\wedge_\infty^m(\mathbb{R}^m)$ into $\wedge_\infty^m(E)$. If

$$\phi_\# : \wedge_\infty^m(E)^* \to \wedge_\infty^m(\mathbb{R}^m)^*$$

is the dual map of $\phi^\# : \wedge_\infty^m(\mathbb{R}^m) \to \wedge_\infty^m(E)$, then

$$\langle \phi_\# T, \omega \rangle := \langle T, \phi^\# \omega \rangle \qquad (4.6.6)$$

for each $T \in \wedge_\infty^m(E)^*$ and each $\omega \in \wedge_\infty^m(\mathbb{R}^m)$.

Every function $g \in L^1(\mathbb{R}^m)$ defines a linear functional

$$\mathbf{E}^m \, \llcorner \, g : a\, dx \mapsto \int_{\mathbb{R}^m} a(x)g(x)\, dx : \wedge_\infty^m(\mathbb{R}^m) \to \mathbb{R},$$

called the *m-current* associated with g [24, Chapter 4]. By definition, $\mathbf{E}^m \, \llcorner \, g$ belongs to $\wedge_\infty^m(\mathbb{R}^m)^*$, and the following proposition shows that if $\phi : \mathbb{R}^m \to \mathbb{R}^m$ is a Lipschitz map, then $\phi_\#(\mathbf{E}^m \, \llcorner \, g)$ is not just an element of $\wedge_\infty^m(\mathbb{R}^m)^*$, but another m-current.

Proposition 4.6.7. *Let $\phi : \mathbb{R}^m \to \mathbb{R}^m$ be a Lipschitz map. If g belongs to $L^1(\mathbb{R}^m)$ or $L_c^1(\mathbb{R}^m)$, then so does, respectively, the function*

$$\deg(\phi, g)(y) := \sum_{x \in \phi^{-1}(y)} g(x) \operatorname{sign} \det D\phi(x)$$

defined for almost all $y \in \mathbb{R}^m$. Moreover,

$$\phi_\#\big(\mathbf{E}^m \, \llcorner \, g\big) = \mathbf{E}^m \, \llcorner \, \deg(\phi, g).$$

PROOF. Since the function $g \operatorname{sign} \det D\phi$ is in $L^1(\mathbb{R}^m)$, the area theorem implies that so is $\deg(\phi, g)$. As $\big\{\deg(\phi, g) \neq 0\big\} \subset \phi(\{g \neq 0\})$, the function $\deg(\phi, g)$ belongs to $L_c^1(\mathbb{R}^m)$ whenever g does. If $T := \mathbf{E}^m \, \llcorner \, g$ and $\omega := a \, dx$ is in $\wedge_\infty^m(\mathbb{R}^m)$, then (4.6.6) and the area theorem yield

$$\langle \phi_\# T, \omega \rangle = \int_{\mathbb{R}^m} a\big[\phi(x)\big] \det D\phi(x) \, g(x) \, dx$$

$$= \int_{\mathbb{R}^m} a\big[\phi(x)\big] g(x) \operatorname{sign} \det D\phi(x) J_\phi(x) \, dx$$

$$= \int_{\mathbb{R}^m} \left[\sum_{x \in \phi^{-1}(y)} a\big[\phi(x)\big] g(x) \operatorname{sign} \det D\phi(x) \right] dy$$

$$= \int_{\mathbb{R}^m} a(y) \left[\sum_{x \in \phi^{-1}(y)} g(x) \operatorname{sign} \det D\phi(x) \right] dy$$

$$= \int_{\mathbb{R}^m} a(y) \deg(\phi, g)(y) \, dy = \big\langle \mathbf{E}^m \, \llcorner \, \deg(\phi, g), \omega \big\rangle.$$

The proposition follows from the arbitrariness of ω. $\qquad\square$

The next theorem is presented without proof; the reader interested in proving it is referred to [24, Sections 4.1 and 4.5].

Theorem 4.6.8. *Let $\phi : \mathbb{R}^m \to \mathbb{R}^m$ be a Lipschitz map. If g belongs to $BV_c(\mathbb{R}^m)$, then so does $\deg(\phi, g)$. In addition,*

$$\big|\deg(\phi, g)\big|_1 \leq c^m |g|_1 \quad and \quad \big\|\deg(\phi, g)\big\| \leq c^{m-1} \|g\|.$$

where c is the Lipschitz constant of ϕ.

Let E be a locally BV set. We denote by $\mathbf{N}_m^{c,\infty}(E)$ the linear space of all m-currents $\mathbf{E}^m \, \llcorner \, g$ with $g \in BV_b^\infty(E)$. The topology $\mathcal{T}$ in $BV_b^\infty(E)$ determines a unique topology in $\mathbf{N}_m^{c,\infty}$, still denoted by $\mathcal{T}$, such that the linear bijection $g \mapsto \mathbf{E}^m \, \llcorner \, g$ becomes a homeomorphism.

Although the map $g \mapsto \mathbf{E}^m \, \llcorner \, g$ is an isomorphism, there is again a conceptual difference between g and $\mathbf{E}^m \, \llcorner \, g$. The reader acquainted with currents will recognize that $\mathbf{N}_m^{c,\infty}(E)$ is a subspace of the space $\mathbf{N}_m^{loc}(\mathbb{R}^m)$ of m-dimensional *normal currents* in $\mathbb{R}^m$ discussed in [24, Section 4.5]. While in this section it would suffice to consider only *integral currents* (see ibid.), we introduced normal currents, since they will be used in Section 5.3 below.

By Kirszbraun's theorem, a Lipschitz map $\phi : E \to \mathbb{R}^m$ has a Lipschitz extension $\psi : \mathbb{R}^m \to \mathbb{R}^m$. Given $g \in BV_b^\infty(E)$, we define

$$\deg(\phi, g) := \deg(\psi, g) \quad \text{and} \quad \phi_\#(\mathbf{E}^m \, \mathsf{L} \, g) := \psi_\#(\mathbf{E}^m \, \mathsf{L} \, g).$$

Note that both $\deg(\phi, g)$ and $\phi_\#(\mathbf{E}^m \, \mathsf{L} \, g) = \mathbf{E}^m \, \mathsf{L} \deg(\phi, g)$ depend only on ϕ and not on the extension ψ; since $\{g \neq 0\} \subset E$.

If A is a bounded BV subset of E, we write $\mathbf{E}^m \, \mathsf{L} \, A$ and $\deg(\phi, A)$ instead of $\mathbf{E}^m \, \mathsf{L} \, \chi_A$ and $\deg(\phi, \chi_A)$, respectively. As

$$\deg(\phi, A)(y) = \sum_{x \in A \cap \phi^{-1}(y)} \text{sign} \det D\phi(x) \tag{4.6.7}$$

for almost all $y \in \mathbb{R}^m$, we see that almost everywhere $\deg(\phi, A)$ is the *oriented multiplicity* of $\phi(A)$.

Proposition 4.6.9. *Let* $\phi : E \to \mathbb{R}^m$ *be a local lipeomorphism of a locally BV set* E. *If g belongs to* $BV_b^\infty(E)$, *then* $\deg(\phi, g)$ *belongs to* $BV_b^\infty[\phi(E)]$, *and the map*

$$g \mapsto \deg(\phi, g) : BV_b^\infty(E) \to BV_b^\infty[\phi(E)]$$

is linear and $\mathfrak{T}$-continuous. In particular, the linear map

$$\phi_\# : \mathbf{E}^m \, \mathsf{L} \, g \mapsto \mathbf{E}^m \, \mathsf{L} \deg(\phi, g) : \mathbf{N}_m^{c,\infty}(E) \to \mathbf{N}_m^{c,\infty}(\mathbb{R}^m)$$

is $\mathfrak{T}$-continuous and $\phi_\#[\mathbf{N}_m^{c,\infty}(E)] \subset \mathbf{N}_m^{c,\infty}[\phi(E)]$.

PROOF. According to Theorem 4.6.8, the function $\deg(\phi, g)$ belongs to $BV_c(\mathbb{R}^m)$. Observation 4.6.5 implies there is an integer $n \geq 1$ such that $\phi \restriction \{g \neq 0\}$ is at most an n-to-one map. Thus $\left|\deg(\phi, g)\right|_\infty \leq n|g|_\infty$, and we conclude that $\deg(\phi, g)$ belongs to $BV_b^\infty[\phi(E)]$.

Employing Observation 4.6.5 and the estimates of Theorem 4.6.8, it is easy to verify the sequence $\{\deg(\phi, g_i)\}$ in $BV_b^\infty[\phi(E)]$ $\mathfrak{T}$-converges to zero whenever the sequence $\{g_i\}$ in $BV_b^\infty(E)$ does. $\square$

Let E be a locally BV set, and let

$$\langle F, \mathbf{E}^m \, \mathsf{L} \, g \rangle := \langle F, g \rangle \tag{4.6.8}$$

for each $\mathbf{E}^m \, \mathsf{L} \, g \in \mathbf{N}_m^{c,\infty}(E)$ and each $F \in CH(E)$. In spite of its formal appearance, formula (4.6.8) serves a useful purpose: as charges become functions of m-currents, they can be pulled back by local lipeomorphisms (see Proposition 4.6.10 below).

By Theorem 4.1.5, for every $F \in CH(E)$, the linear functional

$$\mathbf{E}^m \mathbin{\llcorner} g \mapsto \langle F, \mathbf{E}^m \mathbin{\llcorner} g \rangle : \left(\mathbf{N}_m^{c,\infty}(E), \mathcal{T}\right) \to \mathbb{R}$$

is continuous, and it follows from Theorem 4.3.1 that all continuous linear functionals on $\left(\mathbf{N}_m^{c,\infty}(E), \mathcal{T}\right)$ have this form. Consequently, we identify $CH(E)$ with the space $\mathbf{N}_m^{c,\infty}(E)^*$ dual to $\left(\mathbf{N}_m^{c,\infty}(E), \mathcal{T}\right)$.

Using Example 2.1.3, associate with each form $\omega := a\,dx$ in $\wedge_\infty^m(E)$ a charge $F_\omega := \int a(x)\,dx$ in E. If the function $g \in BV_b^\infty(E)$ is nonnegative, then (4.6.8) and Fubini's theorem yield

$$\langle F_\omega, \mathbf{E}^m \mathbin{\llcorner} g \rangle = \int_0^\infty F_\omega\big(\{g > t\}\big)\,dt = \int_0^\infty \left(\int_{\{g>t\}} a(x)\,dx \right) dt$$

$$= \int_{\mathbb{R}^m} \left(\int_0^{g(x)} a(x)\,dt \right) dx = \int_{\mathbb{R}^m} g(x)a(x)\,dx$$

$$= \langle \mathbf{E}^m \mathbin{\llcorner} g, \omega \rangle .$$

By linearity, the formula

$$\langle F_\omega, \mathbf{E}^m \mathbin{\llcorner} g \rangle = \langle \mathbf{E}^m \mathbin{\llcorner} g, \omega \rangle \tag{4.6.9}$$

holds for each $\omega \in \wedge_\infty^m(E)$ and each $\mathbf{E}^m \mathbin{\llcorner} g \in \mathbf{N}_m^{c,\infty}(E)$. This implies that the map

$$\omega \mapsto F_\omega : \wedge_\infty^m(E) \to CH(E)$$

corresponds to the canonical injection of $\wedge_\infty^m(E)$ into its second dual.

Now let $\phi : E \to \mathbb{R}^m$ be a local lipeomorphism, and let

$$\phi_\# : \mathbf{E}^m \mathbin{\llcorner} g \mapsto \mathbf{E}^m \mathbin{\llcorner} \deg(\phi, g)$$

be the linear map from $\mathbf{N}_m^{c,\infty}(E)$ to $\mathbf{N}_m^{c,\infty}(\mathbb{R}^m)$, whose $\mathcal{T}$-continuity was established in Proposition 4.6.9. The dual map of $\phi_\#$ is the linear map

$$\phi^\# : CH(\mathbb{R}^m) \to CH(E)$$

defined by the identity

$$\langle \phi^\# F, \mathbf{E}^m \mathbin{\llcorner} g \rangle := \langle F, \phi_\#(\mathbf{E}^m \mathbin{\llcorner} g) \rangle \tag{4.6.10}$$

for each $F \in CH(\mathbb{R}^m)$ and each $\mathbf{E}^m \mathbin{\llcorner} g \in \mathbf{N}_m^{c,\infty}(E)$.

A word of caution is in order: since we have denoted by the same symbol $\phi^\#$ two different maps, we must show that this leads to no confusion. If $\omega \in \wedge_\infty^m(\mathbb{R}^m)$ and $T := \mathbf{E}^m \mathbin{\llcorner} g$ is in $\mathbf{N}_m^{c,\infty}(E)$, then

$$\langle \phi^\# F_\omega, T \rangle = \langle F_\omega, \phi_\# T \rangle = \langle \phi_\# T, \omega \rangle = \langle T, \phi^\# \omega \rangle = \langle F_{\phi^\#\omega}, T \rangle$$

by equalities (4.6.10), (4.6.9), and (4.6.6). Since the arbitrariness of T yields $\phi^\# F_\omega = F_{\phi^\# \omega}$, there is a commutative diagram

$$
\begin{array}{ccc}
\wedge_\infty^m(\mathbb{R}^m) & \longrightarrow & CH(\mathbb{R}^m) \\
\phi^\# \downarrow & & \downarrow \phi^\# \\
\wedge_\infty^m(E) & \longrightarrow & CH(E)
\end{array}
$$

where the horizontal maps are the canonical injections $\omega \mapsto F_\omega$. This shows the map $\phi^\#$ on the right is, in a sense, an extension of the map $\phi^\#$ on the left. Consequently denoting both maps by the same symbol is acceptable and, in fact, customary.

Proposition 4.6.10. *Let* $\phi : E \to \mathbb{R}^m$ *be a local lipeomorphism of a locally BV set E. If F is a charge, then*

$$
\phi^\# F(A) = \phi^\# \big[F \, \mathsf{L} \, \phi(E) \big] = \big\langle F, \deg (\phi, A \cap E) \big\rangle
$$

for each bounded BV set A, and the linear map

$$
F \mapsto \phi^\# F : CH(\mathbb{R}^m) \to CH(E)
$$

is $\mathbf{S}$-continuous. If $E \in \mathcal{BV}$ and ϕ is a lipeomorphism, then

$$
\phi^\# F = \phi^* \big[F \, \mathsf{L} \deg (\phi, E) \big] .
$$

PROOF. Let A be a bounded BV set, and let $B = A \cap E$. Equalities (4.6.8) and (4.6.10) together with Proposition 4.6.9 imply

$$
\begin{aligned}
\phi^\# F(A) = \phi^\# F(B) &= \big\langle \phi^\# F, \chi_B \big\rangle = \big\langle \phi^\# F, \mathbf{E}^m \, \mathsf{L} \, B \big\rangle \\
&= \big\langle F, \phi_\#(\mathbf{E}^m \, \mathsf{L} \, B) \big\rangle = \big\langle F, \mathbf{E}^m \, \mathsf{L} \deg (\phi, B) \big\rangle \\
&= \big\langle F, \deg (\phi, B) \big\rangle = \big\langle F, \deg (\phi, A \cap E) \big\rangle .
\end{aligned}
$$

In particular,

$$
\begin{aligned}
\phi^\# \big[F \, \mathsf{L} \, \phi(E) \big](A) &= \big\langle F \, \mathsf{L} \, \phi(E), \deg (\phi, A \cap E) \big\rangle \\
&= \big\langle F, \deg (\phi, A \cap E) \big\rangle = \phi^\# F(A) ,
\end{aligned}
$$

since $\big\{ \deg (\phi, C) \neq 0 \big\} \subset \phi(E)$ for each bounded BV set $C \subset E$.

Fix an integer $p \geq 1$ and a set $A \in \mathcal{BV}_p(\mathbb{R}^m)$. Let

$$
h := \deg(\phi, A \cap E) .
$$

Employing Observation 4.6.5, Theorem 4.6.8, and inequality (4.1.1), it is easy to find a positive integer $n = n(m, p, \phi)$, such that $\Sigma_{h\pm}$ belongs to $\mathcal{BV}_n(\mathbb{R}^{m+1})$. By the first part of the proof,

$$\left|\phi^\# F(A)\right| = \left|\langle F, h \rangle\right| \leq \left|(F \times \mathcal{L}^1)(\Sigma_{h+})\right| + \left|(F \times \mathcal{L}^1)(\Sigma_{h-})\right|$$
$$\leq \|F \times \mathcal{L}^1\|_n.$$

As n does not depend on A, we infer $\|\phi^\# F\|_p \leq \|F \times \mathcal{L}^1\|_n$, which implies the map $F \times \mathcal{L}^1 \mapsto \phi^\# F$ is continuous. The continuity of the map $F \mapsto \phi^\# F$ follows from Theorem 4.1.1.

If $E \in \mathcal{BV}$ and ϕ is a lipeomorphism, then (4.6.7) yields

$$\deg(\phi, A)(y) = \begin{cases} \text{sign} \det D\phi\left[\phi^{-1}(y)\right] & \text{if } y \in \phi(A), \\ 0 & \text{otherwise} \end{cases}$$

for each BV set $A \subset E$ and almost all $y \in \mathbb{R}^m$; in particular, this is true for $A = E$. Consequently for each BV set $A \subset E$,

$$\deg(\phi, A) = \deg(\phi, E)\chi_{\phi(A)}.$$

From this, the first part of the proof, and (4.6.3), the proposition follows by a direct calculation. $\qquad\square$

Let $\phi : E \to \mathbb{R}^m$ be a local lipeomorphism of a locally BV set E. If F is a charge, we can prove directly that the map

$$\phi^\# F : A \mapsto \langle F, \deg(\phi, A \cap E) \rangle : \mathcal{BV}(\mathbb{R}^m) \to \mathbb{R}$$

is a charge in E. Indeed, by (4.6.7),

$$\deg(\phi, A) + \deg(\phi, B) = \deg(\phi, A \cup B)$$

for each disjoint pair A, B of bounded BV subsets of E. Thus $\phi^\# F$ is an additive function on $\mathcal{BV}(\mathbb{R}^m)$ with $\phi^\# F = (\phi^\# F) \mathbin{\mathsf{L}} E$. If $\{A_i\}$ is a sequence of bounded BV sets such that $\{A_i\} \to \emptyset$, then $\{A_i \cap E\} \to \emptyset$ by Proposition 1.8.5. It follows from Theorem 4.6.8 and Observation 4.6.5 that the sequence $\{\deg(\phi, A_i \cap E)\}$ $\mathcal{T}$-converges to zero. Hence $\lim \phi^\# F(A_i) = 0$ according to Theorem 4.1.5, and we conclude $\phi^\# F$ is a charge.

While the above proof may appear simpler than that used in proving Proposition 4.6.10, it gives no motivation for the definition of $\phi^\# F$.

Lemma 4.6.11. *Let $\phi : E \to \mathbb{R}^m$ be a local lipeomorphism of a locally BV set E. If $B \subset E$ is a locally BV set and F is a charge, then*

$$(\phi^\# F) \mathbin{\mathsf{L}} B = (\phi \restriction B)^\# F.$$

PROOF. Let A be a bounded BV set. Since $(\phi \restriction B)^{-1}(y) = B \cap \phi^{-1}(y)$ for each $y \in \mathbb{R}^m$, equality (4.6.7) yields

$$\deg(\phi \restriction B, A \cap B) = \deg(\phi, A \cap B),$$

and the lemma follows from Proposition 4.6.10. $\qquad\square$

Theorem 4.6.12. *Let $\phi : E \to \mathbb{R}^m$ be a local lipeomorphism of a locally BV set E. If F is a charge in $CH_*\big[\phi(E)\big]$, then the charge $\phi^\# F$ belongs to $CH_*(E)$ and*

$$D(\phi^\# F)(x) = DF\big[\phi(x)\big] \det D\phi(x)$$

for almost all $x \in E$.

PROOF. As the space $\mathbb{R}^m$ is hereditarily Lindelöf [21, Theorem 3.8.1], we can cover $\mathrm{cl}\, E$ by open balls $B_1, B_2, \ldots$ so that ϕ is a lipeomorphism on each $B_i \cap E$. For $i = 1, 2, \ldots$, let

$$\phi_i := \phi \restriction (B_i \cap E) \quad \text{and} \quad F_i := F \, \mathsf{L} \, \phi_i(B_i \cap E).$$

By Corollary 3.6.11, the charge F_i belongs to $CH_*\big[\phi_i(B_i \cap E)\big]$, and according to Proposition 4.6.9 and Theorem 4.5.7, so does the charge $F_i \, \mathsf{L} \deg(\phi_i, B_i \cap E)$. It follows from Proposition 4.6.10 and Theorem 4.6.3 that $\phi_i^\# F_i \in CH_*(B_i \cap E)$.

Since $G := \phi^\# F$ is a charge in E, Lemma 4.6.11 and Proposition 4.6.10 imply that $G \, \mathsf{L} \, B_i = \phi_i^\# F_i$. Thus each $G \, \mathsf{L} \, B_i$ is a charge in $CH_*(B_i \cap E)$. Choose a negligible set $N \subset \mathrm{cl}_* E$. Since each B_i is an open set, it is easy to see that

$$V_* G(N \cap B_i) = V_*(G \, \mathsf{L} \, B_i)(N \cap B_i) = 0$$

for $i = 1, 2, \ldots$. As $N \subset \bigcup_i B_i$, we conclude $G \in CH_*(E)$.

Fix a ball B_i, and choose an $x \in B_i \cap E$ so that $DG(x)$ and

$$DF\big[\phi(x)\big] = DF_i\big[\phi_i(x)\big]$$

exist. In view of the first part of the proof, Theorem 3.6.9, and Corollary 2.5.4 almost all $x \in B_i \cap E$ have these properties. As B_i is an open set, $DG(x) = D(G \, \mathsf{L} \, B_i)(x)$. By Proposition 4.6.10 and Theorem 4.6.3,

$$
\begin{aligned}
DG(x) &= D(\phi_i^\# F_i)(x) = D\Big(\phi_i^*\big[F_i \, \mathsf{L} \deg(\phi_i, B_i \cap E)\big]\Big)(x) \\
&= D\big[F_i \, \mathsf{L} \deg(\phi_i, B_i \cap E)\big]\big(\phi_i(x)\big) J_{\phi_i}(x) \\
&= DF_i\big[\phi_i(x)\big] \deg(\phi_i, B_i \cap E)\big(\phi_i(x)\big) J_{\phi_i}(x) \\
&= DF_i\big[\phi(x)\big] \operatorname{sign} \det D\phi_i(x) J_\phi(x) \\
&= DF\big[\phi(x)\big] \det D\phi(x).
\end{aligned}
$$

As $E = \bigcup_i(B_i \cap E)$, the theorem follows. $\square$

For a locally BV set E, the linear map

$$\mathfrak{D} : F \mapsto DF\, dx : CH_{AD}(E) \to \wedge_M^m(E)$$

is called the *geometric derivation*. While superficially there is little difference between derivation and geometric derivation, the reader should keep in mind that the geometric derivate is an m-form, which is a geometric object conceptually different from a measurable function.

Observe that Theorem 4.6.12 is equivalent to the commutativity of the diagram

$$
\begin{array}{ccc}
CH_*\big[\phi(E)\big] & \xrightarrow{\ \mathfrak{D}\ } & \wedge_M^m\big[\phi(E)\big] \\[4pt]
{\scriptstyle\phi^{\#}}\big\downarrow & & \big\downarrow{\scriptstyle\phi^{\#}} \\[4pt]
CH_*(E) & \xrightarrow{\ \mathfrak{D}\ } & \wedge_M^m(E)
\end{array}
\tag{4.6.11}
$$

where the geometric derivation $\mathfrak{D}$ is an injective map. This, in turn, is equivalent to the following statement in the language of categories:

If CH_* and $\wedge_M^m$ denote the obvious contravariant functors from the category $\mathbf{Lip}_{\mathrm{loc}}$ to the category of linear spaces and linear maps, then

$$
\mathfrak{D} : CH_* \to \wedge_M^m
$$

is a natural transformation [78, Chapter 1, Section 2].

It is possible to employ more general maps than lipeomorphisms and local lipeomorphisms. A proper Lipschitz map $\phi : E \to \mathbb{R}^m$ of a set $E \subset \mathbb{R}^m$ is called

- *regular* if it is injective, ess inf $\big\{ J_\phi(x) : x \in E \big\} > 0$, and $\phi^{-1}(T)$ is a thin set whenever $T \subset \mathbb{R}^m$ is a thin set;
- *locally regular* if every point $x \in \operatorname{cl} E$ has a neighborhood U such that the map $\phi : E \cap U \to \mathbb{R}^m$ is regular.

Each lipeomorphism is a regular map, and each local lipeomorphism is a locally regular map. The converse is true if E is a locally BV subset of $\mathbb{R}$, but it is false in general. The next example, due to D. Preiss, shows that in higher dimensions regular maps of bounded BV sets need not be local lipeomorphisms.

Example 4.6.13. Let $t_i = 2^{-i}$ for $i = 1, 2, \ldots$. In $\mathbb{R}^2$, consider the closed disks

$$
B_i = B\big[(t_i, 0), t_i^2/2\big] \quad \text{and} \quad C_i = B\big[(t_i^2, 0), t_i^2/2\big],
$$

and observe $B_i \cap B_j = C_i \cap C_j = \emptyset$ whenever $i \neq j$. The sets $K = \{0\} \cup \bigcup_{i=1}^{\infty} B_i$ and $L = \{0\} \cup \bigcup_{i=1}^{\infty} C_i$ are clearly compact and BV. Define a regular bijection $\phi : K \to L$ by sending 0 to 0 and translating each B_i onto C_i. Since

$$
\Big| \phi^{-1}\big[(t_i^2, 0)\big] - \phi^{-1}\big[(t_j^2, 0)\big] \Big| = |t_i - t_j| = \frac{1}{t_i + t_j} \big|(t_i^2, 0) - (t_j^2, 0)\big|
$$

and $\lim t_i = 0$, we see that ϕ^{-1} is not a Lipschitz map.

Proposition 4.6.14. *Let $\phi : E \to \mathbb{R}^m$ be a regular map of a set $E \subset \mathbb{R}^m$, and let $B \subset E$ be a bounded BV set. Then $\phi(B)$ is a bounded BV set and $r\big[\phi(B)\big] \geq \beta r(B)$ where $\beta = \beta(m, \phi)$ is a positive constant.*

PROOF. It follows from Theorem 4.6.8 that $\deg(\phi, A)$ belongs to $BV_c(\mathbb{R}^m)$ and

$$\|\deg(\phi, B)\| \leq c^{m-1}\|B\|$$

where c is the Lipschitz constant of ϕ. As ϕ is a regular map, equality (4.6.7) implies $\chi_{\phi(B)} = |\deg(\phi, B)|$ almost everywhere. Consequently $\phi(B)$ is a bounded BV set. Using the area theorem, a straightforward calculation reveals

$$r[\phi(B)] \geq \frac{\gamma}{\beta^m}r(B)$$

for any positive $\beta \geq c$ and $\gamma = \text{ess inf}\left\{J_\phi(x) : x \in E\right\}$. $\qquad\square$

Applying Proposition 4.6.14, the reader can easily verify that Proposition 4.6.1 holds for a regular map ϕ. If lipeomorphisms are replaced by regular maps, the conclusions of Proposition 4.6.2 and Theorem 3.7.4 change as follows.

Proposition 4.6.1: $V_*(\phi^* F)(B) \leq V_* F[\bar{\phi}(B)]$.

Theorem 3.7.4: If $F \in CH_*[\phi(E)]$, then $\phi^* F \in CH_*(E)$ and

$$D(\phi^* F)(x) = DF[\phi(x)]J_\phi(x)$$

for almost all $x \in E$.

In view of this, Theorems 5.3.1 and 6.4.8 below are true for regular maps. Moreover, all results of Sections 4.6, 5.3, and 6.4 remain valid when local lipeomorphisms are replaced by locally regular maps.

5

Integration

For a locally BV set E, let $R(E) := D[CH_*(E)]$. It follows from Theorem 3.6.6 the derivation $D : CH_*(E) \to R(E)$ is bijective, and we shall investigate its inverse $I_E : R(E) \to CH_*(E)$. We show that I_E, which has properties analogous to those of the indefinite Lebesgue integral, can be applied to partial derivatives of pointwise Lipschitz functions, and we prove unrestricted versions of the Gauss-Green and Stokes theorems. We also show that an averaging process akin to the classical Riemann integral provides a direct definition of I_E.

5.1. The R-integral

In this section we define the R-integral and prove some of its basic properties; most of them follow readily from the corresponding properties of AC_* charges established in Section 3.6.

Definition 5.1.1. Let E be a locally BV set. A function f defined on E is called *R-integrable* in E if there is an $F \in CH_*(E)$, called the *indefinite R-integral* of f, such that $DF(x) = f(x)$ for almost all $x \in E$.

The family of all R-integrable functions in a locally BV set E is denoted by $R(E)$. It follows from Theorem 3.6.6 that the indefinite R-integral F of a function $f \in R(E)$ is uniquely determined by f, and we denote it by $(R)\int f \, d\mathcal{L}^m$ or $(R)\int f(x) \, dx$. If $A \subset E$ is a bounded BV set, the number $F(A)$ is called the *R-integral* of f over A, denoted by $(R)\int_A f \, d\mathcal{L}^m$ or $(R)\int_A f(x) \, dx$.

From Definition 5.1.1 it is clear the R-integrability and the indefinite integral of a function f defined on a locally BV set E depend only on the equivalence class of f. In particular, if f is an R-integrable function in E, then a function g defined on $\mathrm{cl}_* E$ that coincide with f on $E \cap \mathrm{cl}_* E$ is R-integrable in $\mathrm{cl}_* E$, and the indefinite integrals of f and g are the same. This fact will be used tacitly throughout the remainder of this book.

To make the distinction between the R-integral and Lebesgue integral more apparent, we indicate the Lebesgue integral by employing the symbol $(L) \int$. From Proposition 3.1.7 and Theorems 3.1.9 and 3.6.2 we obtain the following proposition.

Proposition 5.1.2. *If E is a locally BV set, then $L^1_\ell(E) \subset R(E)$ and*

$$(R) \int f(x) \, dx = (L) \int f(x) \, dx$$

for each $f \in L^1_\ell(E)$.

For a locally BV set E, Proposition 5.1.2 says that the map

$$(R) \int : f \mapsto (R) \int f(x) \, dx : R(E) \to CH_*(E)$$

is a bijective extension of the bijection

$$(L) \int : f \mapsto (L) \int f \, dx : L^1_\ell(E) \to CH_{AC}(E).$$

Thus with no danger of confusion, henceforth, we denote both maps $(R) \int$ and $(L) \int$ by the same symbol $\int$. Only occasionally we employ the symbols $(R) \int$ and $(L) \int$ for emphasis.

Proposition 5.1.3. *Let E be a locally BV set. The family $R(E)$ is a linear space and the map*

$$\int : f \mapsto \int f(x) \, dx : R(E) \to CH_*(E)$$

is a linear bijection. If $f \in R(E)$ is a nonnegative function, then so is the indefinite R-integral $\int f(x) \, dx$.

The first claim of Proposition 5.1.3 is a direct consequence of Definition 5.1.1; the second claim follows from Theorem 3.6.6.

Proposition 5.1.4. *Let f be a function defined on a locally BV set E. Then $f \in L^1_\ell(E)$ if and only if both f and $|f|$ belong to $R(E)$.*

PROOF. If $f \in L^1_\ell(E)$ then $|f| \in L^1_\ell(E)$, and so both f and $|f|$ belong to $R(E)$ by Proposition 5.1.2.

Conversely, assuming f and $|f|$ belong to $R(A)$, Proposition 5.1.3 implies $f^{\pm} \in R(E)$ and the indefinite R-integrals $F_{\pm}$ of $f^{\pm}$ are nonnegative. According to Proposition 3.6.5 the charges $F_{\pm}$ are AC. Now it follows from Corollary 3.2.9 that $f^{\pm}$, and hence f, belong to $L_{\ell}^1(E)$. $\square$

Corollary 5.1.5. *Let A be a bounded BV set. If $\{f_i\}$ is a sequence in $R(A)$ and $f = \lim f_i$, then*

$$f \in R(A) \quad \text{and} \quad \int_A f(x)\,dx = \lim \int_A f_i(x)\,dx$$

whenever either of the following conditions is satisfied:

(i) $f_i \le f_{i+1}$ *for* $i = 1, 2, \ldots,$ *and* $\lim \int_A f_i(x)\,dx < \infty$;

(ii) $g \le f_i \le h$ *for* $i = 1, 2, \ldots$ *and* $g, h \in R(A)$.

PROOF. Condition (i) implies $\{f_i - f_1\}$ is an increasing sequence of nonnegative functions, and condition (ii) implies $\{f_i - g\}$ is a sequence of nonnegative functions bounded by a nonnegative function $h - g$. In view of Proposition 5.1.4, the corollary follows from the monotone and dominated convergence theorems for the Lebesgue integral. $\square$

The following proposition is an immediate consequence of Corollaries 3.6.11 and 2.5.4.

Proposition 5.1.6. *Let $B \subset A$ be locally BV sets, and let f be an R-integrable function in A. The restriction $f \restriction B$ is an R-integrable function in B, and if F is the indefinite integral of f, then $F \llcorner B$ is the indefinite integral of $f \restriction B$.*

Proposition 5.1.7. *Let E be a locally BV set. A function f defined on E is R-integrable in E if and only if it is R-integrable in each bounded BV set $B \subset E$, in which case*

$$F : A \mapsto \int_{E \cap A} f(x)\,dx : \mathcal{BV} \to \mathbb{R}$$

is the indefinite R-integral of f.

PROOF. Assume f is R-integrable in each bounded BV set $B \subset E$, and denote by F_k the indefinite integral of the restriction $f \restriction B(k) \cap E$. Since Proposition 5.1.6 implies $F_k = F_{k+1} \llcorner B(k)$, there is a charge F in E such that $F \llcorner B(k) = F_k$ for $k = 1, 2, \ldots$. As $B(k)$ is an open set,

$$DF(x) = DF_k(x) = f(x)$$

for almost all $x \in B(k) \cap E$. Hence $DF(x) = f(x)$ for almost all $x \in E$. Select a negligible set $N \subset \mathrm{cl}_* E$, and observe $N_k = B(k) \cap N$ is a negligible subset of $B(k) \cap \mathrm{cl}_* E \subset \mathrm{cl}_* [B(k) \cap E]$ and

$$V_* F(N_k) = V_* F_k(N_k) = 0$$

for $k = 1, 2, \ldots$; again because $B(k)$ is an open set. Thus $V_* F(N) = 0$, and we infer F is AC_* in E. It follows $f \in R(E)$ and F is the indefinite R-integral of f. In particular,

$$F(A) = F(E \cap A) = \int_{E \cap A} f(x) \, dx$$

for each bounded BV set A. The converse follows directly from Proposition 5.1.6 and Definition 5.1.1. $\qquad\square$

Proposition 5.1.8. *Let E be the disjoint union of locally BV sets E_i, $i = 1, \ldots, k$, and let f be a function defined on E that is R-integrable in every E_i. Suppose that each $\mathrm{cl}_* E_i$ is a relatively closed subset of $\mathrm{cl}_* E$. Then f is R-integrable in E, and if F_i is the indefinite R-integral of $f \upharpoonright E_i$, then $F := \sum_{i=1}^k F_i$ is the indefinite R-integral of f.*

PROOF. By Proposition 3.6.4, (ii), each F_i belongs to $CH_*(E)$ and

$$DF_i(x) = \begin{cases} f(x) & \text{for almost all } x \in E_i, \\ 0 & \text{for almost all } x \in E - E_i. \end{cases}$$

Thus $F \in CH_*(E)$ and $DF(x) = f(x)$ for almost all $x \in E$. $\qquad\square$

We show in Remark 6.1.2, (4) that Proposition 5.1.8 is false without assuming that each $\mathrm{cl}_* E_i$ is a relatively closed subset of E. Fortunately, this apparent deficiency of the R-integral is not serious: a simple extension of the R-integral presented in Chapter 6 has the standard additivity property (Theorem 6.3.10 below).

To indicate the extent to which the R-integral extends the Lebesgue integral, we present without proof a striking result of Buczolich [14, Theorem 2].

Theorem 5.1.9. *In $\mathbb{R}^2$ there is a function f defined on a cell C that is R-integrable in C but not Lebesgue integrable in any nonempty open subset of C. In particular, f is not the sum of a Lebesgue integrable function and a derivate of a charge.*

The *in particular* part of Theorem 5.1.9 is an immediate consequence of the following simple lemma.

Lemma 5.1.10. *Let $C \subset \mathbb{R}^m$ be a nonempty closed set. If a charge F is derivable at each $x \in C$, then there is an open set $U \subset \mathbb{R}^m$ such that $C \cap U \neq \emptyset$ and DF belongs to $L^\infty(C \cap U)$.*

PROOF. In view of Proposition 2.3.5, it suffices to show there is an open set U such that $C \cap U \neq \emptyset$ and DF is bounded on $C \cap U$. Assuming this is not true, observe that C contains no isolated points. Moreover, there is an $x_1 \in C$ with $|DF(x_1)| > 1$, and hence an open ball $B_1 := B(x_1, r_1)$ such that $r_1 < 1$ and $|F(B_1)/|B_1|| > 1$. There is an $x_2 \in C \cap B(x_1, r_1)$ with $|DF(x_2)| > 2$, and hence an open ball $B_2 := B(x_2, r_2)$ such that $r_2 < 1/2$, $B_2 \subset B_1$, and $|F(B_2)/|B_2|| > 2$. Construct inductively open balls $B_i := B(x_i, r_i)$, $i = 1, 2, \ldots$, so that

$$x_i \in C, \quad r_i < \frac{1}{i}, \quad B_{i+1} \subset B_i, \quad \text{and} \quad \left|\frac{F(B_i)}{|B_i|}\right| > i$$

If $\{x\} = \bigcap_{i=1}^{\infty} \mathrm{cl}\, B_i$, then $x \in C$ and a contradiction follows:

$$|DF(x)| = \lim \left|\frac{F(B_i)}{|B_i|}\right| = \infty. \qquad \square$$

In dimension one, the next proposition improves on the *in particular* part of Theorem 5.1.9. For its proof we refer to [7].

Proposition 5.1.11. *If $m = 1$ and $A \subset \mathbb{R}$ is a cell, then*

$$CH_{AC}(A) + CH_{AD}(A) \subsetneqq CH_*(A).$$

From Theorem 3.6.7 and Definitions 2.5.8 and 5.1.1, we obtain an unrestricted version of the Gauss-Green theorem, facilitated by an application of the R-integral (cf. Theorem 3.6.13).

Theorem 5.1.12. *Let E be a locally BV set, and let $v : \mathrm{cl}_* E \to \mathbb{R}^m$ be a charging vector field. Suppose the flux of v is almost derivable at each $x \in \mathrm{cl}_* E - T$ where T is a thin set. Then $\mathfrak{div}\, v(x)$ is defined for almost all $x \in E$, the function $\mathfrak{div}\, v : x \mapsto \mathfrak{div}\, v(x)$ belongs to $R(E)$, and*

$$(R) \int_A \mathfrak{div}\, v \, d\mathcal{L}^m = (L) \int_{\partial_* A} v \cdot \nu_A \, d\mathcal{H}^{m-1}$$

for each bounded BV set $A \subset E$.

The following corollary, which is an immediate consequence of Theorem 5.1.12 and Remark 2.5.9, (i), is a more familiar form of the Gauss-Green theorem — cf. Remark 2.5.9, (4).

Corollary 5.1.13. *Let E be a locally bounded BV set, and let v be a continuous vector field defined on $\mathrm{cl}\, E$. Suppose v is Lipschitz at each $x \in \mathrm{cl}_* E - T$ where T is a thin set. Then $\mathrm{div}_* v(x)$ is defined for almost all $x \in E$, the function $\mathrm{div}_* v : x \mapsto \mathrm{div}_E v(x)$ belongs to $R(E)$, and*

$$(R) \int_A \mathrm{div}_* v \, d\mathcal{L}^m = (L) \int_{\partial_* A} v \cdot \nu_A \, d\mathcal{H}^{m-1}$$

for each bounded BV set $A \subset E$.

Theorem 5.1.12 and Corollary 5.1.13 are our first versions of the Gauss-Green theorem for vector fields whose mean divergence need not be Lebesgue integrable. Note the proof of Theorem 5.1.12 is completely different from the traditional proofs based on Fubini's theorem. In fact, the following example shows Fubini's theorem is false for the R-integral, since it is incompatible with the R-integrability of partial derivatives of differentiable functions.

Example 5.1.14. Define a function $\varphi \in C^\infty(\mathbb{R})$ by the formula

$$\varphi(s) := \begin{cases} \exp\left(\frac{1}{s^2-1}\right) & \text{if } |s| < 1, \\ 0 & \text{if } |s| \geq 1. \end{cases}$$

If $K := [a, b]$ is a one-dimensional cell, let

$$g_K(t) := \begin{cases} c^{-1} \int_{-1}^{u(t)} \varphi(s)\, ds & \text{if } t > a, \\ 0 & \text{if } t \leq a, \end{cases}$$

where $c := \int_{-1}^{1} \varphi(s)\, ds$ and $u(t) := (2t - b - a)/(b - a)$. Note $g_K \in C^\infty(\mathbb{R})$ is an increasing function such that $g_K(t) = 0$ if $t \leq a$ and $g_K(t) = 1$ if $t \geq b$. For $n = 1, 2, \ldots$, let

$$A_n := \left[2^{-n-1}, 2^{-n}\right], \quad K_n := \left[\tfrac{4}{3}2^{-n-1}, \tfrac{5}{3}2^{-n-1}\right], \quad C_n := \left[\tfrac{5}{3}2^{-n-1}, \tfrac{4}{3}2^{-n}\right],$$

and note K_n is the middle third cell in A_n, while C_{n+1} is the middle third cell in $A_n \cup A_{n+1}$. Denote g_{K_n} by g_n, and define a function f on $\mathbb{R}^2$ as follows: given $(s, t) \in \mathbb{R}^2$, let

$$f(s, t) := g_n(t)t^2 \sin(8^n s) + \left[1 - g_n(t)\right]t^2 \sin(8^{n+1} s)$$

whenever $t \in A_n$, $n = 0, 1, \ldots$, and let

$$f(s, t) := \begin{cases} 0 & \text{if } t \leq 0, \\ t^2 \sin s & \text{if } t > 1. \end{cases}$$

Observe f on $\mathbb{R} \times [0, 1]$ is obtained by differentiable deformations of the functions $t^2 \sin(8^n s)$ on $\mathbb{R} \times C_n$ to the functions $t^2 \sin(8^{n+1} s)$ on $\mathbb{R} \times C_{n+1}$. As $|f(s, t)| \leq t^2$ for all $(s, t) \in \mathbb{R}^2$, it is easy to verify that f is differentiable at each point of $\mathbb{R}^2$. Since $h = \partial f/\partial s$ is the divergence of the vector field $v := (f, 0)$, Theorem 5.1.12 implies h is R-integrable in the cell $D := [0, 2\pi] \times [0, 1]$.

It follows from [41, Chapter 1, Theorem 4.1] there is an $\mathcal{L}^1$-negligible subset N of $[0, 2\pi]$ and a strictly increasing sequence $\{n_i\}$ of positive integers having the following property: given $s \in [0, 2\pi] - N$ and $\varepsilon > 0$, we can find integers $k_1, k_2, \ldots$ so that $\left|8^{n_i}s - 2k_i\pi\right| < \varepsilon$ for $i = 1, 2, \ldots$. Choosing $\varepsilon := \pi/24$, we may assume

$$\min\left\{\cos(8^{n_i}s), \cos(8^{n_i+1}s)\right\} > \frac{1}{2}. \qquad (*)$$

Select an $s \in [0, 2\pi] - N$, and assume the function $h_s : t \mapsto h(s, t)$ defined on $\mathbb{R}$ is R-integrable in $[0, 1]$. The indefinite integral of $h_s \restriction [0, 1]$ is a one-dimensional charge H in $[0, 1]$. Since $\{A_{n_i}\}$ $\mathfrak{T}$-converges to the empty set, $\lim H(A_{n_i}) = 0$. As

$h_s \in L^1(A_{n_i})$, the value $H(A_{n_i})$ equals the Lebesgue integral of h_s over A_{n_i}. By $(*)$,

$$H(A_{n_i}) = 8^{n_i}\cos(8^{n_i}s)\int_{A_{n_i}} t^2 g_{n_i}(t)\,dt$$

$$+ 8^{n_i+1}\cos(8^{n_i+1}s)\int_{A_{n_i}} t^2\big[1-g_{n_i}(t)\big]\,dt$$

$$\geq \frac{8^{n_i}}{2}\int_{A_{n_i}} t^2 g_{n_i}(t)\,dt + \frac{8^{n_i+1}}{2}\int_{A_{n_i}} t^2\big[1-g_{n_i}(t)\big]\,dt$$

$$\geq \frac{8^{n_i}}{2}\int_{A_{n_i}} t^2\,dt > \frac{8^{n_i}}{2}2^{-n_i-1}\big(2^{-n_i-1}\big)^2 = 2^{-4},$$

a contradiction. In particular, $h \notin L^1(D)$ on the account of Fubini's theorem.

5.2. Multipliers

A *multiplier* for a family X of functions defined on a set E is a function g defined on E such that $fg \in X$ for each $f \in X$. The collection of all multipliers for X is denoted by MX. If X is a linear space containing constant functions, then MX is a subalgebra of X — for instance, we have $ML^1_{\mathrm{loc}}(\mathbb{R}^m) = L^\infty_{\mathrm{loc}}(\mathbb{R}^m)$.

Lemma 5.2.1. *If E is a locally BV set, then $MR(E) \subset BV^\infty_\ell(E)$.*

PROOF. Select a $g \in MR(E)$, and observe g and g^2 belong to $R(E)$. Thus $g^2 \in L^1_\ell(E)$ by Proposition 5.1.4. It follows $g \in L^1_\ell(E)$, and we claim g is essentially bounded in each bounded subset of E. Suppose not, and find an $r > 0$ so that the set

$$C_i := \big\{x \in E \cap B[r] : |g(x)| \geq 2^i\big\}$$

has positive measure for $i = 1, 2, \dots$. Since

$$h := \sum_{i=1}^\infty \frac{1}{2^i|C_i|}\chi_{C_i}\,\mathrm{sign}\,g$$

belongs to $L^1(E) \subset R(E)$, the function gh belongs to $R(E)$ and, being nonnegative, to $L^1_\ell(E)$. On the other hand, $\lim|C_i| = 0$ and

$$\int_{C_i} g(x)h(x)\,dx \geq \frac{1}{2^i|C_i|}\int_{C_i}|g(x)|\,dx \geq 1$$

for $i = 1, 2, \dots$. This contradiction proves our claim.

Next suppose $g \notin BV^\infty_\ell(E)$, and observe there is an open cube $U := (a,b)^m$ such that $f := \bar{g} \restriction U$ is not in $BV(U)$. For $a < t < b$, let

$$U_{t-} := (a,t) \times (a,b)^{m-1} \quad \text{and} \quad U_{t+} := (t,b) \times (a,b)^{m-1}.$$

As $f \in L^1(U)$ by the first part of the proof, there is a $v_1 \in C^1(\mathbb{R}^m; \mathbb{R}^m)$ with $|v_1|_\infty \leq 1$ and $\int_U f(x) \operatorname{div} v_1(x)\, dx > 4$. Find a $t \in (a, b)$ so that

$$\int_{U_{t-}} f(x) \operatorname{div} v_1(x)\, dx = \int_{U_{t+}} f(x) \operatorname{div} v_1(x)\, dx$$

$$= \frac{1}{2} \int_U f(x) \operatorname{div} v_1(x)\, dx > 2\,,$$

and denote by V_1 the element of $\{U_{t-}, U_{t+}\}$ in which f is not BV; it follows from Theorem 1.7.5 that such a V_1 exists. The other member of $\{U_{t-}, U_{t+}\}$ is denoted by $U_1 = (r_1, s_1) \times (a, b)^{m-1}$. Applying the previous argument to V_1, we obtain disjoint subsets

$$U_2 = (r_2, s_2) \times (a, b)^{m-1} \quad \text{and} \quad V_2 = (t, t') \times (a, b)^{m-1}$$

of V_1 such that f is not BV in V_2 and

$$\int_{U_2} f(x) \operatorname{div} v_2(x)\, dx > 4$$

for a $v_2 \in C^1(\mathbb{R}^m; \mathbb{R}^m)$ with $|v_2|_\infty \leq 1$. Proceeding recursively, find disjoint intervals $U_k = (r_k, s_k) \times (a, b)^{m-1}$ and $v_k \in C^1(\mathbb{R}^m; \mathbb{R}^m)$ with

$$|v_k|_\infty \leq 1 \quad \text{and} \quad \int_{U_k} f(x) \operatorname{div} v_k(x)\, dx > 2^k \quad \text{for} \quad k = 1, 2, \ldots.$$

From the recursion it is clear that $\lim(s_k - r_k) = 0$ and that the set $\{r_k, s_k : k = 1, 2, \ldots\}$ has at most two cluster points α and β. Each U_k contains a cell A_k such that

$$\int_{A_k} f(x) \operatorname{div} v_k(x)\, dx > 2^k\,,$$

and we can modify v_k so that it vanishes outside U_k [65, Lemma 10.4.1]. The vector field $v := \sum_{k=1}^{\infty} 2^{-k} v_k$ is continuous everywhere and differentiable at each point $x \in \mathbb{R}^m$ that is not in the thin set $\{\alpha, \beta\} \times (a, b)^{m-1}$. According to Remark 2.5.9, (iv) and Corollary 5.1.13, the restriction $\varphi = (\operatorname{div} v) \restriction E$ belongs to $R(E)$, and so does the product $g\varphi$. Since

$$g(x)\varphi(x) = 2^{-k} g(x) \operatorname{div} v_k(x)$$

for each $x \in E \cap A_k$, the function $g\varphi$ is in $L^1(A_k \cap E)$. If F is the indefinite R-integral of $g\varphi$, then Propositions 5.1.6 and 5.1.2 imply

$$F(A_k) = F(E \cap A_k) = 2^{-k} \int_{E \cap A_k} g(x) \operatorname{div} v_k(x) \, dx$$

$$= 2^{-k} \int_{A_k} \bar{g}(x) \operatorname{div} v_k(x) \, dx = 2^{-k} \int_{A_k} f(x) \operatorname{div} v_k(x) \, dx \geq 1$$

for $k = 1, 2, \ldots$. As $\lim |A_k| = 0$ and $\sup \|A_k\| \leq \|U\|$, we have a contradiction. $\qquad\square$

Theorem 5.2.2. *Let E be a locally BV set, and let F be the indefinite R-integral of an $f \in R(E)$. If $g \in BV_\ell^\infty(E)$, then $fg \in R(E)$ and*

$$\int f(x)g(x) \, dx = F \mathsf{L} \, g \,.$$

In particular, $MR(E) = BV_\ell^\infty(E)$.

PROOF. If $g \in BV_\ell^\infty(E)$ then $\bar{g} \in BV_{\mathrm{loc}}^\infty$, and the first claim follows from (4.5.2) and Theorem 4.5.7. In particular, $BV_\ell^\infty(E) \subset MR(E)$ and an application of Lemma 5.2.1 completes the proof. $\qquad\square$

Remark 5.2.3. Example 5.1.14 shows Fubini's theorem is false for the R-integral. However, for a bounded BV set A, a function $f \in R(A)$, and a nonnegative function $g \in BV_\ell^\infty(E)$, Theorem 5.2.2 provides a Fubini type transposition

$$\int_A \left(\int_0^{g(x)} f(x) \, dt \right) dx = \int_A f(x)g(x) \, dx = \int_0^c \left(\int_{\{g>t\}} f(x) \, dx \right) dt$$

where $c = |g|_\infty$. In view of Proposition 4.5.10, the affirmative answer to Conjecture 4.5.9 implies the existence of the double integral

$$\int_{\Sigma_g} (f \otimes 1_\mathbb{R}) \, d\mathcal{L}^{m+1} = \int_A fg \, d\mathcal{L}^m \,.$$

The following theorem shows the R-integral facilitates the usual integral representation of continuous linear functionals on $\big(CH_*(A), \mathcal{S}\big)$ when A is a bounded BV set (cf. Theorem 4.3.5).

Theorem 5.2.4. *Let A be a bounded BV set. Then*

$$T_g : F \mapsto (R) \int_A DF(x)g(x) \, dx : \big(CH_*(A), \mathcal{S}\big) \to \mathbb{R}$$

is a continuous linear functional for each $g \in BV_\ell^\infty(A)$. The linear map $T : g \mapsto T_g$ is a bijection from $BV_\ell^\infty(A)$ onto the dual space $CH_*(A)^*$ of $\big(CH_*(A), \mathcal{S}\big)$.

PROOF. Recall the definition of $BV_b^\infty(A)$ from (4.1.3). Since A is bounded, we have a linear bijection

$$\Psi : g \mapsto \overline{g} : BV_\ell^\infty(A) \to BV_b^\infty(A).$$

By Theorem 5.2.2, for each $g \in BV_\ell^\infty(A)$ and each $F \in CH_*(A)$,

$$T_g(F) = (R) \int_A DF(x)g(x)\, dx = (F \, \mathsf{L}\, g)(A)$$

$$= \langle F, \overline{g}\chi_A \rangle = \langle F, \overline{g} \rangle = \Phi_A(\overline{g})(F).$$

Hence $T = \Phi_A \circ \Psi$, and the theorem follows from Theorem 4.3.5.   □

Theorem 5.2.5. *Let A be a bounded BV set, and let k be an integer with $k \geq 1 + 4d(A)^{m-1}$ and $A \subset B[k]$. Let $f \in R(A)$ and $g \in BV_\ell^\infty(A)$, and suppose that $\{f_i\}$ and $\{g_i\}$ are sequences in $R(A)$ and $BV_\ell^\infty(A)$, respectively. Then*

$$\lim \int_A f_i(x)g_i(x)\, dx = \int_A f(x)g(x)\, dx$$

whenever the following conditions are satisfied:

(i) $\lim |g_i - g|_1 = 0$ *and* $\sup(\|\overline{g_i}\| + |g_i|_\infty) < \infty$;

(ii) $\displaystyle\lim_{i\to\infty} \sup_B \left| \int_{A \cap B} [f_i(x) - f(x)]\, dx \right| = 0$ *where* $B \in \boldsymbol{\mathcal{BV}}_k(\mathbb{R}^m)$.

PROOF. Condition (i) says the sequence $\{\overline{g_i}\}$ $\mathcal{T}$-converges to $\overline{g}$. If F_i and F denote, respectively, the indefinite R-integrals of f_i and f, then $\lim \|F_i - F\|_k = 0$ by condition (ii). Proposition 2.2.4 and the choice of k imply that the sequence $\{F_i\}$ $\mathcal{S}$-converges to F. Applying Theorems 5.2.2 and 4.1.5, we conclude

$$\lim \int_A f_i(x)g_i(x)\, dx = \lim \langle F_i, \overline{g_i} \rangle = \langle F, \overline{g} \rangle = \int_A f(x)g(x)\, dx.   \quad □$$

The next result, due to De Pauw [17, Theorem 7.16], generalizes Theorem 1.7.6.

Theorem 5.2.6. *Let Ω be a Lipschitz domain, let $v \in C(\mathrm{cl}\,\Omega; \mathbb{R}^m)$, and let $T \subset \Omega$ be a thin set such that the flux of v is almost derivable at each x in $\Omega - T$. Then*

$$(R) \int_\Omega g \,\mathfrak{div}\, v \, d\mathcal{L}^m = (L) \int_{\partial\Omega} (\mathrm{Tr}\, g)v \cdot \nu_\Omega \, d\mathcal{H}^{m-1} - (L) \int_\Omega v \cdot d(Dg)$$

for each $g \in BV^\infty(\Omega)$.

PROOF. There is a sequence $\{v_i\}$ in $C^1(\mathbb{R}^m; \mathbb{R}^m)$ that converges to v uniformly in cl Ω. According to Theorem 1.7.6, the desired equality holds for each v_i. The dominated convergence theorem yields

$$\lim (L) \int_{\partial\Omega} (\text{Tr } g) v_i \cdot \nu_\Omega \, d\mathcal{H}^{m-1} = (L) \int_{\partial\Omega} (\text{Tr } g) v \cdot \nu_\Omega \, d\mathcal{H}^{m-1},$$

$$\lim (L) \int_\Omega v_i \cdot d(Dg) = (L) \int_\Omega v \cdot d(Dg).$$

According to Theorem 5.1.12,

$$\left| (R) \int_{\Omega\cap B} (\text{div } v_i - \mathfrak{div} \, v) \, d\mathcal{L}^m \right| = \left| (L) \int_{\partial_*(\Omega\cap B)} (v_i - v) \cdot \nu_B \, d\mathcal{H}^{m-1} \right|$$

$$\leq \|\Omega \cap B\| \cdot |v_i - v|_\infty$$

$$\leq \big(\|\Omega\| + \|B\| \big) |v_i - v|_\infty$$

for every bounded BV set B. The theorem follows, since

$$\lim (R) \int_\Omega g(x) \, \text{div } v_i(x) \, dx = (R) \int_\Omega g(x) \, \mathfrak{div} \, v(x) \, dx,$$

by Propositions 5.1.2 and Theorem 5.2.5. $\square$

Corollary 5.2.7. *Let* $U \in \mathcal{BV}$ *be open, and let* $w \in C(\text{cl } U; \mathbb{R}^m)$. *Suppose there is a thin set* T *such that the flux of* w *is almost derivable at each* x *in* $U - T$. *Then*

$$\int_U w(x) \cdot D\varphi(x) \, dx = - \int_U \mathfrak{div} \, w(x) \, \varphi(x) \, dx$$

for each $\varphi \in C_c^1(U)$.

PROOF. Denote by G the flux of w, and choose a $\varphi \in C_c^1(U)$. It is easy to verify that there is a Lipschitz domain $\Omega \subset U$ containing supp φ. Clearly $G \llcorner \Omega$ is the flux of the vector field $v := w \restriction \text{cl } \Omega$. Since Ω is an open set, the charge $G \llcorner \Omega$ is derivable at $x \in \Omega$ whenever G is derivable at x. Now $\text{Tr } \varphi = \varphi = 0$ on $\partial\Omega$, and Theorem 5.2.6 implies

$$\int_U \varphi(x) \, \mathfrak{div} \, w(x) \, dx = \int_\Omega \varphi(x) \, \mathfrak{div} \, v(x) \, dx$$

$$= - \int_\Omega D\varphi(x) \cdot v(x) \, dx$$

$$= - \int_U D\varphi(x) \cdot w(x) \, dx.   \square$$

5.3. Change of variables

We begin with a lemma that is an easy consequence of commutative diagram (4.6.4).

Theorem 5.3.1. *Suppose* $\phi : E \to \mathbb{R}^m$ *be a lipeomorphism of a locally BV set* E, *and* $f \in R[\phi(E)]$. *Then* $(f \circ \phi)J_\phi$ *belongs to* $R(E)$ *and*

$$\int_A f[\phi(x)] J_\phi(x)\, dx = \int_{\phi(A)} f(y)\, dy$$

for each bounded BV set $A \subset E$.

PROOF. Denote by F the indefinite R-integral of f. According to Theorem 4.6.3, the charge $\phi^* F$ is the indefinite R-integral of

$$[(DF) \circ \phi] J_\phi = (f \circ \phi) J_\phi .$$

Since $\phi^* F(A) = F[\phi(A)]$ for each $A \in \mathcal{BV}(E)$, the theorem follows. $\square$

Remark 5.3.2. In view of Theorem 5.3.1, the R-integral can be defined on oriented differentiable manifolds in the usual manner [79, Chapter 5]. Thin sets are also defined on differentiable manifolds in the obvious way [62, Section 7]. If M is a compact oriented m-dimensional differentiable manifold with boundary, then Corolary 5.1.13 yields the following version of the Stokes theorem:

$$(R)\int_M d\omega = (L)\int_{\partial M} \omega$$

for each continuous $(m-1)$-form ω defined on M that is Lipschitz at each point $x \in M - T$ where T is a thin subset of M. Two proofs are available: one uses a partition of unity on M [79, Theorem 5-5], the other relies on a triangulation of M [62, Section 7]. As both of these proofs involve only standard techniques, and neither is particularly difficult, we leave them to the reader. The partition of unity approach can be applied to Lipschitz manifolds as well.

Corollary 5.3.3. *Suppose* $\phi : A \to \mathbb{R}^m$ *is a local lipeomorphism of a bounded BV set* A, *and* $f \in R[\phi(A)]$. *Then* $(f \circ \phi)J_\phi$ *belongs to* $R(A)$, *the function*

$$y \mapsto \mathcal{H}^0[\phi^{-1}(y)] f(y) : \phi(A) \to \mathbb{R}$$

belongs to $R[\phi(A)]$, *and*

$$\int_A f[\phi(x)] J_\phi(x)\, dx = \int_{\phi(A)} \mathcal{H}^0[\phi^{-1}(y)] f(y)\, dy .$$

PROOF. It follows from Definition 4.6.4 that A is the union of disjoint BV sets $A_1, \ldots, A_p$ such that each $\mathrm{cl}_* A_i$ is a relatively closed subset of

$\text{cl}_* A$ and each $\phi : A_i \to \mathbb{R}^m$ is a lipeomorphism. By Proposition 5.1.6 and Theorem 5.3.1,

$$\int_{A_i} f[\phi(x)] J_\phi(x)\, dx = \int_{\phi(A_i)} f(y)\, dy$$

for $i = 1, \ldots, p$. Observe that for each $y \in \Phi(A)$,

$$\sum_{i=1}^{p} f(y)\chi_{\phi(A_i)}(y) = \sum_{\{i : y \in \phi(A_i)\}} f(y) = \mathcal{H}^0\left[\phi^{-1}(y)\right] f(y).$$

According to Theorems 1.8.6 and 5.2.2, the functions $f\chi_{\phi(A_i)}$ belong to $R[\phi(A)]$, and by Proposition 5.1.3, so does $\mathcal{H}^0\left[\phi^{-1}(y)\right] f(y)$. In view of Proposition 5.1.8, the function $(f \circ \phi)J_\phi$ belongs to $R(A)$ and

$$\int_{\phi(A)} \mathcal{H}^0\left[\phi^{-1}(y)\right] f(y)\, dy = \sum_{i=1}^{p} \int_{\phi(A_i)} f(y)\, dy =$$

$$\sum_{i=1}^{p} \int_{A_i} f[\phi(x)] J_\phi(x)\, dx = \int_A f[\phi(x)] J_\phi(x)\, dx. \qquad \square$$

Theorem 5.3.4. *Suppose $\phi : E \to \mathbb{R}^m$ is a local lipeomorphism of a locally BV set E, and $g \in BV_b^\infty(E)$. If $f \in R[\phi(E)]$, then the function $(f \circ \phi)(\det D\phi)\, g$ belongs to $R(E)$; if, in addition, $A \subset E$ is a bounded BV set, then $f\deg(\phi, g\chi_A)$ belongs to $R[\phi(A)]$ and*

$$\int_{\phi(A)} f(y)\deg(\phi, g\chi_A)(y)\, dy = \int_A f[\phi(x)] \det D\phi(x)\, g(x)\, dx.$$

PROOF. Denote by F the indefinite R-integral of f. According to Theorems 4.6.12 and 5.2.2, the charge $\phi^\# F \mathsf{L}\, g$ is the indefinite R-integral of $(f \circ \phi)(\det D\phi)g$. Choose a bounded BV set $A \subset E$, and using Proposition 4.6.9, observe $G := F \mathsf{L} \deg(\phi, g\chi_A)$ is the indefinite R-integral of $f\deg(\phi, g\chi_A)$. As (4.6.8) and (4.6.10) imply

$$\begin{aligned}
\left[(\phi^\# F)\mathsf{L}\, g\right](A) &= \langle \phi^\# F, g\chi_A \rangle = \langle \phi^\# F, \mathbf{E}^m \mathsf{L}\, (g\chi_A) \rangle \\
&= \left\langle F, \phi_\#\left[\mathbf{E}^m \mathsf{L}\, (g\chi_A)\right] \right\rangle = \langle F, \mathbf{E}^m \mathsf{L} \deg(\phi, g\chi_A) \rangle \\
&= \langle F, \deg(\phi, g\chi_A) \rangle = \langle F, \deg(\phi, g\chi_A)\chi_{\phi(A)} \rangle \\
&= G[\phi(A)],
\end{aligned}$$

the theorem follows. $\qquad \square$

Let E be a locally BV set, and let

$$\wedge_L^m(E) := L_\ell^1(E; \wedge^m) \quad \text{and} \quad \wedge_R^m(E) := R(E; \wedge^m).$$

Choose an $\omega = a\,dx$ in $\wedge_R^m(E)$ and $T := \mathbf{E}^m \mathop{\llcorner} g$ in $\mathbf{N}_m^{c,\infty}(E)$. Theorem 5.2.2 implies that the product ag is in $R(E)$. If $A \subset E$ is a bounded BV set containing $\{g \neq 0\}$, let

$$(R)\int_T \omega := (R)\int_A a(x)g(x)\,dx, \qquad (5.3.1)$$

and call this number the *R-integral* of ω over T. The two points below show that the definition of $(R)\int_T \omega$ is nonvacuous and does not depend on the choice of A.

- Since $\{g \neq 0\} \subset E$ is bounded, the set $E \cap B(r)$ is a bounded BV subset of E which contains $\{g \neq 0\}$ whenever $r > 0$ is sufficiently large.
- If A and B are bounded BV subsets of E and each contains $\{g \neq 0\}$, then

$$\int_A a(x)g(x)\,dx = \int_{A \cap B} a(x)g(x)\,dx = \int_B a(x)g(x)\,dx$$

according to Propositions 5.1.6 and 5.1.3.

If ω is in $\wedge_L^m(E) \subset \wedge_R^m(E)$, we may write $(L)\int_T \omega$ instead of $(R)\int_T \omega$, however, when no confusion is possible, the symbol $\int_T \omega$ will denote both $(R)\int_T \omega$ and $(L)\int_T \omega$. Observe

$$\int_T \omega = \langle T, \omega \rangle \qquad (5.3.2)$$

for each $T \in \mathbf{N}_m^{c,\infty}(\mathbb{R}^m)$ and each $\omega \in \wedge_\infty^m(\mathbb{R}^m)$.

The following result, which is a mere translation of Theorem 5.3.4, illuminates the geometry connected with changing variables.

Theorem 5.3.5. *Let $\phi : E \to \mathbb{R}^m$ be a local lipeomorphism of a locally BV set E, and let $\omega \in \wedge_R^m[\phi(E)]$. Then $\phi^\# \omega \in \wedge_R^m(E)$ and*

$$\int_T \phi^\# \omega = \int_{\phi_\# T} \omega$$

for each $T \in \mathbf{N}_m^{c,\infty}(E)$.

PROOF. Let $\omega := a\,dx$, and choose a $T := \mathbf{E}^m \mathop{\llcorner} g$ in $\mathbf{N}_m^{c,\infty}(E)$. By equality (4.6.5) and Proposition 4.6.7,

$$\phi^\# \omega = (a \circ \phi)\det D\phi\,dx \quad \text{and} \quad \phi_\# T = \mathbf{E}^m \mathop{\llcorner} \deg(\phi, g).$$

If $A \in \mathcal{BV}(E)$ contains $\{g \neq 0\}$, then $\phi(A) \in \mathcal{BV}\big[\phi(E)\big]$ according to Proposition 4.6.6, and $\{\deg(\phi, g) \neq 0\} \subset \phi(A)$. As $g\chi_A = g$, an application of Theorem 5.3.4 completes the proof:

$$\int_T \phi^{\#}\omega = \int_A a\big[\phi(x)\big] \det D\phi(x)\, g(x)\, dx$$

$$= \int_{\phi(A)} a(y)\deg(\phi, g)(y)\, dy = \int_{\phi_{\#}T} \omega. \qquad \square$$

Let E be a measurable set. Elements of $\wedge_M^{m-1}(E)$ are forms

$$\omega := \sum_{i=1}^m (-1)^{i-1} w_i\, dx_1 \wedge \cdots \widehat{dx_i} \cdots \wedge dx_m$$

where $w_\omega = (w_1, \ldots, w_m)$ belongs to $M(E; \mathbb{R}^m)$, and $\widehat{dx_i}$ indicates that dx_i has been omitted from the wedge product. We call w_ω the vector field *associated* with ω. Let

$$\wedge_C^{m-1}(E) := C(E; \wedge^{m-1}),$$

$$\wedge^{m-1}(\mathbb{R}^m) := C_c^1(\mathbb{R}^m; \wedge^{m-1}).$$

If $\omega \in \wedge^{m-1}(\mathbb{R}^m)$ then $w_\omega \in C_c^1(\mathbb{R}^m; \mathbb{R}^m)$, and we define an m-form

$$d\omega := \sum_{i=1}^m (-1)^{i-1} \frac{\partial w_i}{\partial x_i}\, dx_i \wedge dx_1 \wedge \cdots \widehat{dx_i} \cdots \wedge dx_m$$

$$= \sum_{i=1}^m \frac{\partial w_i}{\partial x_i}\, dx_1 \wedge \cdots \wedge dx_m = \operatorname{div} w_\omega\, dx \tag{5.3.3}$$

in $\wedge_\infty^m(\mathbb{R}^m)$, called the *exterior derivative* of ω [79, Chapter 4]. The exterior derivative d induces a linear map

$$d : \omega \mapsto d\omega : \wedge^{m-1}(\mathbb{R}^m) \to \wedge_\infty^m(\mathbb{R}^m),$$

called the *exterior differentiation*. The dual map of the exterior differentiation d, called the *boundary operator*, is the map

$$\partial : \wedge_\infty^m(\mathbb{R}^m)^* \to \wedge^{m-1}(\mathbb{R}^m)^*$$

defined by the identity

$$\langle \partial T, \omega \rangle := \langle T, d\omega \rangle \tag{5.3.4}$$

for each $T \in \wedge_\infty^m(\mathbb{R}^m)^*$ and each $\omega \in \wedge^{m-1}(\mathbb{R}^m)$. If T is an m-current, the linear functional ∂T is called the *boundary* of T.

Every vector-valued measure μ defines a linear functional

$$T_\mu : \omega \mapsto \int_{\mathbb{R}^m} w_\omega \cdot d\mu : \wedge^{m-1}(\mathbb{R}^m) \to \mathbb{R}$$

called the $(m-1)$-*current* associated with μ. If $\mathbf{E}^m \mathsf{L}\, g$ is an m-current in $\mathbf{N}_m^{c,\infty}(\mathbb{R}^m)$, then for each $\omega \in \wedge^{m-1}(\mathbb{R}^m)$,

$$
\begin{aligned}
\langle \partial(\mathbf{E}^m \mathsf{L}\, g), \omega \rangle = \langle \mathbf{E}^m \mathsf{L}\, g, d\omega \rangle &= \int_{\mathbb{R}^m} g(x) \mathrm{div}\, w_\omega(x)\, dx \\
&= -\int_{\mathbb{R}^m} w_\omega \cdot d(Dg) = \langle -T_{Dg}, \omega \rangle
\end{aligned}
\tag{5.3.5}
$$

by equalities (5.3.1)–(5.3.4) and (1.7.3). Consequently,

$$\partial(\mathbf{E}^m \mathsf{L}\, g) = -T_{Dg} \tag{5.3.6}$$

for each $\mathbf{E}^m \mathsf{L}\, g$ in $\mathbf{N}_m^{c,\infty}(\mathbb{R}^m)$.

Suppose E is a locally BV set, and w_ω is the vector field associated with $\omega \in \wedge_C^{m-1}(\mathrm{cl}\, E)$. Given a $T := \mathbf{E}^m \mathsf{L}\, g$ in $\mathbf{N}_m^{c,\infty}(E)$, we let

$$\int_{\partial T} \omega := -\int_{\mathrm{cl}\, E} w_\omega \cdot d(Dg), \tag{5.3.7}$$

and call this number the *integral* of ω over ∂T. As $\int_{\partial T} \omega$ is defined by means of the Lebesgue integral, for emphasis, we denote it occasionally by $(L)\int_{\partial T} \omega$. Observe

$$\int_{\partial T} \omega = \langle \partial T, \omega \rangle \tag{5.3.8}$$

for each $T \in \mathbf{N}_m^{c,\infty}(\mathbb{R}^m)$ and each $\omega \in \wedge^{m-1}(\mathbb{R}^m)$ — cf. (5.3.2).

If $T \in \mathbf{N}_m^{c,\infty}(\mathbb{R}^m)$ and $\omega \in \wedge^{m-1}(\mathbb{R}^m)$, then the Stokes theorem

$$\int_T d\omega = \langle T, d\omega \rangle = \langle \partial T, \omega \rangle = \int_{\partial T} \omega \tag{5.3.9}$$

is a direct consequence of the definitions; see (5.3.2), (5.3.4), and (5.3.8). In fact, the definitions were set up so that (5.3.9) holds. We show next that the validity of Stokes' theorem can be extended to a substantially larger class of $(m-1)$-forms than $\wedge^{m-1}(\mathbb{R}^m)$.

Proposition 5.3.6. *If $g \in BV_c^\infty(\mathbb{R}^m)$, then*

$$\int_{\mathrm{cl}\, \{g \neq 0\}} w \cdot d(Dg) = -\int_{\mathbb{R}} \left(\int_{\partial_* \{g > t\}} w \cdot \nu_{\{g > t\}}\, d\mathcal{H}^{m-1} \right) dt$$

for each $w \in C(\mathrm{cl}\, \{g \neq 0\}; \mathbb{R}^m)$.

PROOF. If $w \in C_c^1(\mathbb{R}^m)$, then $\int_{\mathbb{R}^m} \operatorname{div} w(x)\, dx = 0$ by equality (1.8.1). Fubini's theorem implies

$$\int_{\mathbb{R}^m} g^+(x) \operatorname{div} w(x)\, dx = \int_0^\infty \left(\int_{\{g>t\}} \operatorname{div} w(x)\, dx \right) dt\,,$$

$$\int_{\mathbb{R}^m} g^-(x) \operatorname{div} w(x)\, dx = \int_0^\infty \left(\int_{\{-g \geq t\}} \operatorname{div} w(x)\, dx \right) dt$$

$$= \int_0^\infty \left(\int_{\mathbb{R}^m - \{g > -t\}} \operatorname{div} w(x)\, dx \right) dt$$

$$= -\int_0^\infty \left(\int_{\{g > -t\}} \operatorname{div} w(x)\, dx \right) dt$$

$$= -\int_{-\infty}^0 \left(\int_{\{g > t\}} \operatorname{div} w(x)\, dx \right) dt\,,$$

and using (1.7.3) and (1.8.1), we conclude

$$\int_{\mathbb{R}^m} w \cdot d(Dg) = -\int_{\mathbb{R}^m} g(x) \operatorname{div} w(x)\, dx$$

$$= -\int_{\mathbb{R}} \left(\int_{\{g > t\}} \operatorname{div} w(x)\, dx \right)$$

$$= -\int_{\mathbb{R}} \left(\int_{\partial_* \{g > t\}} w \cdot \nu_{\{g > t\}}\, d\mathcal{H}^{m-1} \right) dt\,.$$

Now suppose $w \in C(\operatorname{cl}\{g \neq 0\}; \mathbb{R}^m)$. Extend w to a vector field v in $C_c(\mathbb{R}^m; \mathbb{R}^m)$, and find a sequence $\{v_k\}$ in $C_c^1(\mathbb{R}^m; \mathbb{R}^m)$ that converges uniformly to v. As Dg is a finite measure,

$$\int_{\mathbb{R}^m} v \cdot d(Dg) = \lim \int_{\mathbb{R}^m} v_k \cdot d(Dg) \qquad (*)$$

by the dominated convergence theorem. Let

$$J(t) := \int_{\partial_* \{g > t\}} v \cdot \nu_{\{g > t\}}\, d\mathcal{H}^{m-1}\,,$$

$$J_k(t) := \int_{\partial_* \{g > t\}} v_k \cdot \nu_{\{g > t\}}\, d\mathcal{H}^{m-1}\,,$$

and observe that

$$\left|J(t) - J_k(t)\right| = \left|\int_{\partial_*\{g>t\}} (v - v_k) \cdot \nu_{\{g>0\}} \, d\mathcal{H}^{m-1}\right|$$

$$\leq \left\|\{g > t\}\right\| \cdot |v - v_k|_\infty$$

for $k = 1, 2, \ldots$, and $\mathcal{L}^1$-almost all $t \in \mathbb{R}$. In view of the coarea theorem, another application of the dominated convergence theorem gives

$$\int_\mathbb{R} J(t) \, dt = \lim \int_\mathbb{R} J_k(t) \, dt. \qquad (**)$$

Combining equalities $(*)$ and $(**)$ with the first part of the proof,

$$\int_{\mathbb{R}^m} v \cdot d(Dg) = \int_\mathbb{R} \left(\int_{\partial_*\{g>t\}} v \cdot \nu_{\{g>t\}} \, d\mathcal{H}^{m-1}\right) dt.$$

As both sets $\partial_*\{g > t\}$ and $\operatorname{supp} \|Dg\|$ are contained in $\operatorname{cl}\{g \neq 0\}$, the desired equality follows. □

Proposition 5.3.7. *The map $g \mapsto Dg$ defined on $BV_{\mathrm{loc}}(\mathbb{R}^m)$ is linear.*

PROOF. By equality (1.7.3), the functional

$$L_v : g \mapsto \int_{\mathbb{R}^m} v \cdot d(Dg) : BV(\mathbb{R}^m) \to \mathbb{R}$$

is linear for each $v \in C_c^1(\mathbb{R}^m; \mathbb{R}^m)$. Since $\|Dg\|$ is a Radon measure, and since every $v \in C_c(\mathbb{R}^m; \mathbb{R}^m)$ is uniformly approximable by elements of $C_c^1(\mathbb{R}^m; \mathbb{R}^m)$, the dominated convergence theorem shows that L_v is linear for each $v \in C_c(\mathbb{R}^m; \mathbb{R}^m)$. As for each $g \in BV_{\mathrm{loc}}(\mathbb{R}^m)$, the measure Dg is uniquely determined by the linear functional

$$v \mapsto L_v(g) : C_c(\mathbb{R}^m; \mathbb{R}^m) \to \mathbb{R}$$

[60, Theorem 13.9], the proposition follows. □

Corollary 5.3.8. *If E is a locally BV set, then the pairings*

$$(T, \omega) \mapsto \int_T \omega : \mathbf{N}_m^{c,\infty}(E) \times \wedge_R^m(E) \to \mathbb{R},$$

$$(T, \omega) \mapsto \int_{\partial T} \omega : \mathbf{N}_m^{c,\infty}(E) \times \wedge_C^{m-1}(\operatorname{cl} E) \to \mathbb{R}$$

are bilinear functionals.

PROOF. The bilinearity of the first pairing is obvious, and so is the linearity of the second pairing with respect to ω. Thus the corollary follows from (5.3.7) and Proposition 5.3.7. □

Let E be a locally BV set, and let w_ω be the vector field associated with a form $\omega \in \wedge_C^{m-1}(E)$. We call the *flux* of ω the flux of the vector field $w_\omega \in C(\mathrm{cl}\,E; \mathbb{R}^m)$ associated with ω. If the mean divergence $\mathfrak{div}\, w_\omega(x)$ of w_ω exists at $x \in \mathrm{int}_c E$ (see Definition 2.5.8), we let

$$\mathfrak{d}\omega(x) := \mathfrak{div}\, w_\omega(x)\, dx$$

and call this m-form the *mean exterior derivative* of ω at x. With this notation at hand, we formulate and prove the promised extension of the Stokes theorem.

Theorem 5.3.9. *Let E be a locally BV set, and suppose the flux of $\omega \in \wedge_C^{m-1}(\mathrm{cl}\,E)$ is almost derivable at each $x \in E$ except for points of a thin subset of E. Then $\mathfrak{d}\omega(x)$ is defined for almost all $x \in E$, the form $\mathfrak{d}\omega : x \mapsto \mathfrak{d}\omega(x)$ belongs to $\wedge_R^m(E)$, and*

$$(R)\int_T \mathfrak{d}\omega = (L)\int_{\partial T} \omega$$

for each $T \in \mathbf{N}_m^{c,\infty}(E)$.

PROOF. Let w_ω be the vector field associated with ω, and choose an m-current $T := \mathbf{E}^m \llcorner g$ in $\mathbf{N}_m^{c,\infty}(E)$. In view of Corollary 5.3.8, we may assume that $g \geq 0$. By Theorem 5.1.12, the function $\mathfrak{div}\, w_\omega$, defined almost everywhere in E, is R-integrable in E, which means $\mathfrak{d}\omega$ belongs to $\wedge_R^m(E)$. Denote by F the indefinite R-integral of $\mathfrak{div}\, w_\omega$, and find a bounded BV set $A \subset E$ containing $\{g \neq 0\}$. Theorems 5.2.2 and 5.1.12 together with Proposition 5.3.6 imply

$$\int_T \mathfrak{d}\omega = \int_A g(x)\, \mathfrak{div}\, w_\omega(x)\, dx = (F \llcorner g)(A)$$

$$= \langle F, g \rangle = \int_0^\infty F(\{g > t\})\, dt$$

$$= \int_0^\infty \left(\int_{\{g>t\}} \mathfrak{div}\, w_\omega\, d\mathcal{L}^m \right) dt$$

$$= \int_0^\infty \left(\int_{\partial_*\{g>t\}} v_\omega \cdot \nu_{\{g>t\}}\, d\mathcal{H}^{m-1} \right) dt$$

$$= -\int_{\mathrm{cl}\,\{g \neq 0\}} w_\omega \cdot d(Dg)$$

$$= -\int_{\mathrm{cl}\,E} w_\omega \cdot d(Dg) = \int_{\partial T} \omega\,. \qquad \square$$

5.4. Averaging

We show that the R-integral is an averaging process akin to the classical Riemann integral (hence the letter R). The following theorem is the first step in this direction; a more elaborate demonstration will be given in Section 5.5 below.

Theorem 5.4.1. *Let E be a locally BV set. A function f defined on $\mathrm{cl}_* E$ belongs to $R(E)$ if and only if there is a charge F in E having the following property: given $\varepsilon > 0$, we can find a gage δ on $B(1/\varepsilon) \cap \mathrm{cl}_* E$ so that*

$$\sum_{i=1}^{p} \left| f(x_i)|A_i| - F(A_i) \right| < \varepsilon$$

for each ε-regular δ-fine partition $P = \{(A_1, x_1), \ldots, (A_p, x_p)\}$ with $[P] \subset E$. In this case $F = (R)\int f(x)\,dx$.

PROOF. Suppose $f \in R(E)$, and let $F := \int f(x)\,dx$. There is a negligible set $N \subset \mathrm{cl}_* E$ such that $DF(x) = f(x)$ for each x in $\mathrm{cl}_* E - N$. Choose an $\varepsilon > 0$ and let $B_\varepsilon := B(1/\varepsilon)$. Given x in $\mathrm{cl}_* E - N$, find a $\rho_x > 0$ so that

$$\left| f(x)|B| - F(B) \right| < \frac{\varepsilon}{3|B_\varepsilon|}|B|$$

for each BV set B with $d(B \cup \{x\}) < \rho_x$ and $r(B \cup \{x\}) > \varepsilon$. Making ρ_x smaller, we can assume $B[x, \rho_x] \subset B_\varepsilon$ whenever $x \in B_\varepsilon \cap (\mathrm{cl}_* E - N)$. As $V_* F \llcorner \mathrm{cl}_* E$ is absolutely continuous, there is a gage σ on N with $F(\sigma, \varepsilon) < \varepsilon/3$. Let μ be the Radon measure associated with N and $\varepsilon/3$ according to Lemma 2.6.8. Making σ smaller, assume $|f(x)| \cdot |A| < \mu(A)$ for each $x \in N$ and each set $A \subset B[x, \sigma(x)]$. The formula

$$\delta(x) := \begin{cases} \sigma(x) & \text{if } x \in B_\varepsilon \cap N, \\ \rho_x & \text{if } x \in B_\varepsilon \cap (\mathrm{cl}_* E - N), \end{cases}$$

defines a gage δ on $B_\varepsilon \cap \mathrm{cl}_* E$. Now if $\{(A_1, x_1), \ldots, (A_p, x_p)\}$ is any ε-regular δ-fine partition, then

$$\sum_{i=1}^{p} \left| f(x_i)|A_i| - F(A_i) \right| \le$$

$$\sum_{x_i \in N} \left[\mu(A_i) + |F(A_i)| \right] + \sum_{x_i \notin N} \left| f(x_i)|A_i| - F(A_i) \right| \le$$

$$\mu(\mathbb{R}^m) + F(\sigma, \varepsilon) + \frac{\varepsilon}{3|B_\varepsilon|} \sum_{x_i \notin N} |A_i| < \varepsilon.$$

Conversely, suppose F and f satisfy the condition of the theorem. Choose a negligible set $N \subset \mathrm{cl}_* E$, and a positive $\varepsilon < 1/(2m)$. Let μ be as in the first part of the proof. Find a gage δ on $B_\varepsilon \cap \mathrm{cl}_* E$ so that

$$\sum_{i=1}^{p} \left| f(x_i)|A_i| - F(A_i) \right| < \frac{\varepsilon}{2}$$

for each ε-regular δ-fine partition $P = \{(A_1, x_1), \ldots, (A_p, x_p)\}$ with $[P] \subset E$. Making $\Delta := \delta \restriction (N \cap B_\varepsilon)$ smaller, we may assume that $|f(x)| \cdot |A| < \mu(A)$ for each $x \in N \cap B_\varepsilon$ and each $A \subset B[x, \Delta(x)]$. If $P := \{(A_1, x_1), \ldots, (A_p, x_p)\}$ is an ε-regular Δ-fine partition such that $[P] \subset E$, then

$$\sum_{i=1}^{p} |F(A_i)| \leq \sum_{i=1}^{p} \left| f(x_i)|A_i| - F(A_i) \right| + \sum_{i=1}^{p} |f(x_i)| \cdot |A_i|$$

$$< \frac{\varepsilon}{2} + \sum_{i=1}^{p} \mu(A_i) \leq \frac{\varepsilon}{2} + \mu(\mathbb{R}^m) < \varepsilon,$$

and hence $F(\Delta, \varepsilon; E) \leq \varepsilon$. As ε is arbitrary, this and Proposition 3.6.10 imply $V_* F(N) = 0$. Therefore $F \in CH_*(E)$.

Now denoting by S the set of all points x in $E \cap \mathrm{int}_c E$ at which F is derivable and $DF(x) \neq f(x)$, it suffices to show that S is negligible. To this end, select an integer $n \geq 1$ and let

$$S_n := \left\{ x \in S \cap B(n) : |f(x) - DF(x)| > \tfrac{1}{n} \right\}.$$

In view of Lemma 2.5.2, making δ smaller, we may assume

$$r\left[\left(E \cap B[x, r] \right) \cup \{x\} \right] > \varepsilon$$

whenever x is in $B_\varepsilon \cap (E \cap \mathrm{int}_c E)$ and $0 < r < \delta(x)$. Making δ still smaller, the inequality

$$\left| f(x)|E \cap B[x, r]| - F(E \cap B[x, r]) \right| > \frac{1}{n} |E \cap B[x, r]|$$

holds whenever $x \in B_\varepsilon \cap S_n$ and $0 < r < \delta(x)$. By Vitali's theorem, there are $x_i \in B_\varepsilon \cap S_n$ and $0 < r_i < \delta(x_i)/2$, $i = 1, 2, \ldots$, such that the close balls $A_i := B[x_i, r_i]$ are disjoint and $\left| B_\varepsilon \cap S_n - \bigcup_i A_i \right| = 0$. Since the collections $P_k := \{(E \cap A_1, x_1), \ldots, (E \cap A_k, x_k)\}$ are ε-regular δ-fine partitions with $[P_k] \subset E$, we obtain

$$|B_\varepsilon \cap S_n| \leq \sum_i |E \cap A_i| < n \sum_i \left| f(x_i)|E \cap A_i| - F(E \cap A_i) \right| \leq n\varepsilon.$$

It follows from the arbitrariness of ε that S_n is a negligible set, and so is the union $\bigcup_{n=1}^{\infty} S_n = S$. $\qquad\qquad\qquad\qquad\qquad\qquad\qquad\qquad\square$

Our next task is to characterize the Lebesgue integrability in a way comparable with Theorem 5.4.1. To this end, we need a lemma similar to Lemma 2.6.8.

Lemma 5.4.2. *Let E be a measurable set, and let $\varepsilon > 0$. There is an absolutely continuous Radon measure $\nu < \varepsilon$ such that*

$$\lim \frac{\nu(B_k)}{|B_k|} = \infty$$

whenever $\{B_k\}$ is a sequence of subsets of E with $|B_k| > 0$ for all k and $\lim d\big(B_k \cup \{x\}\big) = 0$ for an x in $\mathbb{R}^m - E$.

PROOF. Let $B = \mathbb{R}^m - E$, and find a decreasing sequence $\{U_i\}$ of open sets containing B so that $|U_i - B| < \varepsilon 2^{-i}$ for $i = 1, 2, \ldots$. Observe

$$\nu = \sum_{i=1}^{\infty} \big[\mathcal{L}^m \, \mathsf{L} \, (U_i - B) \big]$$

is an absolutely continuous Radon measure, and $\nu(\mathbb{R}^m) < \varepsilon$. Choose an $x \in B$ and a sequence $\{B_k\}$ of subsets of E with $\lim d\big(B_k \cup \{x\}\big) = 0$. If $i \geq 1$ is an integer, then $B_k \subset U_i - B$ for all sufficiently large k. Thus $\nu(B_k) \geq i|B_k|$ for all sufficiently large k, and the lemma follows. $\qquad\square$

Theorem 5.4.3. *Let E be a locally BV set. A function f defined on $\operatorname{cl} E$ belongs to $L_\ell^1(E)$ if and only if there is a charge F in E having the following property: given $\varepsilon > 0$, we can find a positive function δ defined on $B(1/\varepsilon) \cap \operatorname{cl} E$ so that*

$$\sum_{i=1}^{p} \Big| f(x_i)|A_i| - F(A_i) \Big| < \varepsilon$$

for each δ-fine partition $P = \{(A_1, x_1), \ldots, (A_p, x_p)\}$ with $[P] \subset E$. In this case $F = (L)\int f(x)\, dx$.

PROOF. Suppose F and f satisfy the condition of the theorem. Then f and F satisfy the condition of Theorem 5.4.1, and hence $F \in CH_*(E)$ and $DF = f$ almost everywhere in E. In view of Proposition 3.6.5 and Corollary 3.2.9, it suffices to show that $VF < \infty$. To this end, we use the idea employed previously in the proof of Proposition 3.6.14. Fix an $x \in \operatorname{cl} E$. Given a positive $\varepsilon < 1/\big(1 + |x|\big)$, find a positive function δ on $B(1/\varepsilon) \cap \operatorname{cl} E$ so that the condition of the theorem is satisfied. Let

$B_x := B\big(x, \delta(x)\big)$. Making $\delta(x)$ smaller, we may assume $\big|f(x)\big| \cdot |B_x| \le 1$. If $A_1, \ldots, A_p$ are disjoint BV subsets of $B_x \cap E$, then

$$\sum_{i=1}^{p} |F(A_i)| \le \sum_{i=1}^{p} \Big| f(x)|A_i| - F(A_i) \Big| + |f(x)| \sum_{i=1}^{p} |A_i|$$
$$< \varepsilon + \big| f(x) \big| \cdot |B_x| < 2,$$

since $P = \big\{(A_1, x), \ldots, (A_p, x)\big\}$ is a δ-fine partition with $[P] \subset E$. Consequently $VF(B_x \cap E) \le 2$. If B is a bounded BV set, then there are $x_i \in \operatorname{cl} E$ such that the open balls $B_{x_1}, \ldots, B_{x_k}$ cover $B \cap E$. Thus $VF(B) \le 2k$, and we conclude that $VF < \infty$.

Conversely, suppose $f \in L^1_\ell(E)$. Let $F := \int f(x)\, dx$, and observe $g = \overline{f \restriction E}$ belongs to $L^1_{\mathrm{loc}}(\mathbb{R}^m)$ with $\int g(x)\, dx = F$. Choose an $\varepsilon > 0$, and use Proposition 3.1.8 to find a positive function σ on $B(1/\varepsilon)$ so that the inequality

$$\sum_{i=1}^{p} \Big| g(x_i)|A_i| - F(A_i) \Big| < \frac{\varepsilon}{2}$$

holds for each σ-fine partition $\big\{(A_1, x_1), \ldots, (A_p, x_p)\big\}$. Let ν be the Radon measure associated with $\varepsilon/2$ and E according to Lemma 5.4.2. Making σ smaller, we can assume

$$\big| f(x) - g(x) \big| \cdot |B| < \nu(B)$$

for each $x \in B(1/\varepsilon) - E$ and $B \subset B\big[x, \sigma(x)\big] \cap E$. Let δ be the restriction of σ to $B(1/\varepsilon) \cap \operatorname{cl} E$, and choose a δ-fine partition

$$P := \big\{(A_1, x_1), \ldots, (A_p, x_p)\big\}$$

with $[P] \subset E$. The following calculation completes the argument:

$$\sum_{i=1}^{p} \Big| f(x_i)|A_i| - F(A_i) \Big| =$$
$$\sum_{x_i \notin E} \Big| f(x_i)|A_i| - F(A_i) \Big| + \sum_{x_i \in E} \Big| g(x_i)|A_i| - F(A_i) \Big| \le$$
$$\sum_{x_i \notin E} \big| f(x_i) - g(x_i) \big| \cdot |A_i| + \sum_{i=1}^{p} \Big| g(x_i)|A_i| - F(A_i) \Big| <$$
$$\sum_{x_i \notin E} \nu(A_i) + \frac{\varepsilon}{2} \le \nu(\mathbb{R}^m) + \frac{\varepsilon}{2} < \varepsilon. \qquad \square$$

Comparing Theorems 5.4.1 and 5.4.3, we see that the regularity of partitions and use of gages are critical components which distinguish the R-integrability from Lebesgue integrability.

Remark 5.4.4. Two comments are in order.

(i) The condition $[P] \subset E$ can be omitted in the statement of Theorem 5.4.1. Indeed, this transpires from the first part of the proof.

(ii) Employing results established in Section 5.7 below, it is easy to see that Theorem 5.4.3 remains valid when the words *"a charge F in E"* are replaced replaced by *"an additive function F on $\mathcal{BV}$ with $F = F \lfloor E$"*.

The formal similarity between Theorems 5.4.1 and 5.4.3 is striking. For the classical *Denjoy-Perron integral* [75, Chapter 8, Section 1] a result analogous to Theorem 5.4.1 was established independently by Henstock [30] and Kurzweil [42]; a comprehensive exposition can be found in [28]. Theorem 5.4.3 was proved by McShane [53] in a more general setting; see also [1].

Replacing cl E by $\mathrm{cl}_* E$ in Theorem 5.4.3 characterizes a linear space of functions defined almost everywhere in E that lies properly in between $L^1_\ell(E)$ and $R(E)$ — cf. Remark 6.1.2, (6) below. Replacing a positive function δ by a gage has a similar effect [8, Proposition 2.8]. In general, by minor modifications of Theorems 5.4.1 and 5.4.3, it is easy to define linear spaces of functions which properly contain $L^1_\ell(E)$. During the past twenty years the whole industry evolved around this idea, but it does not appear any significant discoveries were made. A notable exception is a recent work of Bongiorno, Di Piazza, and Preiss [10]: in the real line, they obtained a *minimal* Riemann type integral which integrates all Lebesgue integrable functions as well as the derivatives of all everywhere differentiable functions.

5.5. The Riemann approach

We refine the results of Section 5.4 by showing that a function f is R-integrable in a bounded BV set A if and only if the Stieltjes sums

$$\sum_{i=1}^{p} f(x_i)|A_i|$$

associated with suitable partitions $P := \big\{(A_1, x_1), \ldots, (A_p, x_p)\big\}$ approximate a real number I; in which case $I = (R)\int_A f(x)\, dx$.

In accordance with our general principle, view $\mathcal{BV}$ as the family of *equivalence classes* of all bounded BV sets, and denote by $\mathfrak{V}$ the smallest topology on $\mathcal{BV}$ for which all charges are continuous. In the topology $\mathfrak{V}$, a neighborhood base at the empty set $\emptyset$ is the family $\mathfrak{V}(\emptyset)$ consisting of all collections

$$\mathcal{U}(F_1,\ldots,F_k;\varepsilon) := \bigcap_{i=1}^{k}\left\{B \in \mathcal{BV} : |F_i(B)| < \varepsilon\right\}$$

where $k \geq 1$ is an integer, $F_1,\ldots,F_k$ are charges, and $\varepsilon > 0$. A neighborhood of $A \in \mathcal{BV}$ in the topology $\mathfrak{V}$ is the family

$$\left\{B \in \mathcal{BV} : A \triangle B \in \mathcal{U}\right\}$$

where $\mathcal{U} \in \mathfrak{V}(\emptyset)$. As $\mathcal{L}^m \restriction \mathcal{BV}$ is a charge, it is easy to see the space $(\mathcal{BV},\mathfrak{V})$ is Hausdorff. The next observation is obvious.

Observation 5.5.1. *If $A \in \mathcal{U}(F_1,\ldots,F_k;\eta)$ and $B \in \mathcal{U}(F_1,\ldots,F_k;\theta)$ do not overlap, then $A \cup B$ belongs to $\mathcal{U}(F_1,\ldots,F_k;\eta + \theta)$.*

If $F_1,\ldots,F_k$ are charges and $\varepsilon > 0$, then

$$\mathcal{U}(F_1,\ldots,F_k;\varepsilon) = \left\{B \in \mathcal{BV} : F(B) < \varepsilon\right\} \tag{5.5.1}$$

where $F = \max\{|F_1|,\ldots,|F_k|\}$. In view of this, the following lemma is a mere translation of Theorem 2.6.7.

Lemma 5.5.2. *Let $A \in \mathcal{BV}$ and $0 < \eta < 1/(2m\sqrt{m})$. Given a gage δ on $\mathrm{cl}_* A$ and $\mathcal{U} \in \mathfrak{V}(\emptyset)$, there is an η-regular δ-fine partition*

$$P := \left\{(A_1,x_1),\ldots,(A_p,x_p)\right\}$$

such that $[P] \subset A$ and $A - [P]$ belongs to $\mathcal{U}$.

Unless specified otherwise, throughout this section we denote by G an *arbitrary* but fixed function defined on $\mathcal{BV}$. Eventually, we assume that G is a charge, and then let $G = \mathcal{L}^m \restriction \mathcal{BV}$. However, in the beginning some nontrivial observations can be made without any assumptions about G.

Given a partition $P := \left\{(A_1,x_1),\ldots,(A_p,x_p)\right\}$ and a function f whose domain contains $\{x_1,\ldots,x_p\}$, we call the number

$$\sigma(f,P;G) := \sum_{i=1}^{p} f(x_i)G(A_i) \tag{5.5.2}$$

the *G-Stieltjes sum* associated with f and P. When no confusion can arise, we usually write $\sigma(P)$ instead of $\sigma(f,P;G)$.

Definition 5.5.3. Let A be a bounded BV set. A function f defined on cl_*A is called *S-integrable* in A with respect to G if there is a real number $I(f, A; G)$ satisfying the following condition: given $\varepsilon > 0$, we can find a gage δ on cl_*A and $\mathfrak{U} \in \mathfrak{V}(\emptyset)$ so that

$$\left| \sigma(f, P; G) - I(f, A; G) \right| < \varepsilon$$

for each ε-regular δ-fine partition P with $[P] \subset A$ and $A - [P]$ in $\mathfrak{U}$.

For a bounded BV set A, we denote by $S(A; G)$ the family of all functions f defined on cl_*A which are S-integrable in A with respect to G. The number $I(f, A; G)$ is uniquely determined by $f \in S(A; G)$. Indeed, if I and J satisfy the condition of Definition 5.5.3, then given $\varepsilon > 0$, we can find a gage δ on cl_*A and $\mathfrak{U} \in \mathfrak{V}(\emptyset)$ so that

$$\left| \sigma(f, P; G) - I \right| < \varepsilon \quad \text{and} \quad \left| \sigma(f, P; G) - J \right| < \varepsilon$$

for each ε-regular δ-fine partition P with $[P] \subset A$ and $A - [P]$ in $\mathfrak{U}$. According to Lemma 5.5.2, such a partition P exists whenever ε is sufficiently small. Thus for each sufficiently small $\varepsilon > 0$,

$$\left| I - J \right| \leq \left| I - \sigma(f, P; G) \right| + \left| \sigma(f, P; G) - J \right| < 2\varepsilon$$

and the equality $I = J$ follows. Clearly,

$$S(A; G) = S(A; G \sqcup A) \quad \text{and} \quad I(f, A; G) = I(f, A; G \sqcup A)$$

for each bounded BV set A and each $f \in S(A; G)$. A routine argument reveals $S(A; G)$ is a linear space and the map $f \mapsto I(f, A; G)$ is a linear functional on $S(A; G)$, which is nonnegative whenever G is nonnegative.

Lemma 5.5.4. *Let A be a bounded BV set. A function f defined on cl_*A belongs to $S(A; G)$ if and only if given $\varepsilon > 0$, we can find a gage δ on cl_*A and $\mathfrak{U} \in \mathfrak{V}(\emptyset)$ so that*

$$\left| \sigma(f, P; G) - \sigma(f, Q; G) \right| < \varepsilon$$

for all ε-regular δ-fine partitions P and Q such that $[P] \cup [Q] \subset A$ and both $A - [P]$ and $A - [Q]$ are in $\mathfrak{U}$.

PROOF. For $\varepsilon_k = 1/(2km\sqrt{m})$, $k = 1, 2, \ldots$, find gages δ_k on cl_*A and $\mathfrak{U}_k$ in $\mathfrak{V}(\emptyset)$ so that the condition of the proposition is satisfied. With no loss of generality, we may assume the sequences $\{\delta_k\}$ and $\{\mathfrak{U}_k\}$ are decreasing. It follows from Lemma 5.5.2 there are ε_k-regular δ_k-fine partitions P_k such that $[P_k] \subset A$ and $A - [P_k]$ belongs to $\mathfrak{U}_k$. Since $\left| \sigma(P_k) - \sigma(P_n) \right| < \varepsilon_k$ whenever $n \geq k \geq 1$, the sequence $\{\sigma(P_k)\}$ is Cauchy. Let $I = \lim \sigma(P_k)$, and choose an $\varepsilon > 0$. There is an integer

$s \geq 1$ for which $\varepsilon_s < \varepsilon/2$ and $|\sigma(P_s) - I| < \varepsilon/2$. If P is an ε-regular δ_s-fine partition with $[P] \subset A$ and $A - [P]$ in $\mathfrak{U}_s$, then

$$|\sigma(P) - I| \leq |\sigma(P) - \sigma(P_s)| + |\sigma(P_s) - I| < \varepsilon.$$

The converse is obvious. □

Proposition 5.5.5. *Let A be a bounded BV set, and let $f \in S(A; G)$. Then f is S-integrable in each BV set $B \subset A$, and*

$$F : B \mapsto I(f, A \cap B; G) : \mathcal{BV} \to \mathbb{R}$$

is a charge in A.

PROOF. Choose a positive $\varepsilon < 1/(2m\sqrt{m})$, and find a gage δ on $\mathrm{cl}_* A$ and $\mathfrak{U} := \mathfrak{U}(F_1, \ldots, F_k; \eta)$ so that

$$\left|\sigma(P) - I(f, A; G)\right| < \frac{\varepsilon}{2}$$

for each ε-regular δ-fine partition P with $[P] \subset A$ and $A - [P]$ in $\mathfrak{U}$.

Given a BV set $B \subset A$, let $C := A - B$ and denote by δ_B and δ_C the restrictions of δ to $\mathrm{cl}_* B$ and $\mathrm{cl}_* C$, respectively. Select ε-regular δ_B-fine partitions Q_1, Q_2 such that $[Q_1] \cup [Q_2] \subset B$ and both $B - [Q_1]$ and $B - [Q_2]$ belong to $\mathfrak{U}(F_1, \ldots, F_k; \eta/2)$. Employing Lemma 5.5.2, find an ε-regular δ_C-fine partition Q for which $[Q] \subset C$ and $C - [Q]$ belongs to $\mathfrak{U}(F_1, \ldots, F_k; \eta/2)$. Observe

$$P_1 = Q_1 \cup Q \quad \text{and} \quad P_2 = Q_2 \cup Q$$

are ε-regular δ-fine partitions with $[P_1] \cup [P_2] \subset A$. Since $A - [P_1]$ is the disjoint union of $B - [Q_1]$ and $C - [Q]$, Observation 5.5.1 shows that $A - [P_1]$ belongs to $\mathfrak{U}$ and, for the same reason, so does $A - [P_2]$. As

$$\left|\sigma(Q_1) - \sigma(Q_2)\right| = \left|\sigma(P_1) - \sigma(P_2)\right| \leq$$
$$\left|\sigma(P_1) - I(f, A; G)\right| + \left|I(f, A; G) - \sigma(P_2)\right| < \varepsilon,$$

$f \in S(B; G)$ according to Lemma 5.5.4.

Define a function F on $\mathcal{BV}$ by letting $F(B) := I(f, A \cap B; G)$ for each $B \in \mathcal{BV}$. Let A be the union of disjoint BV sets A_i, $i = 1, \ldots, p$. Find a neighborhood $\mathfrak{U}(G_1, \ldots, G_n; \theta)$ and gages δ_i on $\mathrm{cl}_* A_i$ so that

$$\left|\sigma(P_i) - F(A_i)\right| < \frac{\varepsilon}{2p}$$

for each ε-regular δ_i-fine partition P_i for which $[P_i] \subset A_i$ and $A_i - [P_i]$ belongs to $\mathfrak{U}(G_1, \ldots, G_k; \theta)$. With no loss of generality, we may assume

$$\mathfrak{U}(G_1, \ldots, G_n; \theta) \subset \mathfrak{U}(F_1, \ldots, F_k; \eta/p)$$

and $\delta_i(x) \leq \delta(x)$ for each $x \in \mathrm{cl}_* A_i$. Using Lemma 5.5.2, select partitions $P_1, \ldots, P_p$ with the above properties, and observe $P = \bigcup_{i=1}^{p} P_i$ is an ε-regular δ-fine partition with $[P] \subset A$. As $A - [P]$ is the union of disjoint sets $A_i - [P_i]$, Observation 5.5.1 shows $A - [P]$ is in $\mathcal{U}$. Thus

$$\left| F(A) - \sum_{i=1}^{p} F(A_i) \right| \leq \left| F(A) - \sigma(P) \right| + \left| \sum_{i=1}^{p} \left[\sigma(P_i) - F(A_i) \right] \right|$$

$$< \frac{\varepsilon}{2} + \sum_{i=1}^{p} \left| \sigma(P_i) - F(A_i) \right| < \varepsilon,$$

and the additivity of F follows from the arbitrariness of ε.

In view of (5.5.1), there is a $\theta > 0$ such that a BV set $D \subset A$ belongs to $\mathcal{U}(F_1, \ldots, F_k; \eta/2)$ whenever $\|D\| < 1/\varepsilon$ and $|D| < \theta$. Choose such a set D and let $E = A - D$. Since $f \in S(E; G)$, there is a gage δ_E on $\mathrm{cl}_* E$ and $\mathcal{U}(H_1, \ldots, H_r; \nu)$ such that

$$\left| \sigma(P_E) - F(E) \right| < \frac{\varepsilon}{2}$$

for each ε-regular δ_E-fine partition P_E for which $[P_E] \subset E$ and $E - [P_E]$ belongs to $\mathcal{U}(H_1, \ldots, H_r; \nu)$. We may assume

$$\mathcal{U}(H_1, \ldots, H_r; \nu) \subset \mathcal{U}(F_1, \ldots, F_k; \eta/2)$$

and $\delta_E(x) \leq \delta(x)$ for each $x \in \mathrm{cl}_* E$. By Lemma 5.5.2, there is a partition P_E with the above properties. As $A - [P_E]$ is the disjoint union of D and $E - [P_E]$, Observation 5.5.1 shows $A - [P_E]$ is in $\mathcal{U}$. Hence

$$\left| F(D) \right| = \left| F(A) - F(E) \right|$$

$$\leq \left| F(A) - \sigma(P_E) \right| + \left| \sigma(P_E) - F(E) \right| < \varepsilon,$$

which shows that F is $\mathfrak{T}$-continuous. $\qquad\qquad\square$

The next proposition is usually referred to as *Henstock's lemma*; the name *Saks-Henstock lemma* is also used [28, Lemma 9.11].

Proposition 5.5.6. *Let A be a bounded BV set. A function f defined on $\mathrm{cl}_* A$ belongs to $S(A; G)$ if and only if there is a charge F in A having the following property: given $\varepsilon > 0$, we can find a gage δ on $\mathrm{cl}_* A$ so that*

$$\sum_{i=1}^{p} \left| f(x_i) G(A_i) - F(A_i) \right| < \varepsilon$$

for each ε-regular δ-fine partition $P = \{(A_1, x_1), \ldots, (A_p, x_p)\}$ with $[P] \subset A$. In this case $F(B) = I(f, B; G)$ for every BV set $B \subset A$.

PROOF. Let F be a charge in A which satisfies the condition of the theorem. Choose an $\varepsilon > 0$ and find a corresponding gage δ on $\mathrm{cl}_* A$. If $P := \{(A_1, x_1), \ldots, (A_p, x_p)\}$ is an ε-regular δ-fine partition such that $[P] \subset A$ and $A - [P]$ belongs to $\mathcal{U}(F; \varepsilon)$, then

$$\left|\sigma(P) - F(A)\right| \leq \sum_{i=1}^{p}\left|f(x_i)G(A_i) - F(A_i)\right| + \left|F\big(A - [P]\big)\right| < 2\varepsilon.$$

It follows $f \in S(A; G)$. Select a BV set $B \subset A$, and find a gage δ_B on $\mathrm{cl}_* B$ and a $\mathcal{U}$ in $\mathfrak{V}(\emptyset)$ so that

$$\left|\sigma(Q) - I(f, B; G)\right| < \varepsilon$$

for each ε-regular δ_B-fine partition Q with $[Q] \subset B$ and $B - [Q]$ in $\mathcal{U}$. We may assume $\mathcal{U} \subset \mathcal{U}(F; \varepsilon)$ and $\delta_B(x) \leq \delta(x)$ for every $x \in \mathrm{cl}_* B$. Using Lemma 5.5.2, choose a partition $Q := \{(B_1, x_1), \ldots, (B_q, x_q)\}$ which satisfies the above conditions. We obtain

$$\left|I(f, B; G) - F(B)\right| \leq \left|I(f, B; G) - \sigma(Q)\right| + \left|\sigma(Q) - F(B)\right| <$$

$$\varepsilon + \sum_{i=1}^{q}\left|f(x_i)G(B_i) - F(B_i)\right| + \left|F\big(B - [Q]\big)\right| < 3\varepsilon,$$

and the arbitrariness of ε implies $F(B) = I(f, B; G)$.

Conversely, assume $f \in S(A; G)$. Proposition 5.5.5 implies

$$F : B \mapsto I(f, A \cap B; G) : \mathcal{BV} \to \mathbb{R}$$

is a charge in A. Select a positive $\varepsilon < 1/(2m\sqrt{m})$ and find a gage δ on $\mathrm{cl}_* A$ and $\mathcal{U} := \mathcal{U}(G_1, \ldots, G_n; \eta)$ so that

$$\left|\sigma(P) - F(A)\right| < \frac{\varepsilon}{3}$$

for each ε-regular δ-fine partition P with $[P] \subset A$ and $A - [P]$ in $\mathcal{U}$. By Lemma 5.5.2, each ε-regular δ-fine partition P with $[P] \subset A$, is a subcollection of an ε-regular δ-fine partition Q such that $[Q] \subset A$ and $A - [Q]$ is in $\mathcal{U}$. Thus it suffices to consider an ε-regular δ-fine partition $P := \{(A_1, x_1), \ldots, (A_p, x_p)\}$ for which $[P] \subset A$ and $A - [P]$ is in $\mathcal{U}$. Let $\delta_i = \delta \restriction \mathrm{cl}_* A_i$, and applying Lemma 5.5.2 again, find an ε-regular δ_i-fine partition P_i so that

$$\left|\sigma(P_i) - F(A_i)\right| < \frac{\varepsilon}{3p},$$

$[P_i] \subset A_i$, and $A_i - [P_i]$ is in $\mathcal{U}(G_1, \ldots, G_n; \eta/p)$. After a suitable reordering, we may assume that

$$\left|f(x_i)G(A_i) - F(A_i)\right| = \begin{cases} f(x_i)G(A_i) - F(A_i) & \text{if } i = 1, \ldots, k, \\ F(A_i) - f(x_i)G(A_i) & \text{if } i = k+1, \ldots, p, \end{cases}$$

where k is an integer with $0 \le k \le p$. The partitions

$$P_+ := \left\{(A_1, x_1), \ldots, (A_k, x_k)\right\} \cup \bigcup_{i=k+1}^{p} P_i,$$

$$P_- := \left\{(A_{k+1}, x_{k+1}), \ldots, (A_p, x_p)\right\} \cup \bigcup_{i=1}^{k} P_i,$$

are ε-regular and δ-fine with $[P_+] \cup [P_-] \subset A$. Since $A - [P_+]$ is the union of disjoint sets $A - [P]$ and $A_i - [P_i]$, $i = k+1, \ldots, p$, Observation 5.5.1 shows that the set $A - [P_+]$ belongs to $\mathcal{U}$ and, by the same argument, so does the set $A - [P_-]$. Therefore

$$\frac{\varepsilon}{3} > \sigma(P_+) - F(A)$$

$$= \sum_{i=1}^{k}\left[f(x_i)G(A_i) - F(A_i)\right] + \sum_{i=k+1}^{p}\left[\sigma(P_i) - F(A_i)\right]$$

$$\ge \sum_{i=1}^{k}\left|f(x_i)G(A_i) - F(A_i)\right| - \frac{\varepsilon}{3} \cdot \frac{p-k}{p}$$

and, by a similar calculation,

$$\frac{\varepsilon}{3} > \sum_{i=k+1}^{p}\left|f(x_i)G(A_i) - F(A_i)\right| - \frac{\varepsilon}{3} \cdot \frac{k}{p}.$$

Adding the previous inequalities, we obtain

$$\sum_{i=1}^{p}\left|f(x_i)G(A_i) - F(A_i)\right| < \varepsilon,$$

which establishes the proposition. $\qquad\square$

Proposition 5.5.7. *Let* $A \in \mathcal{BV}$, *and for* $h \in S(A; G)$, *let*

$$H : B \mapsto I(h, A \cap B; G) : \mathcal{BV} \to \mathbb{R}.$$

A function f *defined on* $\mathrm{cl}_* A$ *belongs to* $S(A; H)$ *if and only if the product* fh *belongs to* $S(A; G)$, *in which case*

$$I(f, A; H) = I(fh, A; G).$$

PROOF. Choose an $\varepsilon > 0$, and for $n = 1, 2, \ldots$, find a gage δ_n on $\mathrm{cl}_* A$ so that

$$\sum_{i=1}^{p} \left| h(x_i) G(A_i) - H(A_i) \right| < \frac{\varepsilon}{n2^n}$$

for each ε-regular δ_n-fine partition $P := \{(A_1, x_1), \ldots, (A_p, x_p)\}$ with $[P] \subset A$. Such gages exist by Proposition 5.5.6. Observe $\mathrm{cl}_* A$ is the disjoint union of the sets

$$E_n := \left\{ x \in \mathrm{cl}_* A : n - 1 \le \left| f(x) \right| < n \right\}, \quad n = 1, 2, \ldots,$$

and define a gage δ on $\mathrm{cl}_* A$ so that $\delta(x) := \delta_n(x)$ whenever $x \in E_n$. If $P := \{(A_1, x_1), \ldots, (A_p, x_p)\}$ is an ε-regular δ-fine partition such that $[P] \subset A$, then

$$\left| \sigma(fh, P; G) - \sigma(f, P; H) \right| \le$$

$$\sum_{i=1}^{p} \left| f(x_i) \right| \cdot \left| h(x_i) G(A_i) - H(A_i) \right| \le$$

$$\sum_{n=1}^{\infty} \sum_{x_i \in E_n} \left| f(x_i) \right| \cdot \left| h(x_i) G(A_i) - H(A_i) \right| < \sum_{n=1}^{\infty} n \frac{\varepsilon}{n2^n} = \varepsilon$$

and the proposition follows.   □

The following theorem demonstrates the Riemannian nature of the R-integral.

Theorem 5.5.8. *Let $G := \mathcal{L}^m \upharpoonright \mathcal{BV}$, and let A be a bounded BV set. A function f defined on $\mathrm{cl}_* A$ belongs to $R(A)$ if and only if it belongs to $S(A; G)$, in which case*

$$I(f, A; G) = \int_A f(x)\, dx.$$

Theorem 5.5.8 is a direct consequence of Proposition 5.5.6 and Theorem 5.4.1. It implies, in particular, that for $G := \mathcal{L}^m \upharpoonright \mathcal{BV}$ the values f takes on $\mathrm{cl}_* A - A$ affect neither the S-integrability of f, nor the number $I(f, A; G)$. Example 5.5.10 below shows this is not true in general.

Corollary 5.5.9. *Let A be a bounded BV set, and let H be the indefinite R-integral of an $h \in R(A)$. Then $BV_\ell^\infty(\mathrm{cl}_* A) \subset S(A; H)$ and*

$$I(g, A; H) = (H \sqcup g)(A)$$

for each $g \in BV_\ell^\infty(\mathrm{cl}_* A)$.

In view of Proposition 5.5.7, the corollary follows immediately from Theorems 5.5.8 and 5.2.2.

Example 5.5.10. Let G be the charge associated with the devil's stair-case s_B in a cell B (see Example 3.2.3 and Remark 2.1.5). Observe that $A := B - C_B$ is a BV set in $\mathbb{R}$ with $\mathrm{cl}_* A = B$. The formula

$$\delta(x) := \begin{cases} \mathrm{dist}(x, C_B) & \text{if } x \in B - C_B, \\ 1 & \text{if } x \in C_B, \end{cases}$$

defines a positive gage on $\mathrm{cl}_* A$, and it is easy to see $\sigma(\chi_A, P; G) = 0$ for each δ-fine partition P. Thus $I(\chi_A, A; G) = 0$. On the other hand, Henstock's lemma implies

$$I(\chi_B, A; G) = G(A) = (G \,\mathsf{L}\, \chi_A)(A)$$

and $G(A) = G(B) = 1$. Note, however, $\chi_B(x)$ is equal to the precise value $\chi_A^*(x)$ of χ_A at each $x \in \mathbb{R} - \partial B$ (Notation 4.4.2).

Question 5.5.11. Let G be a charge, and let A be a bounded BV set.

(i) Does the inclusion $BV_\ell^\infty(\mathrm{cl}_* A) \subset S(A; G)$ hold?

(ii) Assuming the answer to (i) is affirmative, does the equality

$$I(f^*, A; G) = (G \,\mathsf{L}\, f)(A)$$

hold when $f \in BV_\ell^\infty(\mathrm{cl}_* A)$ and f^* is the precise value of $\overline{f}$?

5.6. Charges as distributional derivatives

The concept of S-integrability introduced in Section 5.5 is a convenient vehicle for interpreting charges as distributional derivatives of functions in $L^1_{\mathrm{loc}}(\mathbb{R}^m)$. We begin with a theorem that gives a partial answer to Question 5.5.11.

Theorem 5.6.1. *Let G be a charge, and let A be a bounded BV set. If $\varphi : \mathrm{cl}_* A \to \mathbb{R}$ is Lipschitz, then $\varphi \in S(A; G) \cap BV_\ell^\infty(\mathrm{cl}_* A)$ and*

$$I(\varphi, A; G) = (G \,\mathsf{L}\, \varphi)(A).$$

PROOF. Using Kirszbraun's theorem, extend φ to a bounded Lipschitz function f defined on $\mathbb{R}^m$, and let $c := \mathrm{Lip}(f)$. Then $f \in BV_{\mathrm{loc}}^\infty(\mathbb{R}^m)$ and

$$\|Df\|(E) \leq c|E| \qquad (*)$$

for each set $E \subset \mathbb{R}^m$ [22, Section 4.2.3]; in particular $\varphi \in BV_\ell^\infty(\mathrm{cl}_* A)$. With no loss of generality, we may assume $0 \leq f \leq 1$. Let $F := G \,\mathsf{L}\, f$

and $H := G \times \mathcal{L}^1$. Choose a positive $\varepsilon \leq 1/(1+c)$. By Proposition 2.2.6, there is a $\theta > 0$ such that

$$|H(C)| \leq \theta|C| + \varepsilon^2 (\|C\| + 1) \qquad (**)$$

for every BV set $C \subset A \times [0,1]$. View the number $\delta := \varepsilon/(c\theta)$ as a constant positive gage on $\mathrm{cl}_* A$, and select an ε-regular δ-fine partition $P := \{(A_1, x_1), \ldots, (A_p, x_p)\}$ with $[P] \subset A$. As $\{x_1, \ldots, x_p\} \subset \mathrm{cl}_* A$, we have $f(x_i) = \varphi(x_i)$ for $i = 1, \ldots, p$. The rest of the proof is similar to, but simpler than, that of Proposition 4.5.6. The sets

$$C_i^1 := A_i \times [0, f(x_i)] \quad \text{and} \quad C_i^2 := \Sigma_{f\chi_{A_i}}$$

are contained in $A \times [0,1]$. For $j, k = 1, 2$ and a set $T \subset \{1, \ldots, p\}$, Lemma 4.4.5 and $(*)$ yield

$$\left| \bigcup_{i \in T} (C_i^j - C_i^k) \right| = \sum_{i \in T} |C_i^j - C_i^k| \leq \sum_{i=1}^{p} \int_{A_i} |f(y) - f(x_i)| \, dy$$

$$\leq c \sum_{i=1}^{p} d(A_i \cup \{x_i\})|A_i| < c\delta \sum_{i=1}^{p} |A_i| \leq \frac{\varepsilon}{\theta}|A|,$$

$$\left\| \bigcup_{i \in T} (C_i^j - C_i^k) \right\| \leq \sum_{i \in T} \|C_i^j - C_i^k\|$$

$$\leq \sum_{i=1}^{p} \left[(2 + c)|A_i| + \int_{\partial A_i} |f(y) - f(x_i)| \, d\mathcal{H}^{m-1}(y) \right]$$

$$< (2 + c)|A| + c \sum_{i=1}^{p} d(A_i \cup \{x_i\})\|A_i\|$$

$$< (2 + c)|A| + \frac{c}{\varepsilon} \sum_{i=1}^{p} |A_i| \leq \left(2 + c + \frac{c}{\varepsilon} \right)|A|.$$

From these inequalities and $(**)$, we obtain

$$\sum_{i \in T} H(C_i^j - C_i^k) = H \left[\bigcup_{i \in T} (C_i^j - C_i^k) \right] < \varepsilon\gamma$$

where $\gamma := 1 + (4 + c)|A|$ does not depend on ε. Since T is arbitrary,

$$\sum_{i=1}^{p} \left| H(C_i^1) - H(C_i^2) \right| < 4\varepsilon\gamma.$$

As $H(C_i^1) = \varphi(x_i)G(A_i)$ and $H(C_i^2) = F(A_i)$, we infer

$$\sum_{i=1}^{p} \left| \varphi(x_i)G(A_i) - F(A_i) \right| < 4\varepsilon\gamma.$$

Now $\varphi \in S(A; G)$ according to Henstock's lemma. In view of (4.5.2),

$$I(\varphi, A; G) = F(A) = (G \mathbin{\llcorner} f)(A) = \langle G, f\chi_A \rangle$$
$$= \langle G, \overline{\varphi}\chi_A \rangle = (G \mathbin{\llcorner} \overline{\varphi})(A) = (G \mathbin{\llcorner} \varphi)(A);$$

since $f\chi_A = \overline{\varphi}\chi_A$ almost everywhere by Corollary 1.5.3.   □

The following corollary is a direct consequence of Theorem 5.6.1 and Propositions 5.5.7 and 5.5.5.

Corollary 5.6.2. *Let G be any function defined on $\mathbf{BV}$, and let A be a bounded BV set. Each Lipschitz function defined on $\mathrm{cl}_* A$ is a multiplier for the linear space $S(A; G)$.*

Remark 5.6.3. Observe that in the proof of Theorem 5.6.1 we used only a constant gage. Employing more sophisticated gages, it is not difficult to establish Theorem 5.6.1 for any function $f \in BV_{\mathrm{loc}}^{\infty}(\mathbb{R}^m)$ such that the restriction $f \upharpoonright \mathrm{cl}_* A$ is Lipschitz at each $x \in \mathrm{cl}_* A$ except for points of a thin set [66, Theorem 3.4].

Let G be a charge. If A is a bounded BV set, we let

$$\int_A \varphi \, dG := I(\varphi, A; G)$$

for each Lipschitz function φ defined on $\mathrm{cl}_* A$. This is a legitimate notation, which extends the meaning of $\int_A \varphi \, d\mu$ where μ is either a Borel regular or signed measure. Indeed, if μ is a Borel regular measure for which $G := \mu \upharpoonright \mathbf{BV}$ is a charge, then μ is an absolutely continuous Radon measure. By the Radon-Nikodym theorem, there is a $g \in L_{\mathrm{loc}}^1(\mathbb{R}^m)$ such that $\mu = \mathcal{L}^m \mathbin{\llcorner} g$. Employing Theorem 5.5.8 together with Propositions 5.1.2 and 5.5.7, we obtain

$$\int_A \varphi \, dG = I(\varphi, A; G) = I(\varphi g, A; \mathcal{L}^m \upharpoonright \mathbf{BV})$$

$$= (R)\int_A \varphi g \, d\mathcal{L}^m = (L)\int_A \varphi g \, d\mathcal{L}^m = (L)\int_A \varphi \, d\mu.$$

As on all bounded sets every signed measure is the difference of two Radon measures, the equality $\int_A \varphi \, dG = (L)\int_A \varphi \, d\mu$ remains valid when μ is a signed measure.

If G is a charge and φ is a Lipschitz function on $\mathbb{R}^m$ with compact support, we let

$$\int_{\mathbb{R}^m} \varphi \, dG := \langle G, \varphi \rangle. \tag{5.6.1}$$

Again this is an appropriate notation: if A is a bounded BV set such that $\{\varphi \neq 0\} \subset \operatorname{cl}_* A$, then Theorem 5.6.1 implies

$$\int_{\mathbb{R}^m} \varphi \, dG = \langle G, \varphi \chi_{\operatorname{cl}_* A} \rangle = \langle G, \varphi \chi_A \rangle = (G \, \mathsf{L} \, \varphi)(A) = \int_A \varphi \, dG \, ;$$

since $\varphi \chi_{\operatorname{cl} A} = \varphi \chi_A$ almost everywhere.

With the above definitions at hand it is easy to show that weak derivatives of certain functions are charges. If $G_1, \ldots, G_m$ are charges, we call the map

$$G = (G_1, \ldots, G_m) : \mathcal{BV} \to \mathbb{R}^m$$

a *vector charge*, and for a $\varphi = (\varphi_1, \ldots, \varphi_m)$ in $C_c^1(\mathbb{R}^m; \mathbb{R}^m)$, we let

$$\int_{\mathbb{R}^m} \varphi \cdot dG := \sum_{i=1}^m \int_{\mathbb{R}^m} \varphi_i \, dG_i \, .$$

Denote by $W_{\operatorname{loc}}(\mathbb{R}^m)$ the linear space of all functions $f \in L^1_{\operatorname{loc}}(\mathbb{R}^m)$ whose distributional gradient is a vector charge. Explicitly, a function $f \in L^1_{\operatorname{loc}}(\mathbb{R}^m)$ belongs to $W_{\operatorname{loc}}(\mathbb{R}^m)$ if there is a vector charge

$$\mathcal{D}f = (\mathcal{D}_1 f, \ldots, \mathcal{D}_m f),$$

called the *distributional gradient* of f, such that

$$\int_{\mathbb{R}^m} f(x) \operatorname{div} \varphi(x) \, dx = -\int_{\mathbb{R}^m} \varphi \cdot d(\mathcal{D}f) \tag{5.6.2}$$

for each $\varphi \in C_c^1(\mathbb{R}^m; \mathbb{R}^m)$. If G is a charge and $\int_{\mathbb{R}^m} \varphi \cdot dG = 0$ for every $\varphi \in C_c^1(\mathbb{R}^m)$, then (5.6.1) together with Theorems 1.7.1, (2) and 4.1.5 imply $G = 0$ (cf. the proof of Proposition 5.6.4 below). It follows that each $f \in W_{\operatorname{loc}}(\mathbb{R}^m)$ determines uniquely its distributional gradient $\mathcal{D}f$.

Proposition 5.6.4. *If $f \in BV_{\operatorname{loc}}(\mathbb{R}^m) \cap W_{\operatorname{loc}}(\mathbb{R}^m)$, then*

$$Df(A) = \mathcal{D}f(A)$$

for each bounded BV set A. In particular, $\|Df\|$ is absolutely continuous and $V(\mathcal{D}_i f) < \infty$ for $i = 1, \ldots, m$.

PROOF. According to (1.7.3) and (5.6.2),

$$\int_{\mathbb{R}^m} \varphi \, d(D_1 f) = \int_{\mathbb{R}^m} \varphi \, d(\mathcal{D}_1 f) \qquad (*)$$

for each $\varphi \in C_c^1(\mathbb{R}^m)$. Choose a bounded BV set A, and an open bounded set U containing $\operatorname{cl} A$. Apply Theorem 1.7.1 to find a sequence $\{\varphi_i\}$ in $C_c^1(\mathbb{R}^m)$ so that $\operatorname{supp}\varphi_i \subset U$ for $i = 1, 2, \ldots$,

$$\lim |\varphi_i - \chi_A|_1 = 0, \quad \text{and} \quad \sup \|\varphi_i\| \le \|A\| + 1.$$

Passing to a subsequence, we may assume that $\{\varphi_i\}$ converges to χ_A almost everywhere and $\sup |\varphi_i|_\infty \le 1$; in particular, the sequence $\{\varphi_i\}$ $\mathcal{T}$-converges to χ_A. From the dominated convergence theorem, equalities $(*)$ and (5.6.1), and Theorem 4.1.5, we obtain

$$D_1 f(A) = \lim \int_{\mathbb{R}^m} \varphi_i \, d(D_1 f) = \lim \int_{\mathbb{R}^m} \varphi_i \, d(\mathcal{D}_1 f)$$

$$= \lim \langle \mathcal{D}_1 f, \varphi_i \rangle = \mathcal{D}_1 f(A),$$

and the proposition follows by symmetry. $\qquad \square$

For the definition and properties of the the *Sobolev space* $W_{\mathrm{loc}}^{1,1}(\mathbb{R}^m)$ we refer to [22, Section 4.1]. The next corollary is an immediate consequence of Proposition 5.6.4 and [22, Section 5.1, Example 1].

Corollary 5.6.5. $W_{\mathrm{loc}}(\mathbb{R}^m) \cap BV_{\mathrm{loc}}(\mathbb{R}^m) = W_{\mathrm{loc}}^{1,1}(\mathbb{R}^m)$.

Define the space $W_{\mathrm{loc}}(\mathbb{R}^m; \mathbb{R}^m)$ according to Convention 1.1.1. For a vector field $v = (v_1, \ldots, v_m)$ in $W_{\mathrm{loc}}(\mathbb{R}^m; \mathbb{R}^m)$, let

$$\mathcal{D}\mathrm{iv}\, v = \sum_{i=1}^m \mathcal{D}_i v_i \qquad (5.6.3)$$

and call this charge the *global divergence* of v. From (5.6.2), it is easy to deduce that

$$\int_{\mathbb{R}^m} v(x) \cdot D\varphi(x) \, dx = -\int_{\mathbb{R}^m} \varphi \, d(\mathcal{D}\mathrm{iv}\, v) \qquad (5.6.4)$$

for each $\varphi \in C_c^1(\mathbb{R}^m)$.

Theorem 5.6.6. *If* $v \in W_{\mathrm{loc}}(\mathbb{R}^m; \mathbb{R}^m)$ *is a continuous vector field, then*

$$\mathcal{D}\mathrm{iv}\, v(A) = \int_{\partial_* A} v \cdot \nu \, d\mathcal{H}^{m-1}$$

for each bounded BV set A.

PROOF. By Theorem 1.7.1, (2), there is a sequence $\{\varphi_k\}$ in $C_c^1(\mathbb{R}^m)$ which $\mathcal{T}$-converges to χ_A. As the supports of χ_A and of each $D\varphi_k$ are contained in a fixed compact set, it follows from [22, Section 5.2, Theorem 3] that

$$\lim \int_{\mathbb{R}^m} v(x) \cdot D\varphi_k(x)\, dx = \int_{\mathbb{R}^m} v \cdot d(D\chi_A)\,.$$

In view of this, Theorem 4.1.5 together with (5.6.1) and (5.6.4) imply

$$\mathcal{D}\mathrm{iv}\, v(A) = \langle \mathcal{D}\mathrm{iv}\, v, \chi_A \rangle = \lim \langle \mathcal{D}\mathrm{iv}\, v, \varphi_k \rangle$$

$$= \lim \int_{\mathbb{R}^m} \varphi_k\, d(\mathcal{D}\mathrm{iv}\, v) = -\lim \int_{\mathbb{R}^m} v(x) \cdot D\varphi_k(x)\, dx$$

$$= -\int_{\mathbb{R}^m} v \cdot d(D_i\chi_A) = \int_{\partial_* A} v \cdot \nu_A\, d\mathcal{H}^{m-1}\,;$$

since $D\chi_A = -\mathcal{H}^{m-1} \llcorner \nu_A$ according to Theorem 1.8.2, (3).   □

Whether the space $W_{\mathrm{loc}}(\mathbb{R}^m)$ has some additional useful properties is unclear. Nonetheless, we show that a continuous function defined on $\mathbb{R}^m$ belongs to $W_{\mathrm{loc}}(\mathbb{R}^m)$ whenever it is Lipschitz at each $x \in \mathbb{R}^m$ except for points of a thin set.

Proposition 5.6.7. *Let $v = (f_1, \ldots, f_m)$ be in $C(\mathbb{R}^m; \mathbb{R}^m)$, and let T be a thin set such that the flux of each projected vector field*

$$v_1 = (f_1, 0, \ldots, 0)\,, \ldots, v_m = (0, \ldots, 0, f_m)$$

is almost derivable at all x in $\mathbb{R}^m - T$. Then $v \in W_{\mathrm{loc}}(\mathbb{R}^m; \mathbb{R}^m)$ and

$$\mathcal{D}\mathrm{iv}\, v(A) = (R)\int_A \mathfrak{d}\mathrm{iv}\, v(x)\, dx$$

for every bounded BV set A.

PROOF. Fix integers i and j with $1 \leq i, j \leq m$. By rotation, it follows from our assumptions that the flux of the vector field

$$w_{i,j} := (0, \ldots, \underset{j\text{-th place}}{f_i}, \ldots, 0)$$

is almost derivable at each x in $\mathbb{R}^m - T$. According to Theorem 5.1.12, the function $\mathfrak{d}\mathrm{iv}\, w_{i,j}$ is R-integrable in $\mathbb{R}^m$, and we denote by $F_{i,j}$ the indefinite R-integral of $\mathfrak{d}\mathrm{iv}\, w_{i,j}$. Let $\varphi \in C^1(\mathbb{R}^m)$, and find an open set

$U \in \boldsymbol{BV}$ containing supp φ. Corollary 5.2.7 and Theorem 5.2.2 yield

$$\int_{\mathbb{R}^m} f_i(x) \frac{\partial \varphi}{\partial x_j}(x)\, dx = \int_U w_{i,j}(x) \cdot D\varphi(x)\, dx$$

$$= -\int_U \mathfrak{div}\, w_{i,j}(x)\, \varphi(x)\, dx$$

$$= -(F_{i,j} \, \mathsf{L} \, \varphi)(U) = -\langle F_{i,j}, \varphi \rangle$$

$$= -\int_{\mathbb{R}^m} \varphi\, dF_{i,j}\,.$$

By symmetry, $v \in W_{\mathrm{loc}}(\mathbb{R}^m; \mathbb{R}^m)$ and $\mathcal{D}_j f_i = F_{i,j}$. Therefore,

$$\mathrm{Div}\, v(A) = \sum_{i=1}^m F_{i,i}(A) = \sum_{i=1}^m \int_A \mathfrak{div}\, w_{ii}(x)\, dx = \int_A \mathfrak{div}\, v(x)\, dx$$

for each bounded BV set A. $\square$

Corollary 5.6.8. *Let $f \in C(\mathbb{R}^m)$, and let T be a thin set such that f is Lipschitz at each x in $\mathbb{R}^m - T$. Then $f \in W_{\mathrm{loc}}(\mathbb{R}^m)$ and*

$$\mathcal{D}_i f(A) = (R) \int_A \frac{\partial f}{\partial x_i}(x)\, dx$$

for $i = 1, \ldots, m$ and each bounded BV set A.

PROOF. According to Stepanoff's theorem, f is differentiable almost everywhere in $\mathbb{R}^m$. Fix an integer i with $1 \le i \le m$, and let

$$v := (0, \ldots, \underset{i\text{-th place}}{f}, \ldots, 0)\,.$$

By Example 2.3.2, the vector field v satisfies the assumptions of Proposition 5.6.7, and $\mathfrak{div}\, v(x) = (\partial f/\partial x_i)(x)$ for almost all $x \in \mathbb{R}^m$. The corollary follows from Proposition 5.6.7. $\square$

5.7. The Lebesgue integral

Elaborating on the formal similarity between Theorems 5.4.1 and 5.4.3 we present a Riemannian definition of the Lebesgue integral — an idea discovered and developed by McShane [54, 55]. Claims similar to those established in Section 5.5 are stated without proofs. Throughout this section, we let

$$\sigma(f, P) := \sum_{i=1}^p f(x_i)|A_i| \tag{5.7.1}$$

for each partition $P := \{(A_1, x_1), \ldots, (A_p, x_p)\}$ and each function f whose domain contains $\{x_1, \ldots, x_p\}$. Thus $\sigma(f, P) = \sigma(f, P; \mathcal{L}^m \upharpoonright \boldsymbol{BV})$ in view of (5.5.2).

Definition 5.7.1. Let A be a bounded BV set. A function f defined on cl A is called *L-integrable* in A if there is a real number $J(f, A)$ satisfying the following condition: given $\varepsilon > 0$, there is a positive function δ defined on cl A so that

$$\left| \sigma(f, P) - J(f, A) \right| < \varepsilon$$

for each δ-fine partition P with $[P] = A$.

For a bounded BV set A, we denote by $L(A)$ the family of all functions defined on cl A which are L-integrable in A. If $f \in L(A)$, then using Proposition 2.6.3, it is easy to show the number $J(f, A)$ is unique. Moreover, $L(A)$ is a linear space, and the map $f \mapsto J(f, A)$ is a nonnegative linear functional on $L(A)$.

Lemma 5.7.2. *Let A be a bounded BV set. A function f defined on cl A belongs to $L(A)$ if and only if given $\varepsilon > 0$, there is a function $\delta > 0$ defined on cl A such that*

$$\left| \sigma(f, P) - \sigma(f, Q) \right| < \varepsilon$$

for each δ-fine partitions with $[P] = [Q] = A$.

Proposition 5.7.3. *Let A be a bounded BV set, and let $f \in L(A)$. Then f is L-integrable in each BV set $B \subset A$, and*

$$F : B \mapsto J(f, B) : \mathcal{BV}(A) \to \mathbb{R}$$

is an additive function on $\mathcal{BV}(A)$.

Proposition 5.7.4. *Let A be a bounded BV set. A function f defined on cl A belongs to $L(A)$ if and only if there is an additive function F on $\mathcal{BV}(A)$ having the following property: given $\varepsilon > 0$, we can find a function $\delta > 0$ defined on cl A so that*

$$\sum_{i=1}^{p} \left| f(x_i)|A_i| - F(A_i) \right| < \varepsilon$$

for each δ-fine partition $P := \{(A_1, x_1), \ldots, (A_p, x_p)\}$ with $[P] \subset A$. In this case $F(B) = J(f, B)$ for every BV set $B \subset A$.

Proposition 5.7.4 is a form of Henstock's lemma for the L-integral. Together with Theorem 5.4.3, it implies the following result.

Theorem 5.7.5. *Let A be a bounded BV set, and let f be a function defined on cl A. If $f \in L^1(A)$, then f is L-integrable in A and*

$$J(f, A) = (L) \int_A f(x)\,dx.$$

Corollary 5.7.6. *Let A be a bounded BV set, and let f and g be functions defined on cl A. If $f(x) = g(x)$ for almost all $x \in A$, then f is L-integrable in A if and only if g is, in which case $J(f, A) = J(g, A)$.*

PROOF. Theorem 5.7.5 shows that $f - g$ belongs to $L(A)$ and $J(f - g, A) = 0$. □

If A is a bounded BV set and $f \in L(A)$, then Proposition 5.7.3 implies

$$J(f) : B \mapsto J(f, A \cap B) : \mathcal{BV} \to \mathbb{R}$$

is an additive function on $\mathcal{BV}$ and $J(f) = J(f) \, \mathsf{L} \, A$. We establish the converse of Theorem 5.7.5 by showing that $J(f)$ is a charge. Observe

$$J(f)(B) = J(f\chi_B, A) \tag{5.7.2}$$

for each bounded BV set B. Indeed, by Proposition 5.7.3 and Corollary 5.7.6,

$$J(f\chi_B, A) = J(f\chi_B, A \cap B) + J(f\chi_B, A - B) = J(f, A \cap B) = J(f)(B).$$

Proposition 5.7.7. *Let A be a bounded BV set, and let $\{f_n\}$ be an increasing sequence in $L(A)$ converging to a function f defined on $\operatorname{cl} A$. If $\lim J(f_n, A) < \infty$, then $f \in L(A)$ and $\lim J(f_n, A) = J(f, A)$.*

PROOF. For each BV set $B \subset A$ and $n = 1, 2, \ldots$, let $F_n(B) = J(f_n, B)$. Further let $J = \lim F_n(A)$. Choose an $\varepsilon > 0$, and find functions $\delta_n > 0$ on $\operatorname{cl} A$ so that

$$\sum_{i=1}^{p} \left| f_n(x_i) |A_i| - F_n(A_i) \right| < \varepsilon 2^{-n}$$

for each δ_n-fine partition $P := \{(A_1, x_1), \ldots, (A_p, x_p)\}$ with $[P] \subset A$ (Proposition 5.7.4). Fix an integer n with $F_n(A) > J - \varepsilon$, and for every $x \in \operatorname{cl} A$ select an integer $n(x) \geq n$ for which $f_{n(x)}(x) > f(x) - \varepsilon$. Letting $\delta(x) := \delta_{n(x)}(x)$ for each $x \in \operatorname{cl} A$ defines a positive function on $\operatorname{cl} A$. Let $P := \{(A_1, x_1), \ldots, (A_p, x_p)\}$ be a δ-fine partition with $[P] = A$, and observe

$$\left| \sigma(f, P) - \sum_{i=1}^{p} f_{n(x_i)}(x_i) |A_i| \right| \leq \sum_{i=1}^{p} [f(x_i) - f_{n(x_i)}(x_i)] \cdot |A_i| < \varepsilon |A|. \tag{$*$}$$

If the set $\{n(x_1), \ldots, n(x_p)\}$ consists of integers $k_1 < \cdots < k_s$, then $\{1, \ldots, p\}$ is the union of disjoint sets $T_j := \{i : n(x_i) = k_j\}$ where $j = 1, \ldots, s$. Note

$$A_i \subset B(x_i, \delta(x_i)) = B(x_i, \delta_{n(x_i)}(x_i)) = B(x_i, \delta_{k_j}(x_i))$$

for each $i \in T_j$. Hence $P_j := \{(A_i, x_i) : i \in T_j\}$ is a δ_{k_j}-fine partition with $[P_j] \subset A$. Therefore

$$\left| \sum_{i=1}^{p} f_{n(x_i)}(x_i)|A_i| - \sum_{i=1}^{p} F_{n(x_i)}(A_i) \right| \leq \sum_{j=1}^{s} \sum_{i \in T_j} \left| f_{k_j}(x_i)|A_i| - F_{k_j}(A_i) \right|$$

$$< \varepsilon \sum_{j=1}^{s} 2^{-k_j} < \varepsilon \sum_{j=1}^{\infty} 2^{-j} = \varepsilon. \tag{$**$}$$

The sequence $\{F_n\}$ is increasing. Since $n \leq n(x_i) \leq k_s$ for $i = 1, \ldots, p$,

$$J - \varepsilon < F_n(A) = \sum_{i=1}^{p} F_n(A_i) \leq \sum_{i=1}^{p} F_{n(x_i)}(A_i)$$

$$\leq \sum_{i=1}^{p} F_{k_s}(A_i) = F_{k_s}(A) \leq J < J + \varepsilon.$$

Combining the last inequality with inequalities $(*)$ and $(**)$ yields

$$|\sigma(f, P) - J| < \varepsilon(2 + |A|),$$

which proves the proposition. □

Lemma 5.7.8. *Let A be a bounded BV set. If $f \in L(A)$, then $V_\# J(f)$ is absolutely continuous. In particular, $VJ(f) < \infty$ and $V_\# J(f)$ is locally finite.*

PROOF. Choose a negligible set $N \subset \mathbb{R}^m$ and an $\varepsilon > 0$. In view of Corollary 5.7.6, we may assume $f(x) = 0$ for each $x \in N \cap \operatorname{cl} A$. For $F := J(f)$, Proposition 5.7.4 shows that there is a positive function δ on $\operatorname{cl} A$ such that

$$\sum_{i=1}^{p} \Big| f(x_i)|A_i| - F(A_i) \Big| < \varepsilon$$

for every δ-fine partition $P := \{(A_1, x_1), \ldots, (A_p, x_p)\}$ with $[P] \subset A$. Define a positive function β on N by letting

$$\beta(x) = \begin{cases} \delta(x) & \text{if } x \in N \cap \operatorname{cl} A, \\ \operatorname{dist}(x, \operatorname{cl} A) & \text{if } x \in N - \operatorname{cl} A, \end{cases}$$

and choose a β-fine partition $\{(B_1, y_1), \ldots, (B_q, y_q)\}$. If $y_i \notin \operatorname{cl} A$, then $F(B_i) = 0$ by the choice of β. Since $Q =: \{(A \cap B_i, y_i) : y_i \in \operatorname{cl} A\}$ is a δ-fine partition with $[Q] \subset A$, and since $f(y_i) = 0$ when $y_i \in \operatorname{cl} A$,

$$\sum_{i=1}^{p} |F(B_i)| = \sum_{y_i \in \operatorname{cl} A} |F(A \cap B_i)| = \sum_{y_i \in \operatorname{cl} A} \Big| F(A \cap B_i) - f(y_i)|A \cap B_i| \Big| < \varepsilon.$$

We infer $V_\# F(N) \leq \varepsilon$, and the arbitrariness of ε implies $V_\# F(N) = 0$. The lemma follows from Proposition 3.6.14. $\qquad\Box$

Proposition 5.7.9. *Let A be a bounded BV set. If f belongs to $L(A)$, then so do the fuctions $|f|$, f^+, and f^-.*

PROOF. It suffices to show that $|f| \in L(A)$. To this end choose an $\varepsilon > 0$, and let $F := J(f)$. Since Lemma 5.7.8 implies $VF(A) < \infty$, there is a finite disjoint collection $\{B_1, \ldots, B_q\}$ of BV sets such that $A := \bigcup_{j=1}^{q} B_j$ and

$$\sum_{j=1}^{q} |F(B_j)| > VF(A) - \varepsilon.$$

Using Proposition 5.7.4, find a function $\delta > 0$ defined on $\operatorname{cl} A$ so that

$$\sum_{k=1}^{p} \Big| f(z_k)|C_k| - F(C_k) \Big| < \varepsilon$$

for each δ-fine partition $R := \{(C_1, z_1), \ldots, (C_n, z_n)\}$ with $[R] \subset A$. Select a δ-fine partition $P := \{(A_1, x_1), \ldots, (A_p, x_p)\}$ with $[P] = A$, and let

$$Q := \{(A_i \cap B_j, x_i) : i = 1, \ldots, p; \ j = 1, \ldots, q\}.$$

If $I := \sum_{i=1}^{p} \sum_{j=1}^{q} |F(A_i \cap B_j)|$, then

$$\sum_{j=1}^{q} |F(B_j)| \leq \sum_{j=1}^{q} \Big| \sum_{i=1}^{p} F(A_i \cap B_j) \Big| \leq I \leq VF(A).$$

Since $\sigma(|f|, P) = \sigma(|f|, Q)$, we obtain

$$\left|\sigma(|f|, P) - VF(A)\right| \leq \left|\sigma(|f|, Q) - I\right| + VF(A) - I \leq$$

$$\sum_{i=1}^{p}\sum_{j=1}^{q}\left|\,|f(x_i)| \cdot |A_i \cap B_j| - |F(A_i \cap B_j)|\,\right| + VF(A) - I \leq$$

$$\sum_{i=1}^{p}\sum_{j=1}^{q}\left|f(x_i)|A_i \cap B_j| - F(A_i \cap B_j)\right| + VF(A) - \sum_{j=1}^{q}|F(B_j)| < 2\varepsilon\,.$$

It follows $|f| \in L(A)$ and $J(|f|, A) = VF(A)$. $\qquad\square$

Proposition 5.7.10. *Let A be a bounded BV set. If $f \in L(A)$, then $J(f)$ is an AC charge in A.*

PROOF. As $F := J(f)$ is an additive function on $\mathcal{BV}$ and $F = F \llcorner A$, it suffices to show that F is AC in A (Remark 3.1.3). In view of Proposition 5.7.9, we may assume that f, and hence F, are nonnegative functions. By Lemma 5.7.8 and Proposition 3.6.5, the sharp variation $V_{\#}F$ is absolutely continuous and locally finite. Given $\varepsilon > 0$, Lemma 3.1.6 implies there is an $\eta > 0$ such that $V_{\#}F(E) < \varepsilon$ for each set $E \subset \operatorname{cl} A$ with $|E| < \eta$. In view of (3.6.1), we have $|F(B)| < \varepsilon$ for every BV set $B \subset A$ such that $|\operatorname{cl} B| < \eta$. Select a BV set $B \subset A$ with $|B| < \eta$. It follows from Proposition 1.10.5 that there is an increasing sequence $\{C_n\}$ of essentially closed BV subsets of B such that $\lim |B - C_n| = 0$. Since

$$\left|\operatorname{cl}\left(\operatorname{cl}_* C_n\right)\right| = |\operatorname{cl}_* C_n| = |C_n| \leq |B| < \eta\,,$$

the inequality $F(C_n) < \varepsilon$ holds for $n = 1, 2, \ldots$. If $C := \bigcup_{n=1}^{\infty} C_n$, then

$$F(B) = J(f\chi_B, A) = J(f\chi_C, A) = \lim J(f\chi_{C_n}, A) = \lim F(C_n) \leq \varepsilon$$

by (5.7.2), Corollary 5.7.6 and Propositions 5.7.3 and 5.7.7. $\qquad\square$

The following theorem is a direct consequence of Theorem 5.4.3 and Propositions 5.7.10 and 5.7.4.

Theorem 5.7.11. *Let A be a bounded BV set, and let f be a function defined on $\operatorname{cl}_* A$. Then $f \in L^1(A)$ if and only if $f \in L(A)$.*

The next proposition illuminates the subtle difference between the S-integral and L-integral, and consequently between the Riemannian definitions of the R-integral and Lebesgue integral. Note that positive function and positive gage are interchangable terms.

Proposition 5.7.12. *A function f defined on a bounded BV set A is L-integrable if and only if there is a real number J satisfying the following condition: given $\varepsilon > 0$, we can find a positive gage δ on $\operatorname{cl} A$ and $\mathcal{U} \in \mathfrak{V}(\emptyset)$ so that*

$$\left|\sigma(f, P) - J\right| < \varepsilon$$

for each δ-fine partition P with $[P] \subset A$ and $A - [P]$ in $\mathcal{U}$. In this case $J = J(f, A)$.

PROOF. As the converse follows immediately from Definition 5.7.1, suppose that f is an L-integrable function. Choose an $\varepsilon > 0$ and let $F := J(f)$. Proposition 5.7.10

implies the neighborhood base $\mathfrak{V}(\emptyset)$ contains the family

$$\mathcal{U} = \left\{ B \in \mathcal{BV} : |F(B)| < \tfrac{\varepsilon}{2} \right\}.$$

By Proposition 5.7.4, there is a positive gage δ on cl A such that

$$\sum_{i=1}^{p} \left| f(x_i)|A_i| - F(A_i) \right| < \frac{\varepsilon}{2}$$

for each δ-fine partition $P = \{(A_1, x_1), \ldots, (A_p, x_p)\}$ with $[P] \subset A$. Assuming, in addition, that $A - [P]$ is in $\mathcal{U}$, we conclude

$$\left| \sigma(f, P) - F(A) \right| \leq \sum_{i=1}^{p} \left| f(x_i)|A_i| - F(A_i) \right| + \left| F(A - [P]) \right| < \varepsilon. \qquad \square$$

6

Extending the integral

In Chapter 5 we have demonstrated that the R-integral is an averaging process of appreciable appeal. At the same time, the additivity property of the R-integral is genuinely restricted: cf. Proposition 5.1.8 with Remark 6.1.2, (4) below. To correct this deficiency, we follow ideas of Mařík and extend the R-integral by a transfinite iteration of improper integrals — a process similar to Cauchy's extensions in the constructive definition of the Denjoy integral [75, Chapter 8, Section 4]. We show that the extended integral inherits all desirable properties of the R-integral, and that it is additive in the usual way. The extension is maximal in the sense that the extended integral is closed with respect to further formations of improper integrals.

6.1. Buczolich's example

We present an example, constructed by Buczolich [12], which implies that the assumptions of Propositions 3.6.4, (ii) and 5.1.8 are essential.

Proposition 6.1.1. *Assume $m = 2$, and let $K := [0,1]^2$. There are a vector field $v \in C(\mathbb{R}^m; \mathbb{R}^m)$, whose flux is denoted by H, an open BV set $W \subset K$, and a BV set $A \subset W$ having the following properties.*

(i) *$\mathrm{int}_* W = W$, $\mathrm{cl}_* A = A$, and A is essentially open.*

(ii) *$|\mathrm{cl}\, A - A| = 0$; in particular, $\mathrm{cl}\, A$ is a bounded BV set.*

(iii) *$\{v \neq 0\} \subset W$ and $v \in C^\infty(W; \mathbb{R}^m)$. In particular, H is a charge in K derivable at each $x \in W$, and the function $h : x \mapsto DH(x)$ defined on W is continuous.*

(iv) *If $F := H \, \llcorner \, A$ then $V_* F(\mathrm{cl}\, A - A) > 0$. In particular, F is not AC_* in K.*

Since the proof of Proposition 6.1.1 is quite technical, we precede it by a list of its main consequences.

Remark 6.1.2. Using the notation of Proposition 6.1.1, the following facts are true.

(1) Corollary 2.5.4 shows that $DF(x) = h(x)$ at each $x \in \text{int}_c A$. As $\text{cl}_* A - \text{int}_c A$ is a thin set by (2.5.1) and Theorem 1.8.2, (2), the charge F belongs to $CH_*(A)$ according to Theorem 3.6.7. Since $DF(x) = 0$ for every $x \in K - \text{cl}\, A$, property (iv) implies that the negligible set

$$N_\infty := \left\{ x \in \text{cl}\, A - A : \overline{D}|F|(x) = \infty \right\}$$

is not thin (cf. Remark 2.5.5).

(2) The function $f := h \upharpoonright A$ is R-integrable in A and F is the indefinite R-integral of f. If $g := \overline{f}$ and B is a figure containing A, then g is not R-integrable in B. Indeed, if $g \in R(B)$ and G is the indefinite R-integral of g, then Proposition 5.1.6 implies $G = F$, contrary to property (iv).

(3) By (2) and Theorem 5.2.2, the function f has no R-integrable extension to any figure containing A.

(4) If B is a figure containing A, then g is R-integrable in both A and $B - A$, but not in $B = A \cup (B - A)$.

(5) The set $B - A$ of (4) is essentially closed by property (i). Thus in view of (2), (4), and Proposition 5.1.8, the set $\text{cl}_* A$ is not relatively closed in any figure containing A.

(6) From (3) and Proposition 5.1.2 we see $f \notin L^1(A)$. On the other hand, since f is continuous in $A = \text{cl}_* A$, each $x \in \text{cl}_* A$ has a neighborhood U with $f \in L^1(A \cap U)$ and

$$\lim \frac{F(B_n)}{|B_n|} = \lim \frac{1}{|B_n|} \int_{B_n} f(y)\, dy = f(x)$$

whenever $B_n \subset A \cap U$ and $\lim d(B_n \cup \{x\}) = 0$. From this it is easy to deduce the following statement: given $\varepsilon > 0$, there is a positive function δ on $\text{cl}_* A$ such that

$$\sum_{i=1}^p \Big| f(x_i)|A_i| - F(A_i) \Big| < \varepsilon$$

for each δ-fine partition $P = \{(A_1, x_1), \ldots, (A_p, x_p)\}$ such that $[P] \subset A$ (cf. Theorem 5.4.3).

PROOF OF PROPOSITION 6.1.1. Let $K_{0,1} := K$, and let $V_{0,1}$ be the open square of diameter $d(K)/3$ concentric with $K_{0,1}$. Divide $K_{0,1} - V_{0,1}$ into nonoverlapping closed squares $K_{1,i}$, $i = 1,\ldots,8$, of diameters $d(K)/3$, and denote by $V_{1,i}$ the open square of diameter $d(K)/3^2$ concentric with the square $K_{1,i}$. Next divide

$$\bigcup_{i=1}^{8}(K_{1,i} - V_{1,i})$$

into nonoverlapping closed squares $K_{2,j}$, $j = 1,\ldots,8^2$, of diameters $d(K)/3^2$, and denote by $V_{2,j}$ the open square of diameter $d(K)/3^3$ concentric with $K_{2,j}$. Proceeding inductively, at the n-th step we construct closed nonoverlapping squares $K_{n,i}$, $i = 1,\ldots,8^n$, of diameters $d(K)/3^n$, and denote by $V_{n,i}$ the open square of diameter $d(K)/3^{n+1}$ concentric with $K_{n,i}$.

Denote by $W_{n,i}$ the open square of diameter $d(V_{n,i})/4^{n+1}$ concentric with $V_{n,i}$, and let

$$V := \bigcup_{n=0}^{\infty}\bigcup_{i=1}^{8^n} V_{n,i} \quad \text{and} \quad W := \bigcup_{n=0}^{\infty}\bigcup_{i=1}^{8^n} W_{n,i}.$$

Finally, select an open square $U_{n,i} \subset W_{n,i}$ concentric with $W_{n,i}$, and let

$$U := \bigcup_{n=0}^{\infty}\bigcup_{i=1}^{8^n} U_{n,i} \quad \text{and} \quad A := \bigcup_{n=0}^{\infty}\bigcup_{i=1}^{8^n} \mathrm{cl}\, U_{n,i}.$$

During the proof the open squares $U_{n,i}$, and hence the set A, will be determined more accurately. The sets V, W, and U are open, and since

$$\sum_{n=0}^{\infty}\sum_{i=1}^{8^n} \|U_{n,i}\| = \sum_{n=0}^{\infty}\sum_{i=1}^{8^n} \|\mathrm{cl}\, U_{n,i}\| \leq \sum_{n=0}^{\infty}\sum_{i=1}^{8^n} \|W_{n,i}\| < \infty,$$

it follows from Theorem 1.7.1, (1) that W, U, and A are BV sets. It is easy to verify that $\mathrm{cl}\, A = A \cup (K - V)$, and a direct calculation reveals the set $N := K - V$ is negligible.

Claim 1. The equalities $\mathrm{cl}_* U = \mathrm{cl}_* A = A$ and $\mathrm{int}_* A = \mathrm{int}_* U = U$ hold. In particular, $\mathrm{int}_* W = W$.

Proof. Clearly $A \subset \mathrm{cl}_* U \subset A \cup N$, and we prove $\mathrm{cl}_* U = A$ by showing that no $x \in N$ belongs to $\mathrm{cl}_* U$. Select an x in

$$N = \bigcap_{n=1}^{\infty}\bigcup_{i=1}^{8^n} K_{n,i},$$

and for each integer $n \geq 1$, find a K_{n,i_n} containing x. Since

$$U \cap V_{n,i} \subset W_{n,i} \quad \text{and} \quad U \cap K_{n,i} \subset \bigcup_{k=0}^{\infty} \bigcup_{j=1}^{8^k} W_{n+k,k_j},$$

a direct calculation shows that for each integer $i \geq 1$,

$$\lim_{n \to \infty} \frac{|U \cap V_{n,i}|}{|V_{n,i}|} = \lim_{n \to \infty} \frac{|U \cap K_{n,i}|}{|K_{n,i}|} = 0. \tag{1}$$

Given a positive number $r < 1/3$, find an integer $n = n(r)$ that satisfies the inequality $3^{-n-1} \leq r < 3^{-n}$. Observe

$$K_{n+1,i_{n+1}} \subset K[x,r] \subset V_{n-1,p} \cup \bigcup_{j=1}^{8} K_{n,p_j}$$

where $V_{n-1,p}$ and K_{n,p_j} are the squares adjacent to K_{n,i_n}. In view of these inclusions, $|K[x,r]| \geq |K_{n,p_j}|/9 = |V_{n-1,p}|/9$ and

$$\frac{1}{9} \cdot \frac{|U \cap K[x,r]|}{|K[x,r]|} \leq \frac{|U \cap V_{n-1,p}|}{|V_{n-1,p}|} + \sum_{j=1}^{8} \frac{|U \cap K_{n,p_j}|}{|K_{n,p_j}|}.$$

Since $n \to \infty$ as $r \to 0+$, equality (1) yields

$$\lim_{r \to 0+} \frac{|U \cap K[x,r]|}{|K[x,r]|} = 0.$$

From this it is easy to infer $x \notin \mathrm{cl}_* U$. Now

$$U \subset \mathrm{int}_* U \subset \mathrm{cl}_* U = A,$$

and it is clear that no $x \in A - U$ belongs to $\mathrm{int}_* U$. Thus $\mathrm{int}_* U = U$ and, as the sets A and U are equivalent, the claim follows.

Next we construct the charge H in K. If $Q := (a, a+h) \times (b, b+h)$ is a nonempty open square, define continuous functions φ_Q^n, $n = 1, 2, \dots$, on $\mathbb{R}^2$ by letting

$$\varphi_Q^n(s,t) := \frac{\pi}{4h} \left[1 - \cos \frac{2^n \pi}{h} (s - a) \right] \sin \frac{\pi}{h} (t - b)$$

for $(s,t) \in Q$, and $\varphi_Q^n(s,t) := 0$ for $(s,t) \in \mathbb{R}^2 - Q$. The function φ_Q^n is continuous in $\mathbb{R}^2$ and C^∞ in Q. Moreover,

$$\int_{Q^j} \frac{\partial \varphi_Q^n}{\partial s}(x) \, dx = (-1)^{j-1} \tag{2}$$

for $j = 1, \ldots, 2^n$ and

$$Q^j := \left(a + (j-1)\frac{h}{2^n}, a + j\frac{h}{2^n} \right) \times (b, b+h).$$

The vector field $v_{n,i} := 2^{-4n+1} \left(\varphi^n_{W_{n,i}}, 0 \right)$ belongs to $C(\mathbb{R}^2; \mathbb{R}^2)$, and

$$\sum_{i=1}^{8^n} |v_{n,i}|_\infty \leq \max\{|v_{n,i}|_\infty : i = 1, \ldots, 8^n\} \leq 6\pi \left(\frac{3}{4} \right)^n$$

for $n = 1, 2, \ldots$. It follows the vector field

$$v := \sum_{n=0}^{\infty} \sum_{i=1}^{8^n} v_{n,i},$$

belongs to $C(\mathbb{R}^2; \mathbb{R}^2)$. Clearly $\{v \neq 0\} \subset W$, and $v \upharpoonright W$ belongs to $C^\infty(W; \mathbb{R}^2)$. The flux H of v is a charge in K, and Claim 2 of Example 2.3.2 implies $DH(x) = \operatorname{div} v(x)$ at each $x \in W$. Now assume $\operatorname{cl} U_{n,i} \subset W_{n,i}$, which implies $A \subset W$. As $h := \operatorname{div} v$ is continuous in W, it remains to let $F := H \mathbin{\llcorner} A$ and show $V_* F(\operatorname{cl} A - A) > 0$.

First we choose the open squares $U_{n,i}$ with $\operatorname{cl} U_{n,i} \subset W_{n,i}$ so that

$$F(W_{n,i}^{2j} \cap U_{n,i}) = H(W_{n,i}^{2j} \cap U_{n,i}) < -2^{-4n} \tag{3}$$

for $j = 1, \ldots, 2^n$. This is possible, since (2) and Corollary 5.1.13 imply $H(W_{n,i}^{2j}) = -2^{-4n+1}$. Seeking a contradiction, assume

$$N := \operatorname{cl} A - A = \bigcap_{n=1}^{\infty} \bigcup_{i=1}^{8^n} K_{n,i}$$

is a $V_* F$-negligible set. Under this assumption, for $\varepsilon = 3^{-4}$ there is a gage δ on N such that $F(\delta, \varepsilon) < \varepsilon$.

Given an integer $n \geq 1$, let $\Gamma_n := \{1, \ldots, 8^n\}$ and denote by I_n the set of all $i \in \Gamma_n$ such that $d(K_{n,i}) < \delta(x_{n,i})$ for an $x_{n,i} \in N \cap K_{n,i}$. Therefore $J_n := \Gamma_n - I_n$ consists of all $j \in \Gamma_n$ with $d(K_{n,j}) \geq \delta(x)$ for every $x \in N \cap K_{n,j}$. Consider the closed sets

$$D_n := \bigcap_{k=1}^{n} \bigcup_{j \in J_k} K_{k,j} \quad \text{and} \quad D := \bigcap_{n=1}^{\infty} D_n,$$

and note $D \subset N$. If $x \in N$ and $\delta(x) > 0$, then $d(K_{n,i}) < \delta(x)$ for an integer $n \geq 0$ and all $i \in \Gamma_n$. Since x belongs to one of the squares $K_{n,i}$, it follows x is not in D_n and a fortiori not in D. Thus $D \subset \{\delta = 0\}$, and we obtain a contradiction by showing that D is not a thin set.

Each $V_{n,i} - W_{n,i}$ contains a closed square $C_{n,i}$ whose diameter is equal to $d(K)/3^{n+2}$. For $i \in I_n$ and $n = 1, 2, \ldots$, let

$$B_{n,i} := C_{n,i} \cup \bigcup_{j=1}^{2^{n-1}} \left(W_{n,i}^{2j} \cap U_{n,i} \right).$$

These sets are disjoint and $d\big(B_{n,i} \cup \{x_{n,i}\}\big) \leq d(K_{n,i}) < \delta(x_{n,i})$. A direct calculation shows

$$r\big(B_{n,i} \cup \{x_{n,i}\}\big) \geq \frac{|C_{n,i}|}{d(K_{n,i})\big(\|C_{n,i}\| + 2^{n-1}\|W_{n,i}\|\big)} > \varepsilon,$$

and we conclude the collections

$$P_p := \big\{ (B_{n,i}, x_{n,i}) : i \in I_n \text{ and } n = 1, \ldots, p \big\},$$

$p = 1, 2, \ldots$, are ε-regular δ-fine partitions. For small integers $p \geq 1$ the partitions P_p may be empty, but this is irrelevant. Since $A \cap C_{n,i} = \emptyset$, from inequality (3) we obtain

$$\varepsilon > F(\delta, \varepsilon) \geq \sum_{n=1}^{p} \sum_{i \in I_n} |F(B_{n,i})|$$

$$= -\sum_{n=1}^{p} \sum_{i \in I_n} \sum_{j=1}^{2^{n-1}} F(W_{n,i}^{2j} \cap U_{n,i}) > \frac{1}{2} \sum_{n=1}^{p} 8^{-n} \iota_n \tag{4}$$

where ι_n is the number of elements in I_n.

Let $T_1 = J_1$, and for $n = 2, 3, \ldots$, denote by T_n the set of all $i \in J_n$ for which $K_{n,i} \subset D_{n-1}$. By induction we prove

$$\bigcup_{i \in T_n} K_{n,i} \subset D_n, \quad n = 1, 2, \ldots. \tag{5}$$

Indeed, the case $n = 1$ is clear, and since $D_n = D_{n-1} \cap \bigcup_{i \in J_n} K_{n,i}$ if $n \geq 2$, the inductive step follows. Inclusion (5) implies

$$T_n = \{ j \in J_n : K_{n,j} \subset D_{n-1} \} \supset \left\{ j \in J_n : K_{n,j} \subset \bigcup_{i \in T_{n-1}} K_{n-1,i} \right\}$$

$$= \left\{ j \in \Gamma_n : K_{n,j} \subset \bigcup_{i \in T_{n-1}} K_{n-1,i} \right\} - I_n$$

for $n = 2, 3, \ldots$. Denote by τ_n the number of elements in T_n. Since each square $K_{n-1,i}$ contains 8 distinct squares $K_{n,j}$, the previous inclusion yields $\tau_n \geq 8\tau_{n-1} - \iota_n$ and by induction,

$$\tau_p \geq 8^p - \sum_{n=1}^{p} 8^{p-n}\iota_n = 8^p \left(1 - \sum_{n=1}^{p} 8^{-n}\iota_n\right).$$

This and (4) imply $\tau_p > 8^p(1 - 2\varepsilon) > 8^p/2$.

Let $r := \log 8/\log 3$. In view of [22, Section 2.1, Lemma 2], we complete the proof by showing $\mathcal{H}^r(D) > 0$. For this purpose, we employ the *triadic net measure* [71, Chapter 2, Section 7]. A *triadic square* is the product

$$\left[i3^{-k}, (i+1)3^{-k}\right) \times \left[j3^{-k}, (j+1)3^{-k}\right)$$

where i, j, k are integers and $k \geq 0$. Note each $K_{n,i}$ and $V_{n,i}$ is, respectively, the closure and interior of a triadic square. Two triadic squares are called *adjacent* if their diameters are equal and their closures meet. Cover D by a sequence $\{S_i\}$ of triadic squares, and for $i = 1, 2, \ldots$, denote by $\mathcal{S}_i$ the collection of all triadic squares adjacent to S_i. Note $\mathcal{S}_i$ consists of 9 triadic squares including S_i, and let $S_i^* = \bigcup \mathcal{S}_i$. As the set D is compact, there is an integer $q \geq 1$ such that

$$D \subset \bigcup_{i=1}^{q} \operatorname{int} S_i^* \subset \operatorname{int} \left(\bigcup_{i=1}^{q} S_i^*\right).$$

If $R_1, \ldots, R_p$ are the distinct squares comprising $\bigcup_{i=1}^{q} \mathcal{S}_i$, then

$$\frac{1}{9} \sum_{i=1}^{p} d(R_i)^r \leq \sum_{i=1}^{q} d(S_i)^r \leq \sum_{i=1}^{\infty} d(S_i)^r. \tag{6}$$

If $d(R_i) = 3^{-d_i}\sqrt{2}$, select an integer $n > \max\{d_1, \ldots, d_p\}$ so that

$$D_n \subset \bigcup_{i=1}^{q} S_i^* = \bigcup_{i=1}^{p} R_i. \tag{7}$$

This is possible, since $\{D_k\}$ is a decreasing sequence of compact sets whose intersection is contained in the interior of $\bigcup_{i=1}^{p} R_i$. It is easy to see that each R_j contains no more than 8^{n-d_j} distinct squares $K_{n,i}$. On the other hand, inclusions (5) and (7) imply $\bigcup_{j=1}^{p} R_j$ contains no less than τ_n distinct squares $K_{n,i}$. Hence

$$\sum_{j=1}^{p} 8^{n-d_j} \geq \tau_n > \frac{1}{2}8^n,$$

and consequently $\sum_{j=1}^{p} 8^{-d_j} > 1/2$. As $3^r = 8$, we conclude

$$\sum_{i=1}^{p} d(R_j)^r = \sum_{j=1}^{p} \left(3^{-d_j}\sqrt{2}\right)^r = \left(\sqrt{2}\right)^r \sum_{j=1}^{p} 8^{-d_j} > \frac{1}{2}\left(\sqrt{2}\right)^r.$$

In view of [71, Theorem 49], the last inequality together with inequality (6) imply $\mathcal{H}^r(D) > 0$. $\qquad\square$

6.2. $\mathfrak{T}$-convergence

We begin with a general observation about the topology of sequential spaces.

Proposition 6.2.1. *Let* $(X, \mathfrak{T})$ *be a sequential space, and let* $E \subset X$. *If* $x \in \operatorname{cl}_{\mathfrak{T}} E$, *then there is a countable set* $C \subset E$ *with* $x \in \operatorname{cl}_{\mathfrak{T}} C$.

PROOF. If $\{C_t : t \in T\}$ is the collection of all countable subsets of E, then $D := \bigcup_{t \in T} \operatorname{cl}_{\mathfrak{T}} C_t$ contains E. Choose a sequence $\{x_i\}$ in D $\mathfrak{T}$-converging to an $x \in X$. For $i = 1, 2, \ldots$, find $t_i \in T$ with $x_i \in \operatorname{cl}_{\mathfrak{T}} C_{t_i}$, and note $\bigcup_{i=1}^{\infty} C_{t_i} = C_s$ for an $s \in T$. Since $\bigcup_{i=1}^{\infty} \operatorname{cl}_{\mathfrak{T}} C_{t_i} \subset \operatorname{cl}_{\mathfrak{T}} C_s$, the point x is in $\operatorname{cl}_{\mathfrak{T}} C_s$, and hence in D. As $\mathfrak{T}$ is a sequential topology, this implies the set D is closed. Thus D contains $\operatorname{cl}_{\mathfrak{T}} E$, in particular, D contains x. The proposition follows. $\qquad\square$

Nonsequential topological spaces may also have the property stated in Proposition 6.2.1. Indeed, if $x \in \beta N - N$ where βN is the Stone-Čech compactification of the discrete space N of all positive integers, then $X := N \cup \{x\}$ topologized as a subspace of βN is an example [21, Corollary 3.6.16].

Since the remainder of this chapter is devoted entirely to the sequential space $(\mathcal{BV}, \mathfrak{T})$, we simplify the notation and terminology by writing $\mathcal{E}^-$ instead of $\operatorname{cl}_{\mathfrak{T}} \mathcal{E}$ for each $\mathcal{E} \subset \mathcal{BV}$, and saying a sequence $\{A_i\}$ in $\mathcal{BV}$ *converges* if and only if it $\mathfrak{T}$-converges. In accordance with the notation introduced in Section 1.10, we write $\{A_i\} \to A$ whenever the sequence $\{A_i\}$ converges to a set A, necessarily in $\mathcal{BV}$. If $\{A_i\}$ and $\{B_i\}$ are sequences in $\mathcal{BV}$ converging to A and B, respectively, then Proposition 1.8.4 implies

$$\{A_i \cup B_i\} \to A \cup B \quad \text{and} \quad \{A_i \cap B_i\} \to A \cap B;$$

moreover, it follows from Theorem 1.8.2, (3) and Proposition 1.8.5 that $\{A_i \cap E\} \to A \cap E$ for each locally BV set E. These simple observations will be used tacitly throughout. For subfamilies $\mathcal{B}$ and $\mathcal{C}$ of $\mathcal{BV}$, let

$$\mathcal{B} \vee \mathcal{C} := \{B \cup C : B \in \mathcal{B} \text{ and } C \in \mathcal{C}\},$$
$$\mathcal{B} \wedge \mathcal{C} := \{B \cap C : B \in \mathcal{B} \text{ and } C \in \mathcal{C}\},$$

and let $\mathcal{B} \wedge E := \{B \cap E : B \in \mathcal{B}\}$ whenever E is a locally BV set.

Lemma 6.2.2. *If $\mathcal{B}$ and $\mathcal{C}$ are subfamilies of $\mathcal{BV}$, then*

$$\mathcal{B}^- \vee \mathcal{C}^- \subset (\mathcal{B} \vee \mathcal{C})^- \quad and \quad \mathcal{B}^- \wedge \mathcal{C}^- \subset (\mathcal{B} \wedge \mathcal{C})^-.$$

Moreover, $\mathcal{B}^- \wedge E \subset (\mathcal{B} \wedge E)^-$ for each locally BV set E.

PROOF. Observe the family

$$\mathcal{B}_0 := \left\{ B \in \mathcal{BV} : B \cap E \in (\mathcal{B} \wedge E)^- \right\}$$

is closed and contains $\mathcal{B}$. Thus $\mathcal{B}^- \subset \mathcal{B}_0$, from which it follows that $\mathcal{B}^- \wedge E \subset (\mathcal{B} \wedge E)^-$. In particular,

$$\mathcal{B}^- \wedge C \subset (\mathcal{B} \wedge C)^- \subset (\mathcal{B} \wedge \mathcal{C})^-$$

for each $C \in \mathcal{C}$. From this we infer that the closed family

$$\mathcal{C}_0 := \left\{ C \in \mathcal{BV} : \mathcal{B}^- \wedge C \subset (\mathcal{B} \wedge \mathcal{C})^- \right\}$$

contains $\mathcal{C}$. Thus $\mathcal{C}^- \subset \mathcal{C}_0$, which implies $\mathcal{B}^- \wedge \mathcal{C}^- \subset (\mathcal{B} \wedge \mathcal{C})^-$. The remaining inclusion is proved similarly. □

Lemma 6.2.3. *Let $\mathcal{B} \subset \mathcal{BV}$. There is a $\mathcal{C} \subset \mathcal{BV}$ such that $\mathcal{B}^- \subset \mathcal{C}^-$ and each $C \in \mathcal{C}$ is an essentially closed subset of some $B \in \mathcal{B}$.*

PROOF. By Proposition 1.10.5, for each $B \in \mathcal{BV}$ there is a sequence $\{B_i\}$ of essentially closed BV subsets of B converging to B. Letting

$$\mathcal{C} := \{B_i : B \in \mathcal{B} \text{ and } i = 1, 2, \dots\}$$

yields $\mathcal{B} \subset \mathcal{C}^-$, and the lemma follows. □

Proposition 6.2.4. *If $\mathcal{B} \subset \mathcal{BV}$, then $\left| A - \bigcup \mathcal{B} \right| = 0$ for each $A \in \mathcal{B}^-$.*

PROOF. Since the family $\left\{ A \in \mathcal{BV} : \left| A - \bigcup \mathcal{B} \right| = 0 \right\}$ is closed and contains $\mathcal{B}$, it contains $\mathcal{B}^-$. □

Let $\mathcal{B}$ be a family of bounded BV sets. Proposition 6.2.4 shows the elements of $\mathcal{B}$ cover each $A \in \mathcal{B}^-$ up to a *negligible set*. In the remainder of this section, we prove that the essential interiors of elements of $\mathcal{B}$ cover the essential interior (and hence also the essential closure) of each $A \in \mathcal{B}^-$ up to a *thin set*.

Lemma 6.2.5. *There are positive constants $\theta := \theta(m)$ and $\kappa := \kappa(m)$ such that the essential interior of each bounded BV set A can be covered by a countable collection $\{B_i\}$ of closed balls with*

$$\sum_i |B_i| \leq \theta |A| \qquad \text{and} \qquad \sum_i \|B_i\| \leq \kappa \|A\|.$$

PROOF. If $m = 1$, then $\text{int}_* A$ is the union of finitely many open balls whose closures are disjoint, and the lemma holds. Hence assume $m \geq 2$. Fix an $x \in \text{int}_* A$, and observe the function

$$r \mapsto \frac{|A \cap B[x, r]|}{|B[x, r]|},$$

defined for $r > 0$, is continuous and approaches 1 or 0 as r goes to 0 or to ∞, respectively. Thus there is a positive number R such that for each $x \in \text{int}_* A$ we can find a $B_x := B[x, r_x]$ with $0 < r_x \leq R$ and

$$\frac{1}{2} |B_x| = |B_x \cap A| = |B_x - A|.$$

The relative isoperimetric inequality implies

$$\|B_x\| = m\alpha(m)^{\frac{1}{m}} |B_x|^{\frac{m-1}{m}} \leq \beta \|\mathcal{D}A\|(B_x)$$

where $\beta = \beta(m)$ is a positive constant. According to Theorem 1.4.1, there is a countable set $C \subset \text{int}_* A$ such that the collection $\{B_x : x \in C\}$ is disjoint and $\text{int}_* A \subset \bigcup_{x \in C} B_x^\bullet$ where $B_x^\bullet := B[x, 5r_x]$. Therefore

$$\sum_{x \in C} |B_x^\bullet| = 5^m \sum_{x \in C} |B_x| \leq 2 \cdot 5^m \sum_{x \in C} |A \cap B_x| = 2 \cdot 5^m |A|,$$

$$\sum_{x \in C} \|B_x^\bullet\| = 5^{m-1} \sum_{x \in C} \|B_x\| \leq 5^{m-1} \beta \sum_{x \in C} \|\mathcal{D}A\|(B_x) \leq 5^{m-1} \beta \|A\|,$$

and the lemma follows. $\square$

Lemma 6.2.6. *Let $\{A_n\}$ be a sequence in $\mathcal{BV}$ with $\lim |A_n| = 0$. There is a $\gamma := \gamma(m)$ such that*

$$\mathcal{H}^{m-1}\left(\bigcap_{n=1}^{\infty} \text{int}_* A_n \right) \leq \gamma \liminf \|A_n\|.$$

PROOF. Assume $a := \liminf \|A_n\|$ is a finite number, and choose an $\varepsilon > 0$. Let θ and κ be the constants from Lemma 6.2.5, and let

$$\gamma := \frac{\kappa \, \alpha(m-1)}{m \, \alpha(m)}.$$

There is an integer $p \geq 1$ such that

$$|A_p| < \frac{\alpha(m)}{2^m \theta} \varepsilon^m \quad \text{and} \quad \|A_p\| < a + \frac{\varepsilon}{\gamma}.$$

If $\{B_i\}$ is a countable collection of closed balls associated with A_p according to Lemma 6.2.5, then these balls cover $A := \bigcap_{n=1}^{\infty} \text{int}_* A_n$, and

$$d(B_i) = 2 \left(\frac{|B_i|}{\alpha(m)} \right)^{\frac{1}{m}} \leq 2 \left(\frac{\theta |A_p|}{\alpha(m)} \right)^{\frac{1}{m}} < \varepsilon$$

for $i = 1, 2, \ldots$. Thus

$$\mathcal{H}_{\varepsilon}^{m-1}(A) \leq \sum_i \alpha(m-1) \left[\frac{d(B_i)}{2} \right]^{m-1}$$

$$= \frac{\alpha(m-1)}{m\alpha(m)} \sum_i \|B_i\| \leq \gamma \|A_p\| < \gamma a + \varepsilon,$$

and the lemma follows by letting $\varepsilon \to 0$. $\qquad \square$

Corollary 6.2.7. *Let $\{A_n\}$ be a sequence in $\mathcal{BV}$. If $\{A_n\} \to A$, then*

$$\mathcal{H}^{m-1} \left(\text{int}_* A - \bigcup_{n=1}^{\infty} \text{cl}_* A_n \right) < \infty.$$

PROOF. In view of Corollary 1.5.7,

$$\text{int}_* A - \bigcup_{n=1}^{\infty} \text{cl}_* A_n = \bigcap_{n=1}^{\infty} \text{int}_* (A - A_n).$$

An application of Lemma 6.2.6 completes the proof. $\qquad \square$

Theorem 6.2.8. *If $\mathcal{B}$ is a family of bounded BV sets, then*

$$\text{int}_* A - \bigcup_{B \in \mathcal{B}} \text{int}_* B$$

is a thin set for each $A \in \mathcal{B}^-$.

PROOF. Let $B_0 := \bigcup_{B \in \mathcal{B}} \text{int}_* B$, and let $\mathcal{C}$ be the family of all bounded BV sets C for which $\text{int}_* C - B_0$ is a thin set. If $\{C_n\}$ is a sequence in $\mathcal{C}$ and $\{C_n\} \to C$, let $C_{\text{cl}} := \bigcup_{n=1}^{\infty} \text{cl}_* C_n$ and $C_{\text{int}} := \bigcup_{n=1}^{\infty} \text{int}_* C_n$. Observe

$$\text{int}_* C - B_0 \subset (\text{int}_* C - C_{\text{int}}) \cup (C_{\text{int}} - B_0)$$

$$\subset (\text{int}_* C - C_{\text{cl}}) \cup \left(\bigcup_{n=1}^{\infty} \partial_* C_n \right) \cup \bigcup_{n=1}^{\infty} (\text{int}_* C_n - B_0).$$

By Corollary 6.2.7, Theorem 1.8.2, (2), and our choice of $\mathcal{C}$, the largest set of the previous inclusion is thin. It follows the family $\mathcal{C}$ is closed, and clearly $\mathcal{B} \subset \mathcal{C}$. Therefore $\mathcal{B}^- \subset \mathcal{C}$.   □

6.3. The GR-integral

If f is a function whose domain contains a locally BV set E and F is a charge, denote by $\mathcal{R}(f, F; E)$ the family of all bounded BV sets $B \subset E$ such that $f \restriction B$ belongs to $R(B)$ and the charge $F \, \mathsf{L} \, B$ is the indefinite R-integral of $f \restriction B$.

Remark 6.3.1. The following observations will be used throughout the remainder of this chapter.

(1) If B is a locally BV subset of E, then Proposition 5.1.6 yields $\mathcal{R}(f, F; E) \wedge B \subset \mathcal{R}(f, F; B)$. Consequently,

$$\mathcal{R}(f, F; E) \wedge \mathcal{R}(g, G; E) = \mathcal{R}(f, F; E) \cap \mathcal{R}(g, G; E)$$

for any pairs f, F and g, G for which the symbols are defined.

(2) If $A \in \mathcal{R}(f, F; E)^-$, then

$$\mathcal{BV}(A) \subset \mathcal{R}(f, F; E)^-.$$

Indeed, choose a $B \in \mathcal{BV}(A)$, and observe $B = A \cap B$ belongs to $\mathcal{R}(f, F; E)^- \wedge B$. Since Lemma 6.2.2 and (1) imply

$$\mathcal{R}(f, F; E)^- \wedge B \subset \big[\mathcal{R}(f, F; E) \wedge B\big]^- \subset \mathcal{R}(f, F; E)^-,$$

the set B belongs to $\mathcal{R}(f, F; E)^-$.

(3) Since $\mathcal{BV}(E)$ is a closed family containing $\mathcal{R}(f, F; E)$, we have

$$\mathcal{R}(f, F; E)^- \subset \mathcal{BV}(E).$$

Theorem 6.3.2. *Let f be a function defined on a set $A \in \mathcal{BV}$, and let F be a charge in A. Denote by $\mathcal{B}$ the family of all $B \in \mathcal{R}(f, F; A)$ such that $\mathrm{int}_* B$ is a relatively open subset of $\mathrm{int}_* A$. If $A \in \mathcal{B}^-$, then $f \in R(A)$ and F is the indefinite R-integral of f.*

PROOF. By Proposition 6.2.1, the family $\mathcal{B}$ contains a countable subfamily $\{B_k\}$ such that $A \in \{B_k\}^-$. Choose a negligible set $N \subset \mathrm{cl}_* A$ and an $\varepsilon > 0$. According to our assumption, the measures

$$V_*(F \, \mathsf{L} \, B_k) \, \mathsf{L} \, \mathrm{cl}_* B_k, \quad k = 1, 2, \dots,$$

are absolutely continuous. There are gages δ_k on $N \cap \mathrm{cl}_* B_k$ such that

$$\sum_{i=1}^{p} |F(A_i \cap B_k)| < \varepsilon 2^{-k}$$

for each ε-regular δ_k-fine partition $\{(A_1, x_1), \ldots, (A_p, x_p)\}$. The set

$$S := \partial_* A \cup \left(\mathrm{int}_* A - \bigcup_k \mathrm{int}_* B_k \right) \cup \bigcup_{k=1}^{\infty} \{\delta_k = 0\}$$

is thin by Theorems 1.8.2, (2) and 6.2.8. Observe

$$C_k := \mathrm{int}_* B_k - \bigcup_{j=1}^{k-1} \mathrm{int}_* B_j$$

are disjoint sets whose union contains $\cdot A - S$. Now let $\delta(x) := 0$ for each $x \in S \cap N \cap \mathrm{cl}_* A$. If x is in $N \cap \mathrm{cl}_* A - S$, then x belongs to exactly one set C_k, and we choose a positive number $\delta(x) \leq \delta_k(x)$ so that

$$B[x, \delta(x)] \cap \mathrm{int}_* A \subset \mathrm{int}_* B_k .$$

This choice of $\delta(x)$ is possible, since x belongs to $\mathrm{int}_* B_k$, which is a relatively open subset of $\mathrm{int}_* A$ by our assumptions. Clearly, $\delta : x \mapsto \delta(x)$ is a gage on $N \cap \mathrm{cl}_* A$. If $P := \{(A_1, x_1), \ldots, (A_p, x_p)\}$ is an ε-regular δ-fine partition and $k \geq 1$ is an integer, then $P_k := \{(A_i, x_i) : x_i \in C_k\}$ is an ε-regular δ_k-fine partition such that $[P_k] \cap \mathrm{int}_* A \subset \mathrm{int}_* B_k$ and $P = \bigcup_k P_k$. Since F is a charge in A, we obtain

$$\sum_{i=1}^{p} |F(A_i)| = \sum_{i=1}^{p} |F(A_i \cap \mathrm{int}_* A)| = \sum_{k} \sum_{x_i \in C_k} |F(A_i \cap \mathrm{int}_* B_k)|$$

$$= \sum_{k} \sum_{x_i \in C_k} |F(A_i \cap B_k)| < \varepsilon \sum_{k} 2^{-k} \leq \varepsilon .$$

It follows $V_* F(N) = 0$, and we conclude $F \in CH_*(A)$.

Let $F \llcorner B_k$ be derivable at $x \in \mathrm{int}_c A \cap \mathrm{int}_* B_k$. As $\mathrm{int}_* B_k$ is a relatively open subset of $\mathrm{int}_* A$, it follows from Proposition 2.5.3 that F is derivable at x and $DF(x) = D(F \llcorner B_k)(x)$. Our assumptions yield

$$D(F \llcorner B_k)(x) = f(x)$$

for almost all $x \in \mathrm{int}_c A \cap \mathrm{int}_* B_k$ and $k = 1, 2, \ldots$. By Proposition 6.2.4, the set $\mathrm{int}_c A \cap \bigcup_{k=1}^{\infty} \mathrm{int}_* B_k$ differs from A by a negligible set, and we conclude that $DF(x) = f(x)$ for almost all $x \in A$. $\qquad\square$

The following easy corollary of Theorem 6.3.2 is a version of the classical Hake theorem [75, Chapter 8, Lemma 3.1].

Corollary 6.3.3. *Let f be a function defined on a one-dimensional cell $A := [a, b]$, and suppose $f \in R([a, \beta])$ for each $\beta \in [a, b)$. If a finite limit*

$$\lim_{\beta \to b-} (R) \int_a^\beta f(x)\, dx = I$$

exists, then $f \in R(A)$ and $(R)\int_A f(x)\, dx = I$.

PROOF. By our assumptions, the formula

$$F(x) = \begin{cases} 0 & \text{if } x \le a, \\ \int_a^x f(t)\, dt & \text{if } a < x < b, \\ I & \text{if } x \ge b, \end{cases}$$

defines a continuous function F on $\mathbb{R}$. The function F induces a charge, still denoted by F, such that $A \in \mathcal{R}(f, F; A)^-$. Since the essential interior of each one-dimensional BV set is a finite union of open intervals, the corollary follows from Theorem 6.3.2. $\square$

Theorem 6.3.2 motivates the next definition, introduced first by Mařík in a different context [32, 47, 40, 48]; see also [69, 64].

Definition 6.3.4. A function f defined on a locally BV set E is called *GR-integrable* in E if there is a charge F in E such that

$$\mathcal{R}(f, F; E)^- = \mathcal{BV}(E).$$

The family of all GR-integrable functions in E is denoted by $GR(E)$.

Let E be a locally BV set. A charge F associated with $f \in GR(E)$ according to Definition 6.3.4 is uniquely determined by f. Indeed, if F and G are charges in E such that

$$\mathcal{R}(f, F; E)^- = \mathcal{R}(f, G; E)^- = \mathcal{BV}(E),$$

then Lemma 6.2.2 and Remark 6.3.1, (1) imply that the family

$$\mathcal{R} = \mathcal{R}(f, F; E) \cap \mathcal{R}(f, G; E)$$

is dense in $\mathcal{BV}(E)$. Now F and G are continuous function on $\mathcal{BV}$, and they coincide on $\mathcal{R}$; since for each $B \in \mathcal{R}$, the function $f \upharpoonright B$ has only one indefinite R-integral. Thus F and G coincide on $\mathcal{BV}(E)$, and as both are charges in E, they are equal.

In view of the uniqueness we have established, a charge F associated with a function $f \in GR(E)$ according to Definition 6.3.4 is called the *indefinite GR-integral* of f; we denote it by $(GR)\int f\, d\mathcal{L}^m$ or $(GR)\int f(x)\, dx$. As before, if $A \subset E$ is a bounded BV set, the number $F(A)$ is called the *GR-integral* of f over A, denoted by $(GR)\int_A f\, d\mathcal{L}^m$ or $(GR)\int_A f(x)\, dx$.

If $f \in R(E)$ and $F := (R)\int f(x)\, dx$, then $\mathcal{R}(f, F; E) = \mathcal{BV}(E)$ according to Proposition 5.1.6. Thus $R(E) \subset GR(E)$, and the map

$$(GR)\int : f \mapsto (GR)\int f(x)\, dx : GR(E) \to CH(E)$$

is an injective extension of the bijection

$$(R)\int : f \mapsto (R)\int f(x)\, dx : R(E) \to CH_*(E).$$

Consequently, there is no danger of confusion when we write $\int$ instead of $(GR)\int$. The symbol $(GR)\int$ will be used only occasionally for emphasis.

Proposition 6.3.5. *Let E be a locally BV set. The family $GR(E)$ is a linear space and the map*

$$\int : f \mapsto \int f(x)\, dx : GR(E) \to CH(E)$$

is a linear injection. If $f \in GR(E)$ is a nonnegative function, then so is the indefinite GR-integral $\int f(x)\, dx$.

PROOF. Denote by F and G, respectively, the indefinite GR-integrals of functions f and g in $GR(E)$. By Lemma 6.2.2, Propositions 5.1.3, and Remark 6.3.1, (1) and (3),

$$\begin{aligned}
\mathcal{BV}(E) = \mathcal{R}(f, F; E)^- &\wedge \mathcal{R}(g, G; E)^- \\
&\subset \left[\mathcal{R}(f, F; E) \wedge \mathcal{R}(g, G; E)\right]^- \\
&\subset \left[\mathcal{R}(f, F; E) \cap \mathcal{R}(g, G; E)\right]^- \\
&\subset \mathcal{R}(f + g, F + G; E)^- \subset \mathcal{BV}(E).
\end{aligned}$$

Since Propositions 5.1.3 also implies $\mathcal{R}(f, F; E) \subset \mathcal{R}(cf, cF; E)$ for each $c \in \mathbb{R}$, the linearity part of our claim follows.

If $f \geq 0$, then $F(B) \geq 0$ for each $B \in \mathcal{R}(f, F; E)$ by Propositions 5.1.3. As $F = F\, \llcorner\, E$ is continuous and $\mathcal{R}(f, F; E)$ is dense in $\mathcal{BV}(E)$, we have $F \geq 0$. $\qquad\square$

Proposition 6.3.6. *Let $B \subset A$ be locally BV sets, and let f be a GR-integrable function in A. The restriction $f \upharpoonright B$ is a GR-integrable function in B, and if F is the indefinite GR-integral of f, then $F \mathbin{\mathsf{L}} B$ is the indefinite GR-integral of $f \upharpoonright B$.*

PROOF. According to Lemma 6.2.2 and Proposition 5.1.6,

$$\mathcal{BV}(B) = \mathcal{BV}(A) \wedge B = \mathcal{R}(f, F; A)^- \wedge B$$
$$\subset \left[\mathcal{R}(f, F; A) \wedge B \right]^- = \mathcal{R}(f, F; B)^-$$

and the proposition follows from Remark 6.3.1, (3).   □

Proposition 6.3.7. *Let E be a locally BV set. A function f defined on E is GR integrable in E if and only if it is GR-integrable in each bounded BV set $B \subset E$, in which case*

$$F : A \mapsto \int_{E \cap A} f(x)\, dx : \mathcal{BV} \to \mathbb{R}$$

is the indefinite GR-integral of f.

PROOF. Assume f is GR-integrable in each $A \in \mathcal{BV}(E)$, and denote by F_k the indefinite GR-integral of f restricted to $E_k = B(k) \cap E$. Observe $F_k = F_{k+1} \mathbin{\mathsf{L}} B(k)$ by Proposition 6.3.6. It follows there is a charge F in E such that $F \mathbin{\mathsf{L}} B(k) = F_k$ for $k = 1, 2, \ldots$. Since

$$\mathcal{BV}(E) = \bigcup_{k=1}^{\infty} \mathcal{BV}(E_k) = \bigcup_{k=1}^{\infty} \mathcal{R}(f, F_k; E_k)^-$$
$$\subset \left[\bigcup_{k=1}^{\infty} \mathcal{R}(f, F_k; E_k) \right]^- = \mathcal{R}(f, F; E)^-,$$

we see that $f \in GR(E)$ and F is the indefinite GR-integral of f. The converse follows from Proposition 6.3.6.   □

The following proposition shows that for nonnegative functions there is no difference between the Lebesgue integral, R-integral, and GR-integral.

Proposition 6.3.8. *For a function f defined on a locally BV set E the following is true.*

(i) *If $f \in GR(E)$, then f is measurable.*

(ii) *If both f and $|f|$ belong to $GR(E)$, then $f \in L^1_\ell(E)$.*

PROOF. Claim (i) follows from Propositions 2.3.5, 6.2.1, and 6.2.4. To establish claim (ii), select a bounded BV set $A \subset E$ and denote by F the indefinite GR-integral of $|f| \upharpoonright A$. By Proposition 5.1.4,

$$\mathcal{A} := \left\{ B \in \mathcal{BV}(A) : f \in L^1(B) \right\}$$

is a dense subfamily of $\mathcal{BV}(A)$; since $\mathcal{A} = \mathcal{R}(|f|, F; A)$. Thus it suffices to show that $\mathcal{A}$ is closed. To this end, choose a sequence $\{B_n\}$ in $\mathcal{A}$ converging to a set B, necessarily in $\mathcal{BV}(A)$. As $\{B_n \cap B\}$ is also a sequence in $\mathcal{A}$ converging to B, we may assume that each B_n is a subset of B. Now each $C_k = \bigcup_{n=1}^{k} B_n$ belongs to $\mathcal{A}$, and claim (i) implies

$$(L) \int_B |f(x)| \, dx = \lim (L) \int_{C_k} |f(x)| \, dx = \lim F(C_k) \leq F(A) \,.$$

Since $F(A) < \infty$, the set B belongs to $\mathcal{A}$. $\qquad\square$

Remark 6.3.9. A result completely analogous to Corollary 5.1.5 holds for the GR-integral. We leave its formulation and proof to the reader.

The next theorem and its corollary demonstrate the difference between R-integrability and *GR*-integrability — cf. Propositions 5.1.8 and Remark 6.1.2, (4).

Theorem 6.3.10. *Let E be the union of locally BV sets $E_1, \ldots, E_k$, and let f be a function defined on E that is GR-integrable in every E_i. Then f is GR-integrable in E. If the sets E_i are disjoint and F_i is the indefinite GR-integral of $f \upharpoonright E_i$, then $F := \sum_{i=1}^{k} F_i$ is the indefinite GR-integral of f.*

PROOF. In view of Proposition 6.3.6, it suffices to prove the theorem when the sets E_i are disjoint. Lemma 6.2.3 and Proposition 5.1.6 imply there are families $\mathcal{E}_i \subset \mathcal{R}(f, F_i; E_i)$ consisting of essentially closed sets such that $\mathcal{BV}(E_i) = \mathcal{E}_i^-$ for $i = 1, \ldots, k$. If $A_i \in \mathcal{E}_i$, then it follows from Proposition 5.1.8 that f is R-integrable in $A := \bigcup_{i=1}^{k} A_i$ and

$$F \llcorner A = \sum_{i=1}^{k} (F_i \llcorner A_i)$$

is the indefinite R-integral of $f \upharpoonright A$. Thus $\mathcal{E} := \bigvee_{i=1}^{k} \mathcal{E}_i$ is a subfamily of $\mathcal{R}(f, F; E)$. Extending Lemma 6.2.2 by induction, we obtain

$$\mathcal{BV}(E) = \bigvee_{i=1}^{m} \mathcal{BV}(E_i) = \bigvee_{i=1}^{m} \mathcal{E}_i^- \subset \mathcal{E}^- \subset \mathcal{R}(f, F; E)^-$$

and the theorem follows from Remark 6.3.1, (3). $\qquad\square$

Corollary 6.3.11. *A function f defined on a locally BV set E belongs to $GR(E)$ if and only if $\overline{f}$ belongs to $GR(\mathbb{R}^m)$, in which case the indefinite GR-integrals of f and $\overline{f}$ are the same.*

Proposition 6.3.12. *For a locally BV set E the following is true.*

  (i) *If $m = 1$, then $R(E) = GR(E)$.*

  (ii) *If $m \geq 2$ and $\mathrm{int}\, E \neq \emptyset$, then $R(E) \subsetneqq GR(E)$.*

PROOF. As the essential interior of any one-dimensional BV set is an open set, claim (i) follows from Theorem 6.3.2. Remark 6.1.2, (4) and Theorem 6.3.10 imply claim (ii). $\qquad\qquad\qquad\qquad\qquad\qquad\square$

6.4. Additional properties

In this section we establish the Gauss-Green, multipliers, and change of variable theorems for the GR-integral.

Lemma 6.4.1. *Let A be a bounded BV set, and let $F = \int v \cdot \nu \, d\mathcal{H}^{m-1}$ be the flux of a bounded Borel measurable vector field $v : \mathrm{cl}_* A \to \mathbb{R}^m$. Suppose $\mathcal{B} \subset \mathcal{BV}(A)$ is a family having the following property: for each $B \in \mathcal{B}$, the charge $F_B := F \llcorner B$ is almost derivable at almost all $x \in B$. If $A \in \mathcal{B}^-$, then there is a function $\mathfrak{g}\text{-}\mathfrak{div}\, v$ defined almost everywhere on A such that given $B \in \mathcal{B}$,*

$$\mathfrak{g}\text{-}\mathfrak{div}\, v(x) = \mathfrak{div}\,(v \restriction \mathrm{cl}_* B)(x)$$

for almost all $x \in B$. The function $\mathfrak{g}\text{-}\mathfrak{div}\, v$ is unique up to equivalence and depends only on the vector field v and not on the choice of $\mathcal{B}$.

PROOF. According to Theorem 2.4.3, each $B \in \mathcal{B}$ contains a negligible set N_B such that F_B is derivable at each $x \in B - N_B$. By Corollary 2.5.4, for every pair B, C of sets in $\mathcal{B}$, there is a negligible set $N_{B,C}$ contained in $B \cap C$ such that

$$\begin{aligned}
DF_B(x) = D(F_B \llcorner C)(x) &= D\big[F \llcorner (B \cap C)\big](x) \\
&= D(F_C \llcorner B)(x) = DF_C(x)
\end{aligned} \qquad (*)$$

for each x in $B \cap C - N_{B,C}$. Using Propositions 6.2.1 and 6.2.4, find a sequence $\{B_i\}$ in $\mathcal{B}$ so that $N_0 = A - \bigcup_{i=1}^{\infty} B_i$ is a negligible set. Now

$$N := N_0 \cup \left(\bigcup_{i=1}^{\infty} N_{B_i}\right) \cup \left(\bigcup_{i,j=1}^{\infty} N_{B_i, B_j}\right)$$

is a negligible set, and equality $(*)$ allows us to define a function $\mathfrak{g}\text{-}\mathfrak{div}\, v$ on $A - N$ by letting

$$\mathfrak{g}\text{-}\mathfrak{div}\, v(x) = DF_{B_i}(x)$$

whenever $x \in B_i - N$. Since F_B is the flux of $v \restriction \mathrm{cl}_*B$, we have

$$\mathfrak{g}\text{-}\mathfrak{div}\, v(x) = DF_B(x) = \mathfrak{div}\,(v \restriction \mathrm{cl}_*B)(x)$$

for each $B \in \mathcal{B}$ and all $x \in B - (N \cup N_B)$.

It remains to show that $\mathfrak{g}\text{-}\mathfrak{div}\, v$ does not depend on the choice of the family $\mathcal{B}$. To this end let $\mathcal{B}_k$, $k = 1, 2$, be subfamilies of $\mathcal{BV}(A)$ such that A belongs to both $\mathcal{B}_1^-$ and $\mathcal{B}_2^-$, and let f_k be functions defined almost everywhere on A such that $f_k(x) = \mathfrak{div}\,(v \restriction B)(x)$ for each $B \in \mathcal{B}_k$, $k = 1, 2$, and almost all $x \in B$. Use Propositions 6.2.1 and 6.2.4 to find sequences $\{B_{k,i}\}$ in $\mathcal{B}_k$ so that

$$M := A - \bigcup_{i,j=1}^{\infty} (B_{1,i} \cap B_{2,j}) = \left(A - \bigcup_{i=1}^{\infty} B_{1,i}\right) \cup \left(A - \bigcup_{i=1}^{\infty} B_{2,i}\right)$$

is a negligible set. In view of Corollary 2.5.4,

$$f_1(x) = \mathfrak{div}\,(v \restriction B_{1,i})(x) = \mathfrak{div}\,\left[v \restriction (B_{1,i} \cap B_{2,j})\right](x)$$
$$= \mathfrak{div}\,(v \restriction B_{2,j})(x) = f_2(x)$$

for almost all $x \in B_{1,i} \cap B_{2,j}$ and $i, j = 1, 2, \ldots$. It follows that $f_1 = f_2$ almost everywhere in A. $\qquad\square$

Definition 6.4.2. Let A be a bounded BV set, and let $v : \mathrm{cl}_*A \to \mathbb{R}^m$ be a bounded Borel measurable vector field satisfying the assumptions of Lemma 6.4.1. The function $\mathfrak{g}\text{-}\mathfrak{div}\, v$ defined almost everywhere on A, whose existence and uniqueness have been established in Lemma 6.4.1, is called the *generalized mean divergence* of v.

Calling the function $\mathfrak{g}\text{-}\mathfrak{div}\, v$ the generalized mean divergence of v is justified. Indeed, if $\mathfrak{div}\, v(x)$ is defined for almost all $x \in A$, then the family $\mathcal{A} := \{A\}$ satisfies the conditions of Lemma 6.4.1. Defining $\mathfrak{g}\text{-}\mathfrak{div}\, v$ by means of $\mathcal{A}$ shows that $\mathfrak{g}\text{-}\mathfrak{div}\, v = \mathfrak{div}\, v$.

Remark 6.4.3. Let A be a bounded BV set, and let $v \in C(\mathrm{cl}\, A; \mathbb{R}^m)$. Suppose $\mathcal{B} \subset \mathcal{BV}(A)$ is a family such that for each $B \in \mathcal{B}$, the restriction $v \restriction \mathrm{cl}_*B$ is Lipschitz at almost all $x \in \mathrm{int}_c B$. If $A \in \mathcal{B}^-$, then Lemma 6.4.1 and Remark 2.5.9, (iv) imply that given $B \in \mathcal{B}$,

$$\mathfrak{g}\text{-}\mathfrak{div}\, v(x) = \mathrm{div}_*(v \restriction \mathrm{cl}_*B)(x)$$

for almost all $x \in \mathrm{int}_c B$.

Theorem 6.4.4. *Let E be a locally BV set, and let $v : \mathrm{cl}_* E \to \mathbb{R}^m$ be a charging vector field. Suppose $\mathcal{B} \subset \mathcal{BV}(E)$ is a family having the following property: given $B \in \mathcal{B}$, we can find a thin set T_B such that the flux of $v \restriction \mathrm{cl}_* B$ is almost derivable at each $x \in \mathrm{cl}_* B - T_B$. If $A \in \mathcal{B}^-$, then $\mathfrak{g}\text{-}\mathfrak{div}\, v$ belongs to $GR(A)$, and*

$$(GR) \int_A \mathfrak{g}\text{-}\mathfrak{div}\, v \, d\mathcal{L}^m = (L) \int_{\partial_* A} v \cdot \nu_A \, d\mathcal{H}^{m-1}.$$

PROOF. Denote by F the flux of v, and observe that $F \restriction B$ is the flux of $v \restriction \mathrm{cl}_* B$ for each $B \in \mathcal{B}$. By Lemma 6.4.1 the generalized mean divergence $\mathfrak{g}\text{-}\mathfrak{div}\, v$ is defined almost everywhere on A, and Theorem 5.1.12 implies $\mathcal{B} \subset \mathcal{R}(\mathfrak{g}\text{-}\mathfrak{div}\, v, F; A)$. In particular $A \in \mathcal{R}(\mathfrak{g}\text{-}\mathfrak{div}\, v, F; A)^-$. Now employing Remark 6.3.1, (2) and (3), we obtain

$$\mathcal{BV}(A) = \mathcal{R}(\mathfrak{g}\text{-}\mathfrak{div}\, v, F; A)^-,$$

which proves the theorem. $\square$

The following example shows that Theorem 6.4.4 is a stronger version of the Gauss-Green theorem than Theorem 5.1.12.

Example 6.4.5. Let v, H, h, K, and W be as in Proposition 6.1.1. According to Proposition 1.10.5, there is a sequence $\{B_n\}$ of essentially closed BV subsets of W such that $\{B_n\} \to W$. Since $\mathrm{int}_* B \subset W$, and since $C = K - W$ and $\mathrm{cl}_* B_n$ are closed sets whose intersection is contained in $\partial_* B_n$, it is easy to see that

$$DH(x) = \begin{cases} h(x) & \text{if } x \in \mathrm{cl}_* B_n - \partial_* B_n, \\ 0 & \text{if } x \in \mathrm{cl}_* C - \partial_* B_n. \end{cases}$$

Thus H is derivable at each point of $\mathrm{cl}_*(C \cup B_n) - \partial_* B_n$. As $\partial_* B_n$ is a thin set and $\{C \cup B_n\} \to K$, Lemma 6.4.1 implies

$$\mathfrak{g}\text{-}\mathfrak{div}\, v(x) = \begin{cases} h(x) & \text{for almost all } x \in W, \\ 0 & \text{for almost all } x \in K - W. \end{cases}$$

Moreover, $\mathfrak{g}\text{-}\mathfrak{div}\, v \in GR(K)$ and H is the indefinite GR-integral of $\mathfrak{g}\text{-}\mathfrak{div}\, v$ according to Theorem 6.4.4. On the other hand, Remark 6.1.2, (3) implies that $\mathfrak{g}\text{-}\mathfrak{div}\, v$ is not R-integrable in K.

Note $H \restriction W$ and $H \restriction (K - W)$ are the indefinite R-integrals of $\mathfrak{g}\text{-}\mathfrak{div}\, v \restriction W$ and $\mathfrak{g}\text{-}\mathfrak{div}\, v \restriction (K - W)$, respectively. Thus the equality $H = (GR) \int \mathfrak{g}\text{-}\mathfrak{div}\, v(x)\, dx$ follows also from Theorem 6.3.10.

Theorem 6.4.6. *Let E be a locally BV set. Then*

$$MGR(E) = BV_\ell^\infty(E),$$

and if F is the indefinite GR-integral of $f \in GR(E)$, then $F \mathbin{\mathsf{L}} g$ is the indefinite GR-integral of fg for each $g \in BV_\ell^\infty(E)$.

PROOF. In view of Proposition 6.3.8 and Theorem 6.4.4, the proof of the inclusion $MGR(E) \subset BV_\ell^\infty(E)$ is identical to the proof of Lemma 5.2.1. If F is the indefinite GR-integral of $f \in GR(E)$, then by definition $\mathcal{BV}(E) = \mathcal{R}(f, F; E)^-$. Given $g \in BV_\ell^\infty(E)$, observe that (4.5.2) and Proposition 4.5.3 imply

$$(F \mathbin{\mathsf{L}} g) \mathbin{\mathsf{L}} B = (F \mathbin{\mathsf{L}} \bar{g}) \mathbin{\mathsf{L}} B = (F \mathbin{\mathsf{L}} B) \mathbin{\mathsf{L}} \bar{g} = (F \mathbin{\mathsf{L}} B) \mathbin{\mathsf{L}} (g \upharpoonright B)$$

for each bounded BV set $B \subset E$. Hence

$$\mathcal{R}(fg, F \mathbin{\mathsf{L}} g; E) = \mathcal{R}(f, F; E)$$

according to Theorem 5.2.2, and the theorem follows. $\qquad\square$

Remark 6.4.7. Theorem 6.4.6 implies that Theorem 5.2.5 can be extended verbatim to the GR-integral; in view of Theorem 6.4.4, the same is true about Theorem 5.2.6.

Theorem 6.4.8. *Suppose $\phi : E \to \mathbb{R}^m$ is a lipeomorphism of a locally BV set E, and $f \in GR[\phi(E)]$. Then $(f \circ \phi)J_\phi$ belongs to $GR(E)$ and*

$$\int_A f[\phi(x)] J_\phi(x)\, dx = \int_{\phi(A)} f(y)\, dy$$

for each bounded BV set $A \subset E$.

PROOF. Theorem 1.8.8 implies that the map

$$\phi : B \mapsto \phi(B) : \mathcal{BV}(E) \to \mathcal{BV}[\phi(E)]$$

is a surjective homeomorphism. Thus $\phi(\mathcal{A}^-) = \phi(\mathcal{A})^-$ for each family $\mathcal{A} \subset \mathcal{BV}(E)$. If F is the indefinite GR-integral of f, then

$$\phi[\mathcal{BV}(E)] = \mathcal{BV}[\phi(E)] = \mathcal{R}[f, F; \phi(E)]^-$$

$$= \phi\Big(\mathcal{R}[(f \circ \phi)J_\phi, \phi^* F; E]\Big)^-$$

$$= \phi\Big(\mathcal{R}[(f \circ \phi)J_\phi, \phi^* F; E]^-\Big)$$

by Theorem 5.3.1. Thus $\mathcal{R}[(f \circ \phi)J_\phi, \phi^* F; E]^- = \mathcal{BV}(E)$, which means that $\phi^* F$ is the indefinite GR-integral of $(f \circ \phi)J_\phi$. $\qquad\square$

Theorem 6.4.9. *Suppose* $\phi : A \to \mathbb{R}^m$ *is a local lipeomorphism of a bounded BV set* A, *and* f *is a function defined on* A *such that* $f J_\phi$ *belongs to* $GR(A)$. *Then the function*

$$y \mapsto \sum\left\{f(x) : x \in \phi^{-1}(y)\right\} : \phi(A) \to \mathbb{R}$$

belongs to $GR[\phi(A)]$, *and*

$$\int_A f(x) J_\phi(x)\, dx = \int_{\phi(A)} \sum\left\{f(x) : x \in \phi^{-1}(y)\right\} dy.$$

PROOF. By Definition 4.6.4 the set A is the union of disjoint BV sets $A_1, \ldots, A_p$ such that each $\phi_i := \phi \restriction A_i$ is a lipeomorphism. It follows from Theorem 6.4.8 that $g_i = f \circ \phi_i^{-1}$ belongs to $GR[\phi(A_i)]$ and

$$\int_{\phi(A_i)} g_i(y)\, dy = \int_{A_i} f(x) J_\phi(x)\, dx.$$

Corollary 6.3.11 implies $\overline{g_i} \in GR[\phi(A)]$, and if $g = \sum_{i=1}^p \overline{g_i}$, then

$$\int_{\phi(A)} g(y)\, dy = \sum_{i=1}^p \int_{\phi(A_i)} g_i(y)\, dy = \int_A f(x) J_\phi(x)\, dx.$$

As it is easy to verify that

$$g(y) = \sum\left\{f(x) : x \in \phi^{-1}(y)\right\}$$

for all $y \in \mathbb{R}^m$, the theorem is proved. $\qquad\square$

The following example shows that the GR-integral in Theorem 6.4.9 cannot be replaced by the R-integral.

Example 6.4.10. Let K, W, and h be as in Proposition 6.1.1, and let $W' = (2,0) + W$ be a translation of W. Then the formula

$$\phi(x) := \begin{cases} x & \text{if } x \in K - W, \\ x - 2 & \text{if } x \in W', \end{cases}$$

defines an injective map ϕ from $B = (K - W) \cup W'$ onto K, which is a local lipeomorphism. The function $f = \overline{h} \circ \phi$ is clearly R-integrable in B, and so is $f J_\phi = f$. On the other hand, the function

$$y \mapsto \sum\left\{f(x) : x \in \phi^{-1}(y)\right\}$$

equals $\overline{h}$, which is not R-integrable in K by Remark 6.1.2, (3).

Theorem 6.4.11. *Suppose* $\phi : A \to \mathbb{R}^m$ *is a local lipeomorphism of a bounded BV set A, and* $g \in BV_b^\infty(A)$. *If* $f \in GR[\phi(A)]$, *then* $f\deg(\phi, g)$ *and* $(f \circ \phi)(\det D\phi)g$ *belong to* $GR[\phi(A)]$ *and* $GR(A)$, *respectively, and*

$$\int_{\phi(A)} f(y)\deg(\phi, g)(y)\, dy = \int_A f[\phi(x)]\det D\phi(x)\, g(x)\, dx\,.$$

PROOF. Assume first that ϕ is a lipeomorphism. The map ϕ induces a homeomorphism from $\mathcal{BV}(A)$ onto $\mathcal{BV}[\phi(A)]$, still denoted by ϕ. If F is the indefinite GR-integral of f, then

$$\phi[\mathcal{BV}(A)] = \mathcal{BV}[\phi(A)] = \mathfrak{R}[f, F; \phi(A)]^-$$
$$= \phi\Big(\mathfrak{R}[(f \circ \phi)(\det D\phi)g, \phi^\# F \, \mathsf{L}\, g; A]\Big)^-$$
$$= \phi\Big(\mathfrak{R}[(f \circ \phi)(\det D\phi)g, \phi^\# F \, \mathsf{L}\, g; A]^-\Big)$$

by Theorems 4.6.12 and 4.5.7. Consequently

$$\mathfrak{R}[(f \circ \phi)(\det D\phi)g, \phi^\# F \, \mathsf{L}\, g; A]^- = \mathcal{BV}(A)\,,$$

which means $\phi^\# F \, \mathsf{L}\, g$ is the indefinite GR-integral of $(f \circ \phi)(\det D\phi)g$. From Proposition 4.6.9 and Theorem 6.4.6, we obtain that the charge $G := F \, \mathsf{L}\, \deg(\phi, g)$ is the indefinite GR-integral of $f\deg(\phi, g)$. A calculation identical to that carried out in the proof of Theorem 5.3.4 shows

$$[(\phi^\# F) \, \mathsf{L}\, g](A) = G[\phi(A)]\,,$$

which establishes the theorem when ϕ is a lipeomorphism.

If ϕ is a local lipeomorphism, then A is the union of disjoint BV sets $A_1, \ldots, A_p$ such that each $\phi_i := \phi \restriction A_i$ is a lipeomorphism. By the first part of the proof, for $i = 1, \ldots, p$, the functions $f\deg(\phi_i, g)$ and $(f \circ \phi)(\det D\phi)g$ belong to $GR[\phi(A_i)]$ and $GR(A_i)$, respectively, and

$$\int_{\phi(A_i)} f(y)\deg(\phi_i, g)(y)\, dy = \int_{A_i} f[\phi(x)]\det D\phi(x)\, g(x)\, dx\,.$$

By Theorem 6.3.10, the function $(f \circ \phi)(\det D\phi)g$ belongs to $GR(A)$. As $\deg(\phi_i, g)(y) = 0$ for each y in $\mathbb{R}^m - \phi(A_i)$, Corollary 6.3.11 yields

$$\int_A f[\phi(x)]\det D\phi(x)\, g(x)\, dx = \int_{\phi(A)} f(y)\left[\sum_{i=1}^p \deg(\phi_i, g)(y)\right] dy\,.$$

Now given $y \in \mathbb{R}^m$, the following calculation completes the proof:

$$\sum_{i=1}^{p} \deg(\phi_i, g)(y) = \sum_{i=1}^{p} \sum_{x \in \phi_i^{-1}(y)} g(x)\,\text{sign}\,\det D\phi(x)$$
$$= \sum_{i=1}^{p} \sum_{x \in \phi^{-1}(y) \cap A_i} g(x)\,\text{sign}\,\det D\phi(x)$$
$$= \sum_{x \in \phi^{-1}(y)} g(x)\,\text{sign}\,\det D\phi(x) = \deg(\phi, g)(y). \qquad \square$$

The next proposition shows that iterating Definition 6.3.4 produces no new integrals.

Proposition 6.4.12. *Let f be a function defined on a locally BV set E, and let F be a charge. If $\mathcal{GR}(f, F; E)$ is the family of all bounded BV sets $B \subset E$ such that $f \upharpoonright B$ belongs to $GR(B)$ and $F \llcorner B$ is the indefinite GR-integral of $f \upharpoonright B$, then*

$$\mathcal{GR}(f, F; E)^- = \mathcal{R}(f, F; E)^-.$$

PROOF. The inclusion

$$\mathcal{GR}(f, F; E) \subset \bigcup\{\mathcal{R}(f, F; B)^- : B \in \mathcal{GR}(f, F; E)\}$$
$$\subset \left(\bigcup\{\mathcal{R}(f, F; B) : B \in \mathcal{GR}(f, F; E)\}\right)^-$$
$$\subset \mathcal{R}(f, F; E)^-,$$

yields $\mathcal{GR}(f, F; E)^- \subset \mathcal{R}(f, F; E)^-$. As $\mathcal{R}(f, F; E) \subset \mathcal{GR}(f, F; E)$, the proposition follows. $\qquad \square$

In Remark 3.7.6, we mentioned the possibility of defining the integral by means of figures. Call such an integral the RF-integral, and given a figure A, denote by $RF(A)$ the linear space of all RF-integrable functions in A. For an explicit Riemann type definition of the RF-integral we refer to [65, Chapter 12]. If A is a figure, it follows easily from Lemma 1.10.2, Proposition 2.6.11, and Theorem 3.7.4 that $RF(A) = R(A)$ and that the R-integral and RF-integral of each function $f \in R(A)$ are the same. The following definition extends the RF-integral to functions defined on bounded BV sets.

Definition 6.4.13. *A function f defined on a set $A \in \mathcal{BV}$ is called RF-integrable in A if the zero extension $\bar{f}$ of f is RF-integrable in a figure B containing A.*

The family of all RF-integrable functions in a set $A \in \mathcal{BV}$ is a linear space, still denoted by $RF(A)$. The RF-integral has reasonable properties, including a change of variable theorem similar to Theorem 5.3.1 — see [16]. If $A \in \mathcal{BV}$ is essentially closed, then $RF(A) = R(A)$ by Proposition 5.1.8. Thus defining the linear space $GRF(A)$ analogously to Definition 6.3.4, it follows from Lemma 6.2.3 that

$$GRF(A) = GR(A)$$

for every set $A \in \mathcal{BV}$. On the other hand, Remark 6.1.2, (2) shows that the inclusion $RF(A) \subset R(A)$ is generally proper.

Remark 6.4.14. At the cost of replacing charges by additive functions with weaker continuity properties, further extensions of the GR-integral are possible [61, 64, 39, 57]. The simplest of these, presented in [64, Section 10], follows the pattern of Definition 6.3.4 with respect to the topology in $\mathcal{BV}$ induced by the metric

$$\sigma(A, B) := \|A \triangle B\|.$$

In view of Theorem 1.10.6, given a bounded BV set A, it provides the Gauss-Green formula for a bounded vector field $v : \mathrm{cl}_* A \to \mathbb{R}^m$ that is Lipschitz at each $x \in \mathrm{cl}_* A$ except at the points of a thin set, and continuous outside an $\mathcal{H}^{m-1}$-negligible set [64, Theorem 10.9].

Bibliography

[1] S.I. Ahmed and W.F. Pfeffer, *A Riemann integral in a locally compact Hausdorff space*, J. Austral. Math. Soc. **41, Series A** (1986), 115–137.

[2] L. Ambrosio, *A compact theorem for a new class of functions of bounded variation*, Boll. Un. Mat. Ital. **3-B** (1989), 857–881.

[3] T. Bagby and W.P. Ziemer, *Pointwise differentiability and absolute continuity*, Trans. Amer. Math. Soc. **191** (1974), 129–148.

[4] H. Bauer, *Der Perronsche Integralbegriff und seine Beziehung zum Lebesgueshen*, Monatshefte Math. Phys. **26** (1915), 153–198.

[5] A.S. Besicovitch, *On sufficient conditions for a function to be analytic, and behaviour of analytic functions in the neighbourhood of non-isolated singular points*, Proc. London Math. Soc. **32** (1931), 1–9.

[6] B. Bongiorno, *Essential variation*, Measure Theory Oberwolfach, Lecture Notes in Math. no. 945, Springer-Verlag, New York, 1981, pp. 187–193.

[7] B. Bongiorno, U. Darji, and W.F. Pfeffer, *On indefinite BV-integrals*, Comment. Math. Univ. Carolinae **41** (2000), 843–853.

[8] B. Bongiorno, M. Giertz, and W.F. Pfeffer, *Some nonabsolutely convergent integrals in the real line*, Boll. Un. Mat. Ital. **6-B** (1992), 371–402.

[9] B. Bongiorno, L. Di Piazza, and D. Preiss, *Infinite variations and derivatives in $\mathbb{R}^m$*, J. Math. Anal. Appl. **224** (1998), 22–33.

[10] _____, *A constructive minimal integral which includes Lebesgue integrable functions and derivatives*, J. London Math. Soc. **62** (2000), 117–126.

[11] B. Bongiorno and P. Vetro, *Su un teorema di F. Riesz*, Atti Acc. Sci. Lettere Arti Palermo (IV) **37** (1979), 3–13.

[12] Z. Buczolich, *Functions with all singular sets of Hausdorff dimension bigger than one*, Real Anal. Exchange **15(1)** (1989–90), 299–306.

[13] _____, *Density points and bi-Lipschitz finctions in $\mathbb{R}^m$*, Proc. American Math. Soc. **116** (1992), 53–56.

[14] _____, *A v-integrable function which is not Lebesgue integrable on any portion of the unit square*, Acta Math. Hung. **59** (1992), 383–393.

[15] _____, *The g-integral is not rotation invariant*, Real Anal. Exchange **18(2)** (1992–93), 437–447.

[16] _____, *Lipeomorphisms, sets of bounded variation and integrals*, Acta Math. Hung. **87** (2000), 243–265.

[17] Z. Buczolich, T. De Pauw, and W.F. Pfeffer, *Charges, BV functions, and mul-tipliers for generalized Riemann integrals*, Indiana Univ. Math. J. **48** (1999), 1471–1511.

[18] Z. Buczolich and W.F. Pfeffer, *Variations of additive functions*, Czechoslovak Math. J. **47** (1997), 525–555.

[19] ———, *On absolute continuity*, J. Math. Anal. Appl. **222** (1998), 64–78.

[20] G. Congedo and I. Tamanini, *Note sulla regolarità dei minimi di funzionali del tipo dell'area*, Rend. Acad. Sci. XL, Mem. Mat. **106 (XII, 17)** (1988), 239–257.

[21] R. Engelking, *General Topology*, PWN, Warsaw, 1977.

[22] L.C. Evans and R.F. Gariepy, *Measure Theory and Fine Properties of Func-tions*, CRP Press, Boca Raton, 1992.

[23] K.J. Falconer, *The Geometry of Fractal Sets*, Cambridge Univ. Press, Cam-bridge, 1985.

[24] H. Federer, *Geometric Measure Theory*, Springer-Verlag, New York, 1969.

[25] H. Federer and W.H. Fleming, *Normal and integral currents*, Ann. of Math. **72** (1960), 458–520.

[26] E. Giusti, *Minimal Surfaces and Functions of Bounded Variation*, Birkhäuser, Basel, 1984.

[27] C. Goffman, T. Nishiura, and D. Waterman, *Homeomorphisms in Analysis*, Amer. Math. Soc., Providence, 1997.

[28] R.A. Gordon, *The Integrals of Lebesgue, Denjoy, Perron, and Henstock*, Amer. Math. Soc., Providence, 1994.

[29] P.R. Halmos, *Measure Theory*, Van Nostrand, New York, 1950.

[30] R. Henstock, *Definitions of Riemann type of the variational integrals*, Proc. London Math. Soc. **11** (1961), 79–87.

[31] I.N. Herstein, *Topics in Algebra*, Blaisdell, London, 1964.

[32] J. Holec and J. Mařík, *Continuous additive mappings*, Czechoslovak Math. J. **14** (1965), 237–243.

[33] J. Horváth, *Topological Vector Spaces and Distributions*, vol. 1, Addison-Wesley, London, 1966.

[34] E.J. Howard, *Analyticity of almost everywhere differentiable functions*, Proc. American Math. Soc. **110** (1990), 745–753.

[35] J. Jarník and J. Kurzweil, *A nonabsolutely convergent integral which admits transformation and can be used for integration on manifolds*, Czechoslovak Math. J. **35** (1986), 116–139.

[36] T. Jech, *Set Theory*, Academic Press, New York, 1978.

[37] R.L. Jeffery, *The Theory of Functions of a Real Variable*, Dover, New York, 1985.

[38] W.B. Jurkat, *The divergence theorem and Perron integration with exceptional sets*, Czechoslovak Math. J. **43** (1993), 27–45.

[39] W.B. Jurkat and D.J.F. Nonnenmacher, *A generalized n-dimensional Riemann integral and the divergence theorem with singularities*, Acta Sci. Math. Szeged **59** (1994), 241–256.

[40] K. Karták and J. Mařík, *A non-absolutely convergent integral in E_m and the theorem of Gauss*, Czechoslovak Math. J. **15** (1965), 253–260.

[41] L. Kuipers and H. Niederreiter, *Uniform Distribution of Sequences*, John Wiley, New York, 1974.

[42] J. Kurzweil, *Generalized ordinary differential equations and continuous dependence on a parameter*, Czechoslovak Math. J. **82** (1957), 418–446.

[43] _____, *Nichtabsolut Konvergente Integrale*, Taubinger, Leipzig, 1980.

[44] J. Kurzweil, J. Mawhin, and W.F. Pfeffer, *An integral defined by approximating BV partitions of unity*, Czechoslovak Math. J. **41** (1991), 695–712.

[45] U. Massari and M. Miranda, *Minimal Surfaces of Codimension One*, North-Holland, Amsterdam, 1984.

[46] P. Mattila, *Geometry of Sets and Measures in Euclidean Spaces*, Cambridge Univ. Press, Cambridge, 1995.

[47] J. Mařík, *Extensions of additive mappings*, Czechoslovak Math. J. **15** (1965), 244–252.

[48] J. Mařík and J. Matyska, *On a generalization of the Lebesgue integral in E_m*, Czechoslovak Math. J. **15** (1965), 261–269.

[49] J. Mawhin, *Generalized multiple Perron integrals and the Green-Goursat theorem for differentiable vector fields*, Czechoslovak J. Math. **31** (1981), 614–632.

[50] _____, *Generalized Riemann integrals and the divergence theorem for differentiable vector fields*, E.B. Christoffel (Basel), Birkhäuser, 1981, pp. 704–714.

[51] V.G. Maz'ja, *Sobolev Spaces*, Springer-Verlag, New York, 1985.

[52] R.M. McLeod, *The Generalized Riemann Integral*, Math. Asso. Amer., Washington, D.C., 1980.

[53] E.J. McShane, *A Riemann-type Integral that Includes Lebesgue-Stieltjes, Bochner and Stochastic Integrals*, Mem. Amer. Math. Soc., 88, Providence, 1969.

[54] _____, *A unified theory of integration*, Amer. Math. Monthly **80** (1973), 349–359.

[55] _____, *Unified Integration*, Academic Press, New York, 1983.

[56] F. Morgan, *Geometric Measure Theory*, Academic Press, New York, 1988.

[57] D.J.F. Nonnenmacher, *Sets of finite perimeter and the Gauss-Green theorem*, J. London Math. Soc. **52** (1995), 335–344.

[58] J.C. Oxtoby, *Measure and Category*, Springer-Verlag, New York, 1971.

[59] T. De Pauw, *Topologies for the space of BV-integrable functions in $\mathbb{R}^m$*, J. Func. Anal. **144** (1997), 190–231.

[60] W.F. Pfeffer, *Integrals and measures*, Marcel Dekker, New York, 1977.

[61] _____, *The divergence theorem*, Trans. Amer. Math. Soc. **295** (1986), 665–685.

[62] _____, *The multidimensional fundamental theorem of calculus*, J. Australian Math. Soc. **43** (1987), 143–170.

[63] _____, *A descriptive definition of a variational integral and applications*, Indiana Univ. Math. J. **40** (1991), 259–270.

[64] _____, *The Gauss-Green theorem*, Adv. Math. **87** (1991), 93–147.

[65] _____, *The Riemann Approach to Integration*, Cambridge Univ. Press, New York, 1993.

[66] _____, *The generalized Riemann-Stieltjes integral*, Real Anal. Exchange **21(2)** (1995–96), 521–547.

[67] _____, *The Lebesgue and Denjoy-Perron integrals from a descriptive point of view*, Ricerche Mat. **48** (1999), 211–223.

[68] W.F. Pfeffer and B.S. Thomson, *Measures defined by gages*, Canadian J. Math. **44** (1992), 1306–1316.

[69] W.F. Pfeffer and Wei-Chi Yang, *The multidimensional variational integral and its extensions*, Real Anal. Exchange **15(1)** (1989–90), 111–169.

[70] K.P.S. Bhaskara Rao and M. Bhaskara Rao, *Theory of Charges*, Acad. Press, New York, 1983.

[71] C.A. Rogers, *Hausdorff measures*, Cambridge Univ. Press, Cambridge, 1970.

[72] P. Romanovski, *Intégrale de Denjoy dans l'espace á n dimensions*, Math. Sbornik **51** (1941), 281–307.

[73] W. Rudin, *Real and Complex Analysis*, McGraw-Hill, New York, 1987.

[74] _____, *Functional Analysis*, McGraw-Hill, New York, 1991.

[75] S. Saks, *Theory of the Integral*, Dover, New York, 1964.

[76] V.L. Shapiro, *The divergence theorem for discontinuous vector fields*, Ann. Math. **68** (1958), 604–624.

[77] L. Simon, *Lectures on Geometric Measure Theory*, Proc. C.M.A. 3, Australian Natl. Univ., Cambera, 1983.

[78] E.H. Spanier, *Algebraic Topology*, McGraw-Hill, New York, 1966.

[79] M. Spivak, *Calculus on Manifolds*, Benjamin, London, 1965.

[80] E.M. Stein, *Singular Integrals and Differentiability Properties of Functions*, Princeton Univ. Press, Princeton, 1970.

[81] I. Tamanini and C. Giacomelli, *Approximation of Caccioppoli sets, with applications to problems in image segmentation*, Ann. Univ. Ferrara **VII (N.S.) 35** (1989), 187–213.

[82] _____, *Un tipo di approssimazione "dall'interno" degli insiemi di perimetro finito*, Rend. Mat. Acc. Lincei **(9) 1** (1990), 181–187.

[83] B.S. Thomson, *Spaces of conditionally integrable functions*, J. London Math. Soc. **2** (1970), 358–360.

[84] _____, *Derivatives of Interval Functions*, Mem. Amer. Math. Soc., 452, Providence, 1991.

[85] _____, *σ-finite Borel measures on the real line*, Real Anal. Exch. **23** (1997-98), 185–192.

[86] A.I. Volpert, *The spaces BV and quasilinear equations*, Math. USSR–Sbornik **2** (1967), 225–267.

[87] H. Whitney, *Geometric Integration Theory*, Princeton Univ. Press, Princeton, 1957.

[88] W.P. Ziemer, *Weakly Differentiable Functions*, Springer-Verlag, New York, 1989.

List of symbols

Euclidean spaces

$\mathbb{R}$, $\mathbb{R}^m$, 1

$x \cdot y$, $|x|$, 1

$B(x,r)$, $B[x,r]$, 2

$B(r)$, $B[r]$, 2

$d(E)$, 2

cl E, int E, ∂E, 2

dist (A,B), dist (x,A), 2

$K[x,r]$, $K[r]$, 2

$C^\star$, 2

$B^\bullet$, 16

$\mathbf{e}_i$, 39

E^y, 39

$\Pi_\mathbf{e}$, 40

$\mathfrak{Q}(n)$, $\mathfrak{Q}_n$, 110

Functions

$f \restriction B$, 3

$\{f = 0\}$, $\{f \neq 0\}$, 3

supp f, 3

f^+, f^-, 3

χ_E, 3

$\overline{f}$, 3

$f^{-1}(B)$, 12

$\{f > t\}$, $\{f < t\}$, 37

Σ_g, 138

$f * g$, 143

η_ε, 143

$(g)_{x,r}$, $g^*(x)$, 154

$g \otimes h$, 164

1_E, 164

Topology

$\mathrm{cl}_{\mathcal{T}} E$, 4

$\mathcal{T} \otimes \mathcal{S}$, 4

$\mathcal{T}$, 33

$\mathfrak{T}$, 42

$\mathcal{S}$, 57

$\mathfrak{V}$, 209

$\mathfrak{V}(\emptyset)$, 210

$\mathfrak{U}(F_1, \ldots, F_k; \varepsilon)$, 210

$\mathcal{E}^-$, 236

Measures and related concepts

$\mu \llcorner E$, 11

$\mathfrak{M}_\mu$, 11

supp μ, 11

$\mu \llcorner h$, 12

259

Forms and currents

Other symbols

Index